The butterflyfishes: success on the coral reef

Developments in environmental biology of fishes 9

Series Editor
EUGENE K. BALON

The butterflyfishes:
success on the coral reef

Guest editor:
PHILIP J. MOTTA

Reprinted from *Environmental biology of fishes* 25 (1–3), 1989
with addition of species and subject index

KLUWER ACADEMIC PUBLISHERS
DORDRECHT / BOSTON / LONDON

The Butterflyfishes : success on the coral reef / guest editor, Philip J. Motta.
  p.   cm. — (Developments in environmental biology of fishes ; 9)
  Papers presented at a symposium held at the University of Michigan, Ann Arbor, 6/24–29/88.
  'Reprinted from environmental biology of fishes 25 (1–3), 1989 with addition of species and subject index.'
  Includes index.
  ISBN 0-7923-0168-4
  1. Chaetondontidae-Congresses. I. Motta, Philip Jay.
II. Series.
QL638.C48B88   1989
597'.58-dc 19   89-2409

Published by Kluwer Academic Publishers,
P.O. Box 17, 3300 AA Dordrecht, The Netherlands.

Kluwer Academic Publishers incorporates
the publishing programmes of
Martinus Nijhoff, Dr. W. Junk, D. Reidel and MTP Press.

Sold and distributed in the U.S.A. and Canada
by Kluwer Academic Publishers,
101 Philip Drive, Norwell, MA 02061, U.S.A.

In all countries, sold and distributed
by Kluwer Academic Publishers Group,
P.O. Box 322, 3300 AH Dordrecht, The Netherlands.

Printed on acid free paper

All Rights Reserved
© 1989 by Kluwer Academic Publishers
No part of the material protected by this copyright notice may be reproduced or utilized in any form or by any means, electronic or mechanical including photocopying, recording or by any information storage and retrieval system, without written permission from the copyright owner

Printed in the Netherlands

To Diane, Dominic and Danielle

Contents

Preface, by P.J. Motta 7
Biogeography of the Chaetodontidae: an analysis of allopatry among closely related species, by S.D. Blum 9
Circumtropical patterns in butterflyfish communities, by J.S. Findley & M.T. Findley 33
Correlations between chaetodontid fishes and coral communities of the Gulf of Aqaba (Red Sea), by Y. Bouchon-Navaro & C. Bouchon 47
Environmental determinants of butterflyfish social systems, by T.F. Hourigan 61
Orientation behavior of butterflyfishes (family Chaetodontidae) on coral reefs: spacial learning of route specific landmarks and cognitive maps, by E.S. Reese 79
Larval biology of butterflyfishes (Pisces, Chaetodontidae): what do we really know? by J.M. Leis . 87
Implications of feeding specialization on the recruitment processes and community structure of butterflyfishes, by M.L. Harmelin-Vivien 101
Sexual differentiation, gonad development, and spawning seasonality of the Hawaiian butterflyfish, *Chaetodon multicinctus*, by T.C. Tricas & J.T. Hiramoto 111
Spawning behavior of *Chaetodon multicinctus* (Chaetodontidae); pairs and intruders, by P.S. Lobel 125
Aspects of the spawning of western Atlantic butterflyfishes (Pisces: Chaetodontidae), by P.L. Colin 131
Eye camouflage and false eyespots: chaetodontid responses to predators, by S. Neudecker 143
Dentition patterns among Pacific and Western Atlantic butterflyfishes (Perciformes, Chaetodontidae): relationship to feeding ecology and evolutionary history, by P.J. Motta 159
Prey selection by coral-feeding butterflyfishes: strategies to maximize the profit, by T.C. Tricas 171
Temporal and areal feeding behavior of the butterflyfish, *Chaetodon trifascialis*, at Johnston Atoll, by D.K. Irons 187
Feeding habits of Japanese butterflyfishes (Chaetodontidae), by M. Sano 195
The brain organization of butterflyfishes, by R. Bauchot, J.-M. Ridet & M.-L. Bauchot 205
The eye muscles and their innervation in *Chaetodon trifasciatus* (Pisces, Teleostei, Chaetodontidae), by R. Bauchot, A. Thomot & M.-L. Bauchot 221
The membranous labyrinth and its innervation in *Chaetodon trifasciatus* (Pisces, Teleostei, Chaetodontidae), by R. Bauchot, A. Thomot & M.-L. Bauchot 235
Strengths and weaknesses in butterflyfish research: concluding remarks, by P.J. Motta 243
Species and subject index 251

Preface

Butterflyfishes of the family Chaetodontidae are conspicuous members of almost all tropical reefs. These colorful fishes have attracted a great deal of attention from both the scientific community and especially the aquarium fish industry.

At first one is tempted to say that butterflyfishes are abundant worldwide, but the evidence does not support this statement. The biomass of chaetodontids on reefs may range from 0.02–0.80%, and in terms of numbers they comprise only 0.04–0.61% of the individuals on the reef. Yet in spite of these relatively small numbers they have been extensively studied. A quick census shows some 170 articles on or about butterflyfishes, with 78% of them being published since the 1970's. Along with the cichlids and damselfishes they might be one of the most studied and well published family of tropical fishes.

Why then have chaetodontids attracted so much attention? The butterflyfishes are mostly shallow water inhabitants that are approachable and easily recognizable, making their study very feasible. Their bright coloration has provoked many hypotheses but has posed more questions about coloration than it has provided answers. And despite their apparent overall morphological similarity, their highly structured and varied social systems have made them an ideal model for such studies. The reasons for choosing these organisms are indeed as diverse as the studies themselves.

What then have most of the studies of the group concentrated on? There have been two general categories of scientific publication on the group. Books by Burgess, Allen, Thresher and Steene discuss the classification, identification, and reproduction of butterflyfishes. There is also an extensive background of primary literature, which, not surprisingly, has centered on coloration, their social behavior, reproduction, and feeding. Most of these studies would be classified as being behavioral or ecological. However, as witnessed by this symposium, the burgeoning research base has expanded to encompass systematics and evolution, morphology, neuroanatomy, early ontogeny, and biogeography.

Many of these research programs are scattered throughout the world in geographically isolated camps in the continental United States, Australia, China, Japan, France, Hawaii, and the West Indies, and the field sites are just as diverse. While this has made for a rich comparative base, there has been no overall coordination among members and little scientific contact other than through the scientific literature.

In 1987 I decided to convene as many butterflyfish researchers as possible in order to organize an international symposium on the group. From the very onset this symposium was perhaps not typical of many symposia. The presentations were never intended to be exhaustive reviews of the literature and they were not to be presentations of previously published ideas and data, although such ideas and data could be the building blocks for new thoughts. Invitations and solicitations therefore went out to those only currently involved in chaetodontid research; naturally not all researchers were reached in this effort. The invitations had the desired effect in many cases; it sent researchers into the field to complete data collection. Some papers also tend to be more speculative than traditional research papers as the authors were encouraged to be more speculative in order to expose new ideas and hypotheses. Furthermore, I attempted to include research from as many differing disciplines as possible. One notes that certain areas are represented by many more papers than others, and other disciplines are totally lacking. This in itself indicates strengths and weaknesses in our understanding of the group.

The goals of this symposium were therefore: (1) to present the latest research on butterflyfishes encompassing as many disciplines as possible, (2) to bring these researchers together under one roof for the first time to coordinate efforts, discuss findings, and increase communication, (3) to investigate defficient areas of study and suggest avenues for future research on the group, (4) to relate the findings to other fish groups, (5) to stimulate new ideas and hypotheses not only for chaetodontid research but hopefully for fish biology in general, and (6) to present these ideas to the widest possible audience. The latter could be best served by having the symposium at the joint meeting of the American Society of Ichthyologists and Herpetologists, The Early Life History Section, and the American Elasmobranch Society, and by publishing the proceedings in a widely distributed international journal, the Environmental Biology of Fishes.

A symposium such as this does not occur without the cooperation and effort of many. The American Society of Ichthyologists and Herpetologists and its officers as well as Gerald Smith and the hosting University of Michigan worked with us and supported us from the start. The publishers of Environmental Biology of Fishes and specifically their editor Eugene Balon put in an extraordinary effort into its publication. Our authors worked against many deadlines, pulling together as a group and supporting me from the beginning. The other three chairpersons, Ernst Reese, Stanley Blum, and Timothy Tricas contributed not only their papers but their time. The University of Montana, University of the Virgin Islands, and now the University of South Florida provided support throughout the project. Last but by no means least the reviewers listed at the close of the journal volume put in an outstanding effort in critically evaluating every manuscript. My profound thanks to all.

Tampa, September 1988   Philip J. Motta

Biogeography of the Chaetodontidae: an analysis of allopatry among closely related species

Stanley D. Blum
Department of Vertebrate Paleontology, American Museum of Natural History, Central Park West at 79th Street, New York, NY 10024, U.S.A.

Received 2.8.1988   Accepted 15.11.1988

Key words: Butterflyfishes, Vicariance, Cladistics, Allopatric speciation, Tethys, Indo-Pacific

Synopsis

A recent survey of chaetodontid osteology has produced a hypothesis of relationships among 22 osteologically distinct genera and subgenera. Fourteen supra-specific taxa have distributions that are Indo-Pacific or larger. Most sister taxa inferred by osteology are broadly sympatric. The basal dichotomy within the large genus *Chaetodon* contrasts monophyletic groups centered in the Atlantic and Indo-Pacific with little overlap. Divergence of Atlantic and Indo-Pacific distributions is correlated with the closing of the Tethys seaway 18–13 million years ago. Distributional data of Burgess (1978) and Allen (1980) are reevaluated in the context of putative species pairs and complexes. Species in nearly two thirds of these complexes (18 of 31) are distributed allopatrically. Eight complexes are examined in more detail. Five of these eight contain at least one peripherally isolated species. Distributions of species in four complexes indicate that previously wide-spread species were cleaved more symmetrically. Sympatric distributions within two species pairs indicate that the more narrowly distributed species in each pair arose through central isolation within a broadly distributed ancestor. The area of central isolation corresponds to the classical center of origin. A new hypothesis of vicariance followed by dispersal may partially explain the diversity gradient so prominently featured in dispersal-oriented tropical marine biogeography.

Introduction

Classical works on tropical marine biogeography have primarily concerned the demarcation of faunal provinces, and the decline in diversity that occurs from west to east in the Pacific, and, less evenly, from east to west in the Indian Ocean (e.g. Ekman 1953, Schilder 1961, Briggs 1974). Classical biogeographic methods usually treat species as the units of inference; species ranges are superimposed to demonstrate areas of endemism, or the numbers of species are tabulated to compare diversities, or the numbers of shared species are tabulated to show similarities among regions. Little attention is devoted to the phylogenetic histories of species. Perhaps consequently, the biogeographic patterns elucidated by classical methods (above) are not easily integrated with the most general mechanism of biological diversification, allopatric speciation (Mayr 1963, Futuyama & Meyer 1980, Lande 1980). Note that Mayr (1954, p. 16) concluded from his review of tropical echinoid distributions that 'geographic speciation is the principal, if not exclusive, speciation mechanism in most marine animals'.

In contrast, the methods of vicariance biogeography proceed directly from the prediction of allopatric speciation, that most closely related species are initially allopatric (e.g. Rosen 1975, Nelson & Platnick 1981, Cracraft & Prum 1988). Vicariance methods seek to identify geographic events of the past by combining the distributions of species with the most fundamental component of biological history, the phylogenetic relationships among taxa (cladograms). The barriers to dispersal (and gene flow) that have fostered allopatric speciation are inferred to be located between allopatric sister taxa, and the temporal sequence among barriers may be inferred by the relative recency of common ancestry (i.e. the sequence of cladistic relationships). The methods of vicariance (cladistic) biogeography are applied here in an analysis of chaetodontid species.

The Chaetodontidae are a diverse family of marine fishes that, by virtue of their bright color patterns, have attracted a great deal of scientific and popular attention. The most widely followed taxonomy of the family was erected by Burgess (1978). Using classical systemic methods, he recognized 114 species in 10 genera, nine of which were not subdivided further. He partitioned the largest genus, *Chaetodon* (90 species), into 13 subgenera, to yield a total of 22 higher taxa. He also recognized species pairs and complexes within larger taxa, but did not consider them to be sufficiently differentiated to warrant formal recognition. Allen (1980) followed Burgess's taxonomy for the sake of stability (his disagreements were apparently minor), and indicated that he also agreed with most of Burgess's informally expressed species complexes. An analysis of chaetodontid osteology (Blum 1988) has shown that these species complexes either correspond to groups that can be diagnosed by osteological synapomorphies, or are contained within larger monophyletic groups (that cannot be resolved further for a lack of osteological differentiation; Fig. 1).

In his discussion of chaetodontid biogeography, Burgess (1978) reported that patterns of geographic variation within species are repeated, and that distributions of species within closely related groups (as he perceived them) often conform to the same patterns. Although he did not use terms such as distributional track, vicariance, or allopatric speciation, he recognized evidence for these phenomena (Burgess 1978, pp. 763–764):

'The trend is therefore to a broad Indo-West Pacific distribution for species, species pairs, or even closely related species groups. There is also a trend for some differentiation between the Indian Ocean and Pacific Ocean (including the East Indies) forms at a specific or subspecific level, as well as differentiation between the Indo-West Pacific forms and the Red Sea form, usually however, at the specific level. (. . .) There are several groups of [allopatric] species, sometimes equivalent to an entire subgenus, that include closely related species which, when considering their combined ranges, possibly indicate the range of an ancestral species (or series of species) that has fragmented into the currently accepted species.'

In his brief treatment of chaetodontid biogeography, Allen (1980) followed the classical paradigm, tabulating distributions of species and diversities without regard for the relationships among species. Allen also provided a table of the presence or absence of each species in 38 geographic regions. Because all taxa were listed alphabetically in this table, the distributional patterns among closely related species (noted by Burgess 1978) were not readily apparent. In Table 1, the distributional data of Allen (1980) are updated and the order of species rearranged to reflect genera, subgenera, and putative species complexes.

Almost all of the sister groups inferred by osteology (Fig. 1) are broadly sympatric (see Table 1 for members of terminal groups and their distributions). Thus, most of the geography associated with early chaetodontid evolution has been obscured by subsequent dispersal. The only exception to the pattern of broad sympatry is the dichotomy (node 8, Fig. 1) between *Chaetodon* sensu stricto and the monophyletic group containing all other species of *Chaetodon* sensu lato. The species of *Chaetodon* sensu stricto all have American (including the east Pacific) or west African distributions, except *C. marleyi,* which is found off South Africa from Delagoa Bay to Lamberts Bay (Burgess 1978). The unnamed sister group (node 9, Fig. 1) is predominantly Indo-Pacific. Only two members are found

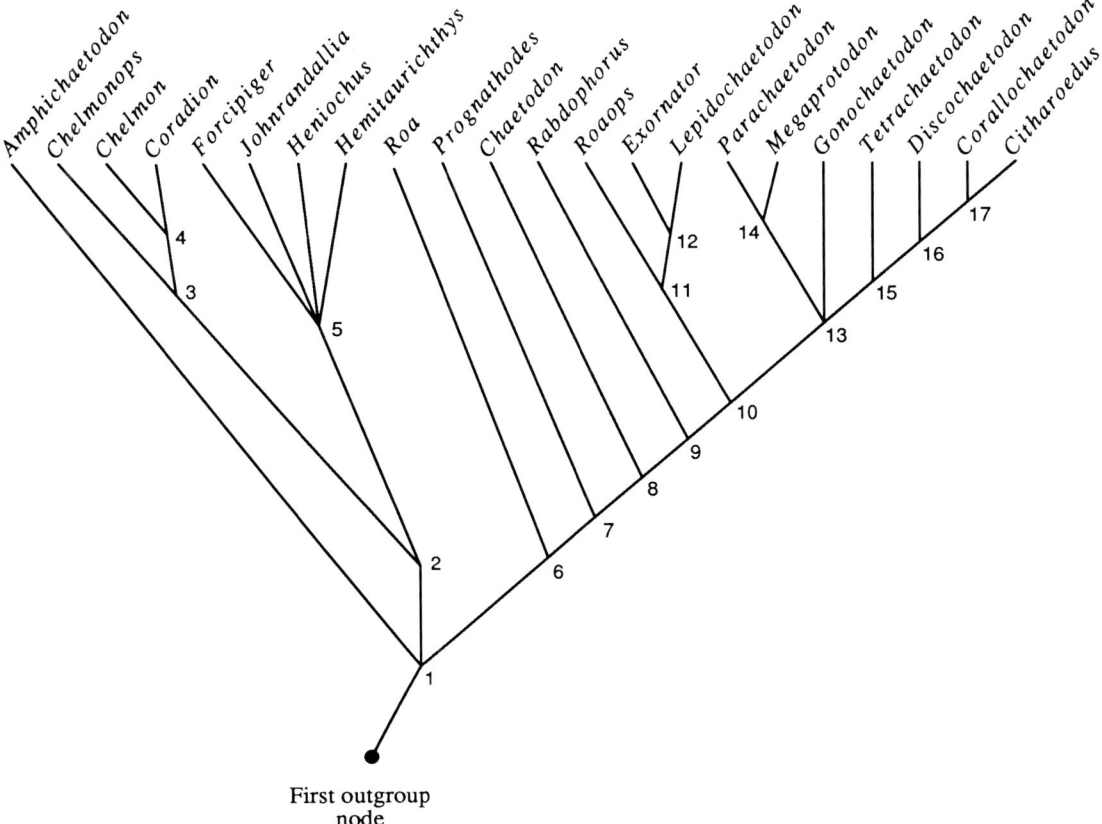

Fig. 1. Phylogenetic relationships among the valid genera and subgenera of the Chaetodontidae (after Blum 1988). Relationships were derived from a cladistic (parsimony) analysis of 34 variable features of osteology and internal soft anatomy (comprising 50 binary and two multistate unordered characters). Characters were polarized by outgroup comparison (Maddison et al. 1984; relationships among outgroups taxa given in Blum 1988). Twelve distinct trees are equally parsimonious (consistency index = 0.659). The relationships shown here are common to all twelve trees. The numbered internal nodes refer to un-named monophyletic groups.

in the Atlantic, *C. sanctaehelenae* (from St. Helena and Ascension Islands) and *C. sedentarius* (from the Caribbean), and both are members of a species complex within a relatively derived subgroup (*Exornator*, Fig. 1). The geological event most commonly associated with the splitting of Atlantic and Indo-Pacific sister-groups is the closing of the Tethyan (or Paratethyan) seaway, which separated the Mediterranean (and thereby the Atlantic) and Indian Ocean. Age estimates of this event range from 18 to 13 million years ago (mya) (early to middle Miocene; Berggren & Hollister 1974, Steininger & Rögl 1979, Bizon 1985, Müller 1985). If this cause and effect are correctly correlated, a minimum age may be inferred for all speciation events below node 8 in the cladogram (Fig. 1). A strong latitudinal temperature gradient has limited exchange between the tropical faunas of the Indian and Atlantic Oceans by way of southern Africa since the development of the circum-Antarctic current 38 mya (Eocene-Oligocene boundary, Kennett 1978). The three species that create sympatry between the two larger clades are among the more deep-dwelling and cold-tolerant chaetodontids. These habits probably facilitated their dispersal (or dispersion) around the southern tip of Africa.

Table 1 also shows that a significant amount of biogeographic information is present within the 31 putative species complexes (including species pairs). (Monotypic taxa and species listed as in-

Table 1. Distributions of chaetodontid species (after Allen 1980). Species are listed alphabetically within complexes, higher taxa are arranged phylogenetically (after Blum 1988, see Fig. 1). Species pairs and complexes are numbered. An uppercase letter in parentheses after each numbered group indicates whether the contained species show an allopatric, sympatric, or mixed pattern of distribution. Under the regional columns, an X indicates the presence of a species. Other letters refer to subregions. Regions: ArG – Arabian Gulf, Ber – Bermuda, CVI – Cape Verde Islands, EAf – East Africa, EPc – Eastern Pacific, EsI – Easter Island, Fij – Fiji, Fla – Florida, GBR – Great Barrier Reef, GMa – (G = Guam and Marianas, M = Marshall Islands, P = Palau, X = all), GMx – Gulf of Mexico, Haw – Hawaiian Islands, Ind – India, LHI – Lord Howe Island, Maq – Marquesas, MIn – Malaysia and Indonesia, Mld – Maldives, MMR – Madagascar Mauritius and Reunion, NCb – Northern Caribbean, NGi – New Guinea, Phi – Philippines, PiR – (P = Pitcairn, R = Rapa, X = both), RdS – Red Sea, Ryu – Ryukyu Islands, SAf – South Africa, Sam – Samoa, SAs – Southern Australia (W = West, E = East), SCb – Southern Caribbean, Sey – Seychelles, SAP – (S = Saint Helena Island, A = Ascencion Island, P = Saint Paul's Rocks), SJp – Southern Japan, SLa – Sri Lanka, SmN – Solomon Islands and New Caledonia, STu – (S = Society Islands, T = Tuamotus, X = both), Taw – Taiwan, Thi – Thailand, WAf – West Africa, WAs – Western Australia.

Taxon	SAf	EAf	MMR	Sey	RdS	ArG	Mld	SLa	Ind	Thi	WAs	MIn	Phi	Taw	Ryu	SJp	NGi	GBR	LHI	SAs	Sam	Fij	STu	SmN	PiR	EsI	Haw	GMa	Maq	EPc	GMx	Fla	Ber	NCb	SCb	CVI	WAf
(1) Genus *Amphichaetodon* (A)																																					
Amphichaetodon howensis																		X	X																		
Amphichaetodon melbae																											X										
(2) Genus *Chelmonops* (A)																																					
Chelmonops curiosus[1]																				W																	
Chelmonops trucatus																				E																	
(3) Genus *Chelmon* (S)																																					
Chelmon marginalis													X					X																			
Chelmon mulleri																		X																			
Chelmon rostratus		X	X						X	X		X	X	X	X		X	X		X																	
(4) Genus *Coradion* (S)																																					
Coradion altivelis														X		X																					
Coradion chrysozonus											X	X	X	X	X			X																			
Coradion melanopus												X						X																			
(5) Genus *Forcipiger* (S)																																					
Forcipiger flavissimus	X	X	X	X	X		X	X	X	X	X	X	X	X	X	X	X	X			X	X	X	X		X	X	X	X	X							
Forcipiger longirostris				X										X	X						X	X			X	P			X	X							
Genus *Johnrandallia*																																					
Johnrandallia nigrirostris																														X							
(6) Genus *Heniochus* (S)																																					
Heniochus acuminatus	X	X	X	X			X	X	X	X	X	X	X	X	X	X	X	X			X	X	X	X				X									
Heniochus chrysostomus									X	X	X	X	X	X	X		X	X			X	X	X	X	P		X										
Heniochus diphreutes[2]	X			X		X	X			X			X			X					X						X										
Heniochus intermedius					X																																
Heniochus monocerus	X	X	X	X				X	X			X	X	X	X	X	X	X			X	X	X	X			X										
Heniochus pleurotaenia				X	X	X	X																														
Heniochus singularis											X	X	X	X	X	X	X	X						X													
Heniochus varius												X	X	X	X	X	X	X			X	X	X														
Genus *Hemitaurichthys*																																					
(7) *H. polylepis* spp. grp. (A)																																					
Hemitaurichthys polylepis												X		X	X	X	X	X	X	X	X	X	X		X		X	X	X	P				X			
Hemitaurichthys zoster	X	X	X	X			X	X	X																												
(8) *H. thompsoni* spp. grp. (A)																																					
Hemitaurichthys multispinosus																									P												
Hemitaurichthys thompsoni																											X										
(9) Genus *Prognathodes* (A)																																	X	X		X	
Prognathodes aculeatus																																	X	X		X	
Prognathodes aya																															X	X		X			
Prognathodes dichrous																																				S	
Prognathodes falcifer																														X							
Prognathodes guezei		X																																			
Prognathodes guyanensis																															X	X					
Prognathodes guyotensis[3]										X				X																							
Prognathodes marcellae																																			X	X	
Prognathodes obliquus[4]																																				P	

Table 1. (Continued).

Taxon	Region
	SAf EAf MedR SRy RSG Ad Md Sa Ini Tha Iwu Map Pi TRI RjR SGI NBj GHm Lau SMa SPM Eel GCx HAr MbP EbI GCC FAV BCC WAf
(10) Genus *Roa*[5] (A)	
Roa excelsa	. . . . X X . . . . . .
Roa jayakari	. . . X X . . X . . . .
Roa modestus	. . . . X . X X X . . . .
Genus *Chaetodon*	
(11) Subgenus *Chaetodon* (M)	
Chaetodon capistratus	. . . . X X X X X . . .
Chaetodon hoefleri	 X
Chaetodon humeralis	. . . . X . . . .
Chaetodon marleyi	X . . . .
Chaetodon ocellatus	. . . . X X X X X . . .
Chaetodon robustus	. . . . X X
Chaetodon striatus	. . . . X X X X X . . .
Subgenus *Rabdophorus*	
incertae sedis	
Chaetodon mesoleucos	. . . . X . . . .
Chaetodon nigropunctatus	. . . . . X . . . .
Chaetodon rafflesi	. . . . X X X . . X X X X X X X . . X X X X . . . . .
Chaetodon semeion	. . . . X . . . X X X X . . X X . X X X X . X . . . .
(12) *Chaetodontops* spp. grp. (S)	
Chaetodon adiergastos	. . . . X X X X X . . . .
Chaetodon auripes	. . . . X X X X . . . .
Chaetodon collare	. . . . . . X X X X . . X X . . . .
Chaetodon flavirostris	. . . . X X . X X X . X . . . . . .
Chaetodon semilarvatus	. . . . X . . . .
Chaetodon wiebeli	. . . . X . X X X X X . . . .
(13) *C. auriga* spp. grp. (S)	
Chaetodon auriga	X X . X X X X X X X X X X X X X X X X X . X X X X R . X X X
Chaetodon decussatus	. . . . . . X X X X . X . . . .
Chaetodon vagabundus	X X X X X . . X X . X . X X X X X X X X . X X X X . X . . .
(14) *C. ephippium* spp. grp. (A)	
Chaetodon ephippium	. . . . X X X X X X X X X X X . . . X X X X R . X X X
Chaetodon xanthocephalus	X X X X . . X X . . . .
(15) *C. falcula* spp. grp. (A)	
Chaetodon falcula	. X X X . . X X X . . . .
Chaetodon ulietensis	. . . . X X X X X X X X X X X . . X X X X . X . . .
(16) *C. lineolatus* spp. grp. (S)	
Chaetodon lineolatus	. X X X X . . X X X X X X X X X X X X . . X X X X . X . . .
Chaetodon oxycephalus	. . . . X X . . X X . . . . X . . . .
(17) *C. lunula* spp. grp. (A)	
Chaetodon fasciatus	. . . . X . . . .
Chaetodon lunula	X X X X . . X X X X X X X X X X X X X X . X X X X . X X X .
(18) *C. melannotus* spp. grp. (S)	
Chaetodon melannotus	. X X X X . X X X X . X X X X X X X X . . X X X X . X . . .
Chaetodon ocellicaudus	. . . . X X . . X . . . X . . . X . . . .
(19) *C. selene* spp. grp. (A)	
Chaetodon gardneri	. . . . . X . X . . . .
Chaetodon leucopleura	. X . X X . . . .
Chaetodon selene	. . . . X X X X X X . . . .
Subgenus *Roaops*[6]	
incertae sedis	
Chaetodon nippon	. . . . X . . . .
(20) *C. tinkeri* spp. grp. (A)	
Chaetodon burgessi	. . . . P . . . .
Chaetodon declivis	. . . . X . . . .
Chaetodon flavocoronatus[7]	. . . . G . . . .
Chaetodon mitratus	. . X . . . X . . . .
Chaetodon tinkeri	. . . . M X . . . .

Table 1. (Continued).

```
Taxon                          Region
                               S E M S R A M S I T W M P T R S N G L S S F S S P E G H M E G F B N S S C W
                               A A M e d r l L n h A I h a y J G B H A m i a T i s M a a P M l e C C A V A
                               f f R y S G d a d i s n i w u p i R I s N j m u R I a w q c x a r b b P I f

Subgenus Exornator
  incertae sedis
    Chaetodon citrinellus      . X X X . . X X X X X X X X X X X X X X X . . X X X X . . . X X X . . . . .
    Chaetodon daedalma         . . . . . . . . . . . . X X . . . . . . . . . . . . . . . . . . . . . . . .
    Chaetodon litus            . . . . . . . . . . . . . . . . . . . . . . . . . X . . . . . . . . . . . .
    Chaetodon quadrimaculatus  . . . . . . . . . . . . X . . . . . . . . . . . X X P . X X . . . . . . . .
    Chaetodon smithi           . . . . . . . . . . . . . . . . . . . . . . . . . X . . . . . . . . . . . .
(21) Rhombochaetodon spp.
     grp. (A)
    Chaetodon argentatus       . . . . . . . . . . . X X X X . . . . . . . . . . . . . . . . . . . . . . .
    Chaetodon madagascariensi  X X X X . . X X . . . . . . . . . . . . . . . . . . . . . . . . . . . . . .
    Chaetodon mertensii        . . . . . . . . . . X X X X . . X X X . X X X X R . X . . . . . . . . . . .
    Chaetodon paucifasciatus   . . . . X . . . . . . . . . . . . . . . . . . . . . . . . . . . . . . . . .
    Chaetodon xanthurus        . . . . . . . . . . . X X X X . . . . . . . . . . . . . . . . . . . . . . .
(22) C. fremblii spp. grp. (A)
    Chaetodon blackburnii      X X X . . . . . . . . . . . . . . . . . . . . . . . . . . . . . . . . . . .
    Chaetodon fremblii         . . . . . . . . . . . . . . . . . . . . . . . . . . . X . . . . . . . . . .
(23) C. miliaris spp. grp. (A)
    Chaetodon assarius         . . . . . . . . . X . . . . . . . . . . . . . . . . . . . . . . . . . . . .
    Chaetodon dolosus          X X X . . . . . . . . . . . . . . . . . . . . . . . . . . . . . . . . . . .
    Chaetodon guntheri         . . . . . . . . . . . X . X X X X X X X . . . . . . . . . . . . . . . . . .
    Chaetodon miliaris         . . . . . . . . . . . . . . . . . . . . . . . . . . . X . . . . . . . . . .
    Chaetodon sanctaehelenae   . . . . . . . . . . . . . . . . . . . . . . . . . . . . . . . . . . X . . .
    Chaetodon sedentarius      . . . . . . . . . . . . . . . . . . . . . . . . . . X X . X X . . . . . . .
(24) C. punctatofasciatus spp
     grp. (A)
    Chaetodon guttatissimus    X X X X . . X X X X . . . . . . . . . . . . . . . . . . . . . . . . . . . .
    Chaetodon multicinctus     . . . . . . . . . . . . . . . . . . . . . . . . . . . X . . . . . . . . . .
    Chaetodon pelewensis       . . . . . . . . . . . . . X X X . X X X X X . . . X . . . . . . . . . . . .
    Chaetodon punctatofasciatus. . . . . . . . . X X X X X X X X X . . X . . . X . . . . . . . . . . . . .
Subgenus Lepidochaetodon
  incertae sedis
    Chaetodon unimaculatus     X X X X . . X X X X X X X X X X X X X X X X . X X X X . . . X X X . . . . .
(25) C. kleinii spp. grp. (A)
    Chaetodon kleinii          X X X X . . X X X X X X X X X X X X X X X X . X X X T . . X X . . . . . . .
    Chaetodon trichrous        . . . . . . . . . . . . . . . . . . . . . S . . . . . . X . . . . . . . . .
(26) Subgenus Megaprotodon
     (S)
    Chaetodon oligacanthus⁸    . . . . . X X X X X X X . . . X X . . . . X X . . . . . . . . . . . . . . .
    Chaetodon trifascialis     X X X X X . X X X X . X X X X X X X X X X X . X X X X R . X X . . . . . . .
(27) Subgenus Gonochaetodon
     (A)
    Chaetodon baronessa        . . . . . . . . . X X X X X . . X X . . X X . . . . . . . . . . . . . . . .
    Chaetodon larvatus         . . . . X . . . . . . . . . . . . . . . . . . . . . . . . . . . . . . . . .
    Chaetodon triangulum       . X . X . . X X X X . . . X . . . . . . . . . . . . . . . . . . . . . . . .
(28) Subgenus Tetrachaetodon
     (S)
    Chaetodon bennetti         X X . X . . X X X X X . X X X X X X X X X X . X X X X R . X . . . . . . . .
    Chaetodon plebeius         . . . . . . . X . X X X X X X X X X X X . X X . . . . . . . . . . . . . . .
    Chaetodon speculum         . . . . . . . . . X X X X X X X X X . . . . . . . . . . . . . . . . . . . .
    Chaetodon zanzibariensis   X X X . . . . . . . . . . . . . . . . . . . . . . . . . . . . . . . . . . .
(29) Subgenus Discochaetodon
     (S)
    Chaetodon aureofasciatus   . . . . . . . . . X . . X . . . . X . . . . . . . . . . . . . . . . . . . .
    Chaetodon rainfordi        . . . . . . . . . . . . . . . . . X X X . . . . . . . . . . . . . . . . . .
    Chaetodon octofasciatus    . . . . . . . . . X X X . X X X X X . . . . . . . . . . . . . . . . . . . .
    Chaetodon tricinctus       . . . . . . . . . . . . . . . . . . X . . . . . . . . . . . . . . . . . . .
(30) Subgenus Corallochaeto-
     don (A)
    Chaetodon austriacus       . . . . X . . . . . . . . . . . . . . . . . . . . . . . . . . . . . . . . .
    Chaetodon melapterus       . . . . X . . . . . . . . . . . . . . . . . . . . . . . . . . . . . . . . .
    Chaetodon trifasciatus     . X X X . . X X X X X X X X X X X X X X X X . X X X X R . X X . . . . . . .
```

Table 1. (Continued).

Taxon	Region
	S E M S R A M S I T W M P T R S N G L S S F S S P E G H M E G F B N S S C W
	A A M e d r l L n h A I h a y J G B H A m i a T i s M a a P M l e C C A V A
	f f R y S G d a d i s n i w u p i R I s N j m u R I a w q c x a r b b P I f

(31) Subgenus *Chitharoedus* (S)

    Chaetodon meyeri   X X . X . . X X X X X X X X X . X X . . . X . . . . . X X . . . .

    Chaetodon ornatissimus   . . . . X . . X X X X X X X X X . X X X X . . X X X . . . .

    Chaetodon reticulatus   . . . . X X X X X X . X X X X P . X X . . . .

[1] Kuiter (1986).
[2] Allen & Kuiter (1978).
[3] Species described by Yamamoto & Tameka, in Okamura et al. (1982). Range extended by Randall & de Bruin (1988).
[4] Lubbock & Edwards (1980).
[5] The generic rank of *Roa* follows Blum (1988). The recognition of *Roa* (= *Chaetodon*) *excelsa* Jordan, and *R. jayakari* (Norman) as distinct from *R. modestus* (Schlegel), and their respective distributions follow Klausewitz & Fricke 1985.
[6] Maugé & Bauchot (1984).
[7] Meyers (1980).
[8] *Chaetodon oligacanthus* Bleeker 1850 = *Parachaetodon ocellatus* (Cuvier) 1831, after Blum (1988).

certae sedis were not considered to be members of species complexes). In 18 of the 31 complexes, the contained species are either allopatric or only marginally sympatric (59 species). In 12 of the remaining 13 complexes the species are broadly sympatric (42 species). The 7 species of *Chaetodon* sensu stricto show a mixed pattern; two are isolated endemics, three are sympatric in the Caribbean, and two are sympatric in west Africa. These data suggest that the barriers responsible for many of the recent speciation events within the Chaetodontidae are still intact, and can be located by comparing the distributions of species that are thought to be closely related.

Table 1 is, however, only a crude representation of species distributions. Similarly comprehensive, but more precise data have not been published. Neither Burgess (1978) nor Allen (1980) presented species distributions as 'spot-maps', and neither provided a sufficient number of locality records to completely verify the distributions presented in verbal or tabular format. Moreover, the putative species complexes that are presented in Table 1 have not been tested by cladistic methods. The purposes of this study are thus to provide more reliable estimates of chaetodontid species distributions, and to interpret these within the context of cladistically supported relationships. However, confirming distributions from positively identified material can be done only as one has the opportunities to examine museum collections. At present, I have been able to examine the collections in only three major museums. Consequently, I consider the distributions of species within only 6 of the 18 allopatric complexes to be sufficiently known to warrant further discussion. (Two pairs of sympatric species are also discussed below as their distributions appear to have resulted from the same geographic events.)

Methods

Locality data were derived primarily from specimens held in the United States Natural History Museum (USNM), the Academy of Natural Sciences, Philadelphia (ANSP), and the American Museum of Natural History, New York (AMNH). Identification of specimens from odd or critical localities was confirmed by the author or a curator aware of the diagnostic characters in closely related species. These data were supplemented by recent checklists, where a high degree of confidence can be placed in specimen identifications (Russell 1983, Gloerfelt-Tarp & Kailola 1985, Randell et al. 1985), and published photographs that were accompanied by specific locality information (Allen 1980, Amesbury & Meyers 1982, Masuda 1984).

Species distributions are presented as spot-maps. A complete list of localities is available from the author.

Species pairs and complexes were inferred to be monophyletic by cladistic analyses of osteology (Blum 1988) and color pattern (see Burgess 1978, Steene 1979, and Allen 1980 for color pictures of all species discussed below). Apomorphic (derived) features are reviewed for each complex whose geography is discussed.

In most cases, the minor differences in color pattern that exist among species within a complex cannot be polarized. Therefore, the sister group relationships within these complexes are unknown. Cladistic biogeography can identify a barrier as that which caused or promoted allopatric speciation (whether allopatry arose by vicariance or dispersal) only when adjacent taxa are shown to be sister taxa. (My usage of 'adjacent' in this discussion is not intended to mean parapatric, but rather nearest neighbor in a given direction, regardless of the distance between taxa.) If the relationships within a monophyletic group are unresolved, it is possible that the actual relationships are such that some sister taxa are not geographically adjacent, but separated by one or more species of the same complex that have a more distant cladistic relationship (Fig. 2, case 4). However, the only scenarios that generate such distributions require either extinction of one or both sister taxa in the area occupied by the geographically intermediate species, or 'leap-frog' dispersal (across the intermediate area without the establishment of populations there). Simple extinctions in areas between sister taxa are not assumed to be rare, and are included within the broader concept of barrier formation. All scenarios that produce non-adjacent sister taxa are more complicated (less parsimonious) than those requiring only the development or existence of barriers between populations, and evolution in situ. The inferences drawn here are based on the assumptions that such coordinated extinctions and 'leap-frog' dispersal are rare phenomena.

It could be argued that the second scenario depicted in case 4 (Fig. 2), involving step-wise dispersal and coordinated extinction, is predicted by competitive exclusion, and may not be rare. It should be noted, however, that this scenario involves a number of assumptions beyond the operation of competitive exclusion. First, the barriers that separated the initial two species must become permeable to AC (the ancestor of A and C), but must remain impermeable to B. Additionally, competitive exclusion must not operate during this phase so that AC can colonize the more distant unoccupied habitat (by step-wise dispersal). Then the barriers must again become impermeable to AC, and competitive exclusion must operate such that AC goes extinct, but not B. Moreover, the large number of regionally sympatric and closely related butterflyfishes (cf. Table 1) implies that if competitive exclusion operates among these species at all, it does so at level of habitats within regions, and not among regions.

If 'leap-frog' dispersal and coordinated extinctions are rare, it may be inferred that geographically adjacent species are sister taxa (or members of sister taxa), and that the barriers currently separating these species are those responsible for the initial prevention or interruption of gene flow.

Results

Corallochaetodon Burgess 1978

A sister group relationship between the monophyletic subgenera *Corallochaetodon* and *Citharoedus* Kaup 1860 (Fig. 1, node 17) is indicated by their derived jaw and tooth morphologies (Motta 1985, 1988, Blum 1988). *Corallochaetodon* is more derived than *Citharoedus* in having a palatine base that is higher and shorter. The three species of *Corallochaetodon* share a background color that is yellow to orange and overlaid by a series of 14–16 dark (blue-black) stripes that are nearly horizontal (sightly inclined posteriorly). The more dorsal and ventral stripes are convex towards the periphery. A generally similar striping pattern occurs in two species of *Tetrachaetodon* (the third outgroup to *Corallochaetodon*), *Chaetodon plebeius* and *C. zanzibariensis*. In these fishes the stripes are more numerous, more horizontal, straight, and have jagged edges (formed by rows of contiguous dots).

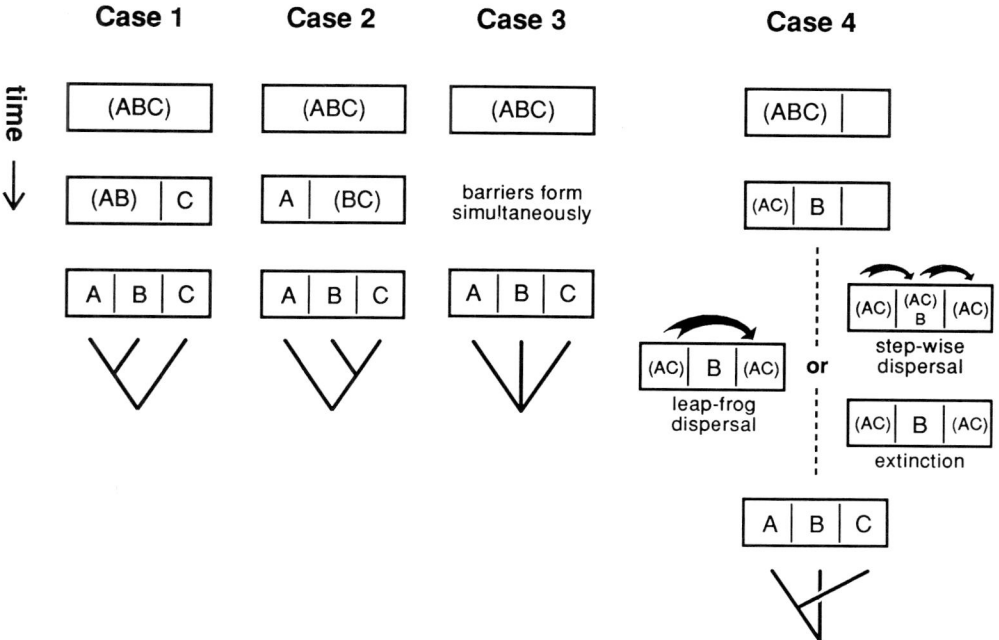

Fig. 2. A comparison of scenarios that generate allopatry among three species descended from a common ancestor. Suitable habitat is enclosed within the rectangle. Barriers are represented as vertical lines within rectangles. Ancestors are represented as their descendants in parentheses. For example, (AB) represents the ancestor of species A and B. If the relationships among species are unknown, any one of four possibilities may represent the true cladistic relationships among species. Scenarios that involve only the development of barriers and evolution in situ always generate sister taxa that are adjacent. Scenarios that generate non-adjacent sister taxa (case 4) require either 'leap-frog' dispersal, or step-wise dispersal followed by extinction in the intermediate area.

The dissimilarities and the relative distance of relationship indicate that the two striping patterns are convergent. Thus, the pattern in *Corallachaetodon* is uniquely derived.

The type species of *Corallachaetodon*, *Chaetodon trifasciatus* Park 1797, has narrow, posteriorly expanding, black markings along the bases of the dorsal and anal fins, and a vertical black bar in the caudal fin. Similar features in *C. meyeri* Schneider 1801 and *C. ornatissimus* Cuvier 1831 (members of *Citharoedus*) indicate that these attributes are plesiomorphic in Corallochaetodon. (The color patterns in the third species of *Citharoedus*, *Chaetodon reticulatus* Cuvier 1831, and the species of *Discochaetodon* are not comparable.) The other two species of *Corallochaetodon*, *Chaetodon austriacus* Rüpell 1835 and *C. melapterus* Guichenot 1862, have anal and caudal fins that are completely black except for fringing white highlights. This single derived feature suggests that these two species are more closely related to each other than either is to *C. trifasciatus*.

Chaetodon trifasciatus is widely distributed in the Indo-Pacific, *C. austriacus* is endemic to the northern half of the Red Sea, and *C. melapterus* is endemic to the Persian Gulf and southern Arabian peninsula, including the Gulf of Aden (Fig. 3). Burgess (1978, p. 491) reported that *C. melapterus* might be sympatric with *C. austriacus* in the southern Red Sea, and with *C. trifasciatus* in the western Indian Ocean. However, I have not yet seen records that confirm this sympatry.

Chaetodon kleinii complex: *C. kleinii* Bloch 1790, and *C. trichrous* Günther 1874

Derived morphologies of the first dorsal pterygiophore, vomer, and hypurapophysis infer that *C. kleinii*, *C. trichrous*, and *C. unimaculatus* Bloch 1787 form the monophyletic subgenus, *Lepidochaetodon* Bleeker 1876 (Blum 1988). Color pat-

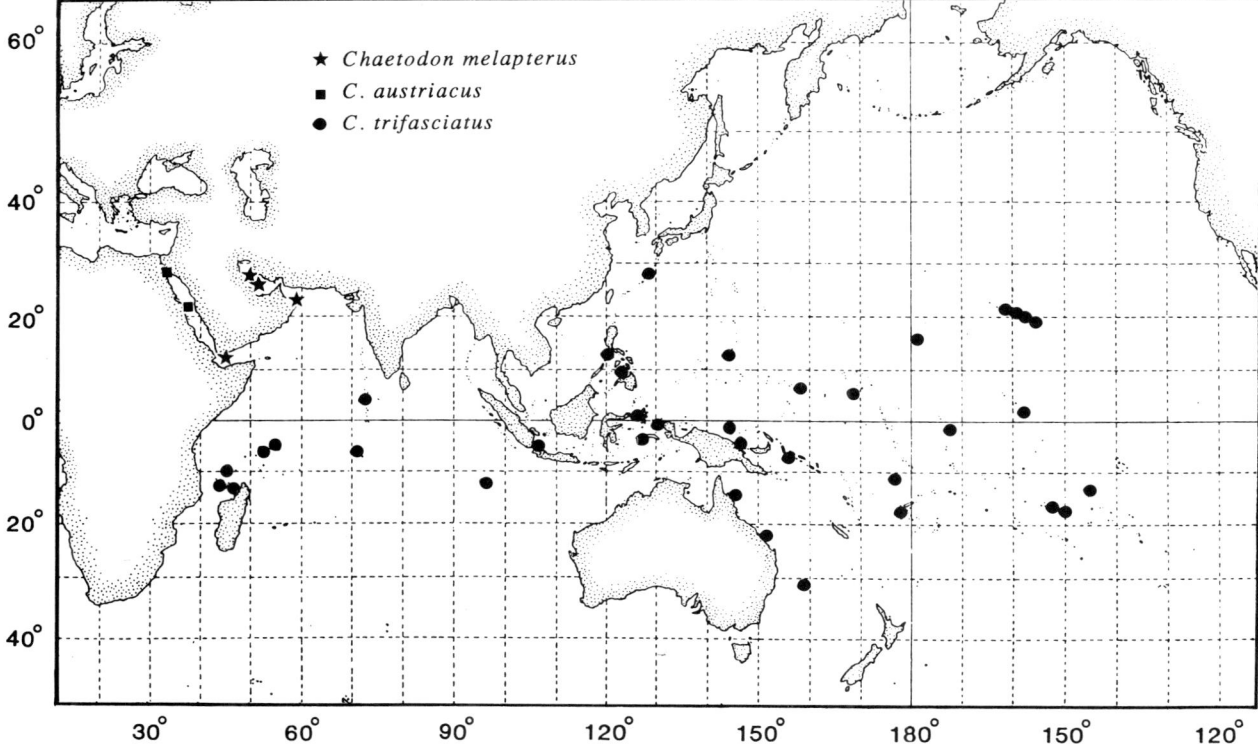

Fig. 3. Distributions of species comprising the subgenus *Corallochaetodon*.

tern indicates that *C. kleinii* and *C. trichrous* form a pair of allopatric sister species. They are perhaps the most drably colored chaetodontids. The body is a dull cream color from the eyebar posteriorly to the pectoral fin, then changes quickly, but without sharp contrast, to an orange-brown. (*Chaetodon kleinii* differs from *C. trichrous* in having a second vertical band of lighter coloring between the end of the pectoral fin and the origin of the anal fin.) Both species are further characterized by a narrow white band on the caudal peduncle, and an eyebar that curves posteriorly below the eye and extends to the origin of the pelvic fins. In all species closely related to the *kleinii* complex (*C. unimaculatus*, and the species of *Exornator* Nalbant 1971, and *Roaops* Maugé & Bauchot 1984), the eyebar terminates anterior to the pelvic fins.

Chaetodon kleinii is widely distributed throughout the Indo-Pacific, but does not occur in the Society Islands or the Marquesas (Fig. 4). *Chaetodon trichrous* was previously considered to be endemic to the Society Islands, but has been collected in the Marquesas (Bernice Pauahi Bishop Museum, lots 10991 and 11743). The boundary between these two species is confused by a record of *C. kleinii* from Makemo lagoon in the Tuamotus (USNM 133829). If this old specimen (1899) was labelled correctly, and the record accurately indicates the range of *C. kleinii*, then polygons circumscribing these distributions would overlap. This would indicate that one (or both) has dispersed since their differentiation (although the two species are not sympatric in any one island group). However, *C. kleinii* has not been collected subsequently in the Tuamotus. The odd USNM specimen may thus represent a waif or labelling error in the field.

Chaetodon falcula complex: *C. falcula* Bloch 1793, and *C. ulietensis* Cuvier 1831

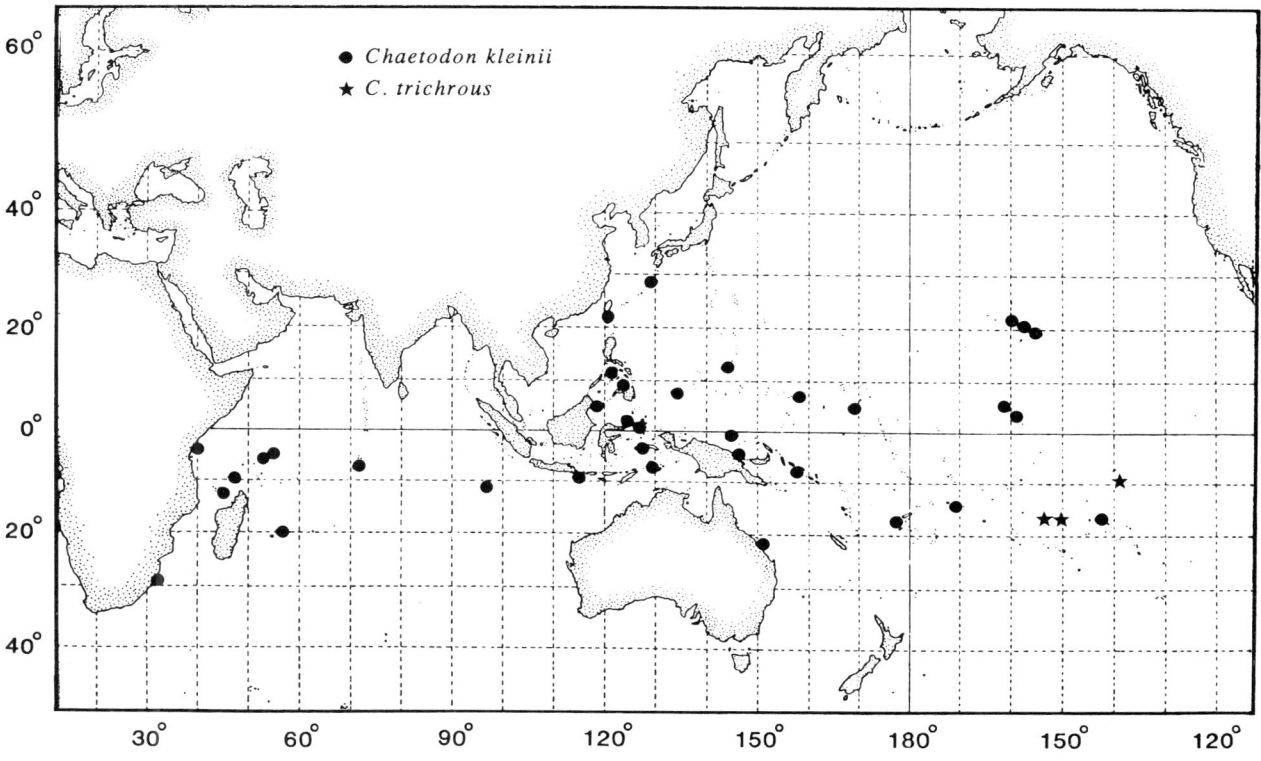

Fig. 4. Distributions of species in the *Chaetodon kleinii* complex.

The skeletal morphologies of *Chaetodon falcula* and *C. ulietensis* show that they are members of a large, possibly paraphyletic (lacking osteological apomorphies), but phenetically homogeneous group of species. This group encompasses two of Burgess's (1978) subgenera, *Rabdophorus* Swainson 1839 and *Chaetodontops* Bleeker 1876. Several species within this group have silvery-white bodies and at least partially yellow dorsal, anal, and caudal fins. In four species, *Chaetodon falcula, C. ulietensis,* and two others (discussed below), these characteristics are combined with a series of 15 to 18 thin vertical black lines. *Chaetodon falcula* and *C. ulietensis* are united as sister species by the presence of two black, V-shaped, dorsal saddles, which are unique among chaetodontids.

Chaetodon falcula is endemic to the western Indian Ocean, whereas *C. ulietensis* occurs widely throughout the central and western Pacific (Fig. 5). Significantly (see discussion), *C. ulietensis* also occurs in the eastern Indian Ocean at Cocos-Keeling and Christmas Islands, whereas *C. falcula* reaches its eastern-most limit at the Chagos and Laccadive Islands.

Chaetodon punctatofasciatus complex: *C. punctatofasciatus* Cuvier 1831, *C. guttatissimus* Bennett 1828, *C. multicinctus* Garrett 1863, and *C. pelewensis* Kner 1868

The *punctatofasciatus* complex is inferred to be part of the large subgenus *Exornator* by three osteological characters. The two, plesiomorphically separate, predorsal bones are fused distally, the first erector dorsalis muscle passes through a foramen in the first dorsal pterygiophore, and the number of branchiostegal rays is reduced from six to five. Members of the *punctatofasciatus* complex are distinguished from other species of *Exornator* by longer, straighter jaw teeth, and by the absence of teeth on the descending process of the premaxilla.

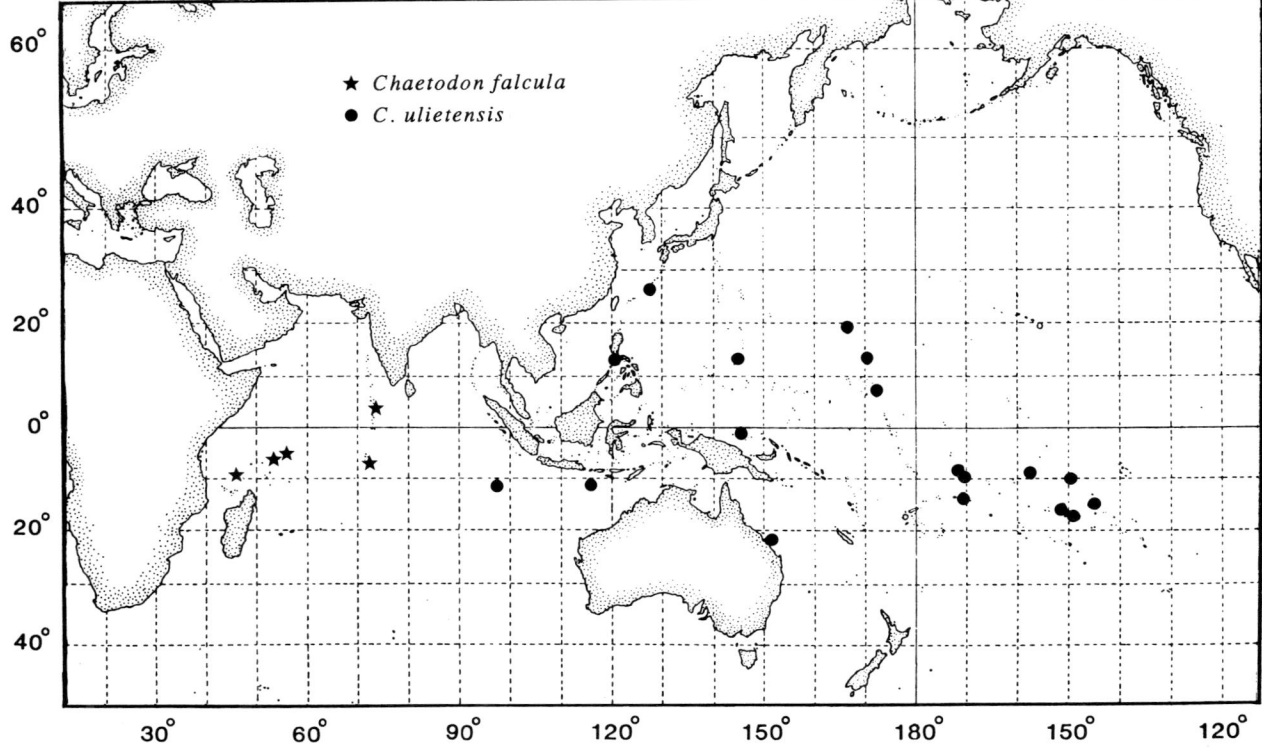

Fig. 5. Distributions of species in the *Chaetodon falcula* complex.

The *punctatofasciatus* complex can also be diagnosed by color pattern. In most species of the more inclusive spotted butterflyfish complex (Ahl 1923, i.e. *C. citrinellus* Cuvier 1831, all members of the *punctatofasciatus* complex, and most members of the *miliaris* complex) there are vertical or diagonal rows of small dark-brown dots, on a lighter (yellow or silvery-white) background. In the *punctatofasciatus* complex, the background color is a light olive-green (except in *C. multicinctus*), and the sizes of dots alternate among rows such that in rows of larger dots, the dots run together (or background color approaches dot color) forming solid dark bands, particularly in the dorsal half of the body (in *C. guttatissimus*, the blending of dots into bars is only partial). All members of the complex also have a lens-shaped black bar covering the proximal region of the caudal fin. On the caudal peduncle, anterior to the bar, the coloring is distinctly lighter than on the body (except in *C. multicinctus*, in which the body is already light), and posterior to the bar, the caudal fin is hyaline (lacking pigment). Colored or dark caudal fin bars are relatively common among chaetodontids. However, such bars are usually arc-shaped. Moreover, the cream-black-hyaline sequence is unique to the *punctatofasciatus* complex. Relationships among the four species are not resolved.

The *punctatofasciatus* complex includes *C. guttatissimus* from the Indian Ocean, *C. punctatofasciatus*, from the northwestern tropical Pacific, *C. pelewensis*, from the southern tropical Pacific, and *C. multicinctus*, from the Hawaiian and Johnston Islands (Fig. 6). Steene (1979) and Allen (1980) reported that *C. punctatofasciatus* and *C. pelewensis* are sympatric in New Guinea, the Solomon Islands and the northern Great Barrier Reef, but my preliminary survey has not produced specific locality records that document this sympatry. In contrast to the situation in the *falcula* complex, the range of

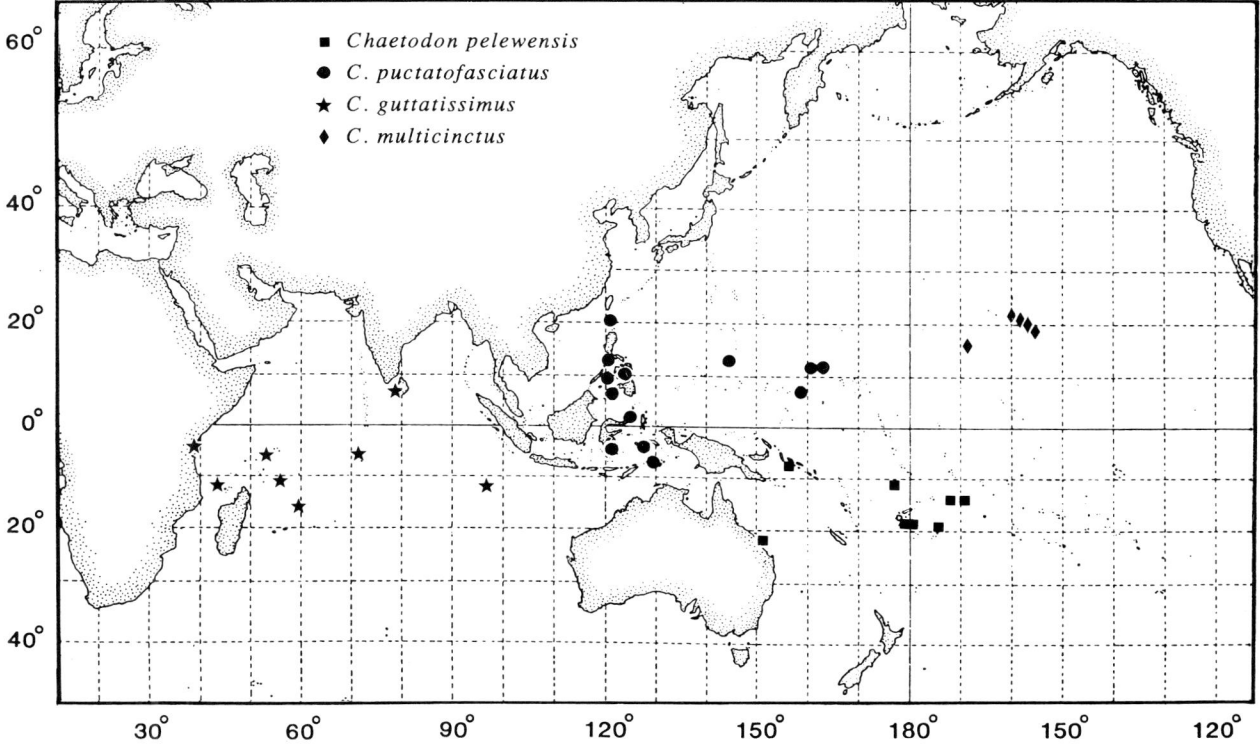

Fig. 6. Distributions of species in the *Chaetodon punctatofasciatus* complex.

the Indian Ocean member in the *punctatofasciatus* complex includes Cocos-Keeling and Christmas Island (Allen 1980, p. 189).

Rhombochaetodon Burgess 1978
The *Rhombochaetodon* species complex (or subgenus) is inferred to be a member of the larger group *Exornator* by the osteological features discussed under the *punctatofasciatus* complex (Blum 1988). The monophyly of these five species is shown by a reticulated or chevron pattern of dark lines that outline the exceptionally large, diamond-shaped scales. Additionally, all species have an identical pattern of intense yellow, orange, red, or black pigment on the soft dorsal fin rays, posterior body, and more posterior anal fin rays. The shapes of the pigmented areas are not duplicated elsewhere in the family. The caudal fin bears a vertical bar of the same color present on the posterior portion of the body.

The relationships among the five species of the *Rhombochaetodon* complex were not resolved. The distributions of these species are as follow: *Chaetodon mertensii* Cuvier 1831, from the oceanic islands of the Pacific (excluding Hawaii), New Caledonia, and northeastern Australia; *C. xanthurus* Bleeker 1857, from the Philippines and Sulawesi; *C. argentatus* Smith & Radcliffe 1911, from the northern most islands of the Philippines to southern Japan; *C. madagascariensis* Ahl 1923 from the Indian Ocean (including Cocos-Keeling); *C. paucifasciatus* Ahl 1923, from the northern Red Sea (i.e. Gulf of Aqaba, Fig. 7). Allen's table (1980) indicates that both *C. argentatus* and *C. mertensii* are sympatric with *C. xanthurus* in the Philippines, but in the text (p. 236) he states that there appears to be little overlap between *C. xanthurus* and *C. mertensii*. The extensive Philippine collections at the USNM contain only *C. xanthurus*. Individuals that have been observed outside the exclusive species

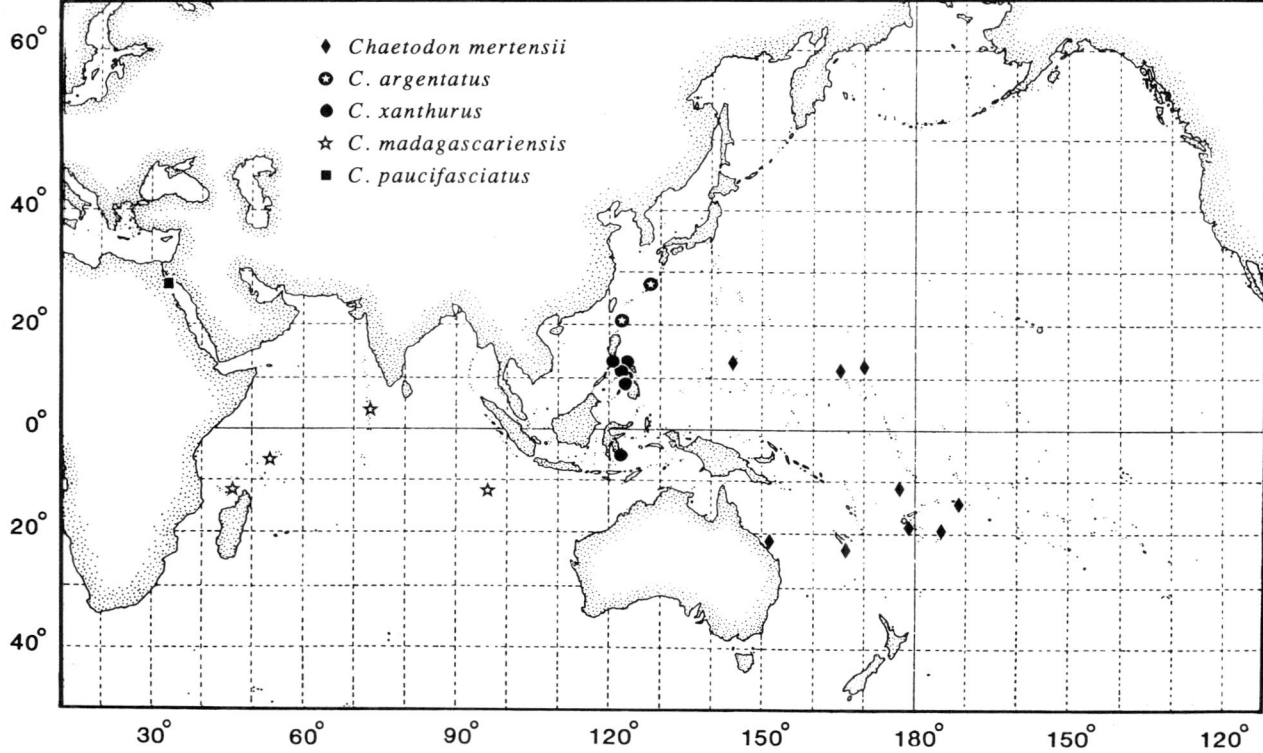

Fig. 7. Distributions of species in the *Rhombochaetodon* species complex.

range (centers) may represent waifs rather than established breeding populations.

Chaetodon miliaris 'complex': *C. miliaris* Quoy & Gaimard 1824, *C. guntheri* Ahl 1923, *C. assarius* Waite 1905, *C. dolosus* Ahl 1923, *C. sactaehalenae* Günther 1868, *C. sedentarius* Poey 1858 Burgess (1978, p. 657, 682, 684, 695, 765) and Allen (1980, p. 181, 204), both referred to a close relationship among the species of the *miliaris* complex. All of these species except *C. sanctaehelenae*, are characterized by vertical rows of small dark or black dots on the body. However, I have interpreted this feature to be plesiomorphic relative to that in the *punctatofasciatus* complex. (Similar patterns of dots also occur in *Roaops* species where the body bears no other pigment.) All species of the *miliaris* complex, except *C. miliaris,* have a silvery-white body color and yellow or dusky umber pigment on the soft dorsal and anal fin rays (*C. miliaris* is usually completely yellow, but some populations have lighter bodies). However, increasing color intensity (particularly yellow) on the median fins is one of the most general attributes of color pattern among chaetodontids. Thus, at present, no morphological features are known to infer that these species form a monophyletic group exclusive of *C. citrinellus, C. quadrimaculatus* Gray 1831, and species of the *punctatofasciatus* complex. A single ecological character, however, suggests that this complex may be monophyletic. All of these species have distributions that are either completely restricted to, or extend into deep water (ca. 100 m). *Amphichaetodon* Burgess 1978, *Prognathodes* Gill 1862, *Roa* Jordan 1923, and the *tinkeri* complex (part of the subgenus *Roaops*) have similar habits, but the phylogenetic positions of these groups indicate that this characteristic was acquired independently in the *miliaris* complex. The relationships among species of the *miliaris* complex are not resolved.

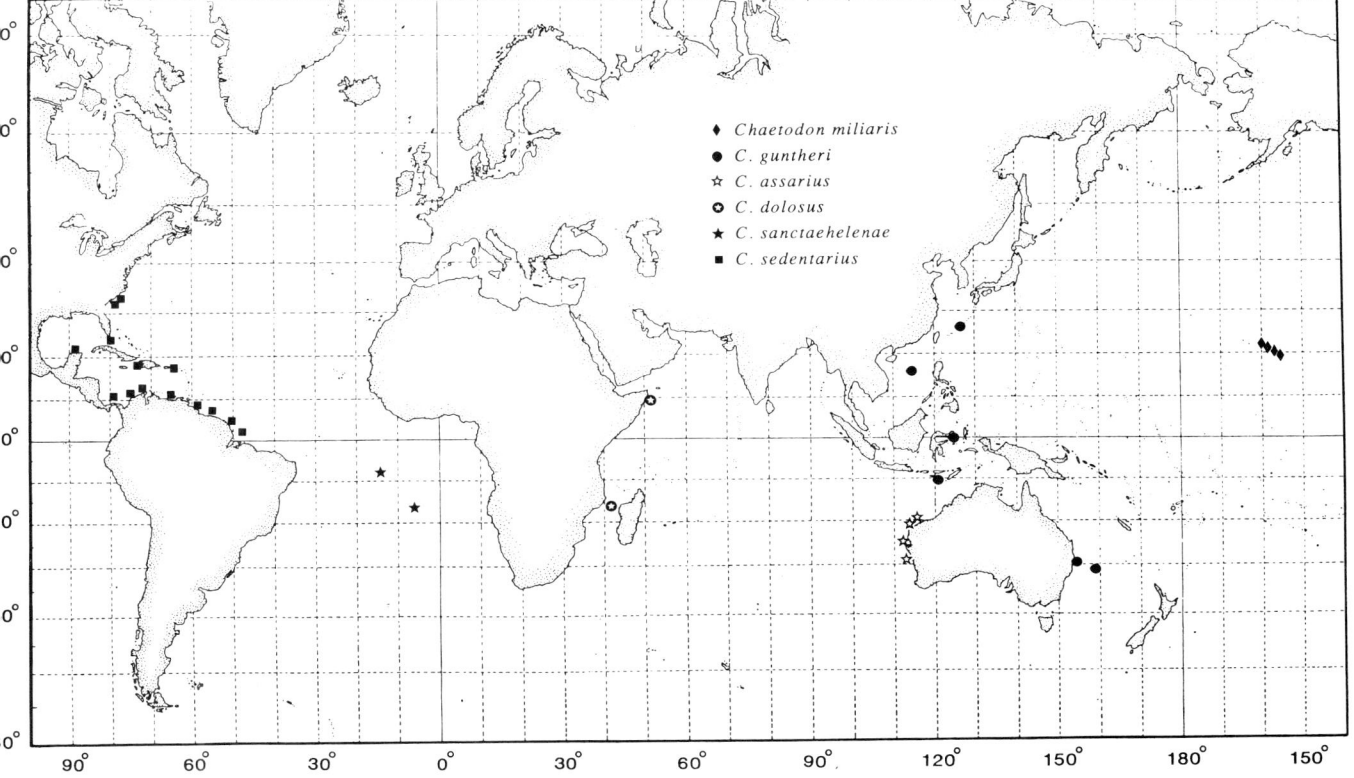

Fig. 8. Distributions of species in the *Chaetodon miliaris* complex.

This complex includes *C. miliaris* from the Hawaiian Islands, *C. guntheri* from the western Pacific, *C. assarius* from western Australia, *C. dolosus* from east Africa, *C. sanctaehalenae* from St. Helena and Ascension Islands, and *C. sedentarius* from the northern tropical Atlantic and Caribbean (Fig. 8). This is the only chaetodontid species complex that extends from the central Pacific to the Caribbean.

Chaetodon lineolatus complex: *C. lineolatus* Cuvier 1831, and *C. oxycephalus* Bleeker 1853
Chaetodon lineolatus and *C. oxycephalus* are characterized by 16–18 thin vertical black lines on a white body, a scythe-shaped black marking along the base of the dorsal fin, and a very wide eyebar (width greater than eye diameter). Two of these features are seen or approximated in other species; the same vertical black lines occur in the *C. falcula* complex, and a black stripe runs along the base of one or both median fins in *C. selene*, *C. vagabundus* and *C. lunula*. However, the shape and orientation of the black scythe are unique to *C. lineolatus* and *C. oxycephalus* (thick and rounded antero-dorsally, narrowing to a point postero-ventrally), and the wide eyebar is not seen in any of the species that have the other two characteristics. These unique features of color pattern infer that these species are sister taxa.

This complex is unlike all others discussed above in that its two species are broadly sympatric. *C. lineolatus* is widely distributed throughout the indo-Pacific from the red Sea to Reunion Island, from the Ryukyus to Lord Howe Island, and from the Hawaiian Islands to the southeastern end of the Tuamotus (Fig. 9). (The lack of records from the Laccadive Islands, Sri Lanka, and the southwestern coast of Indonesia probably reflects collecting

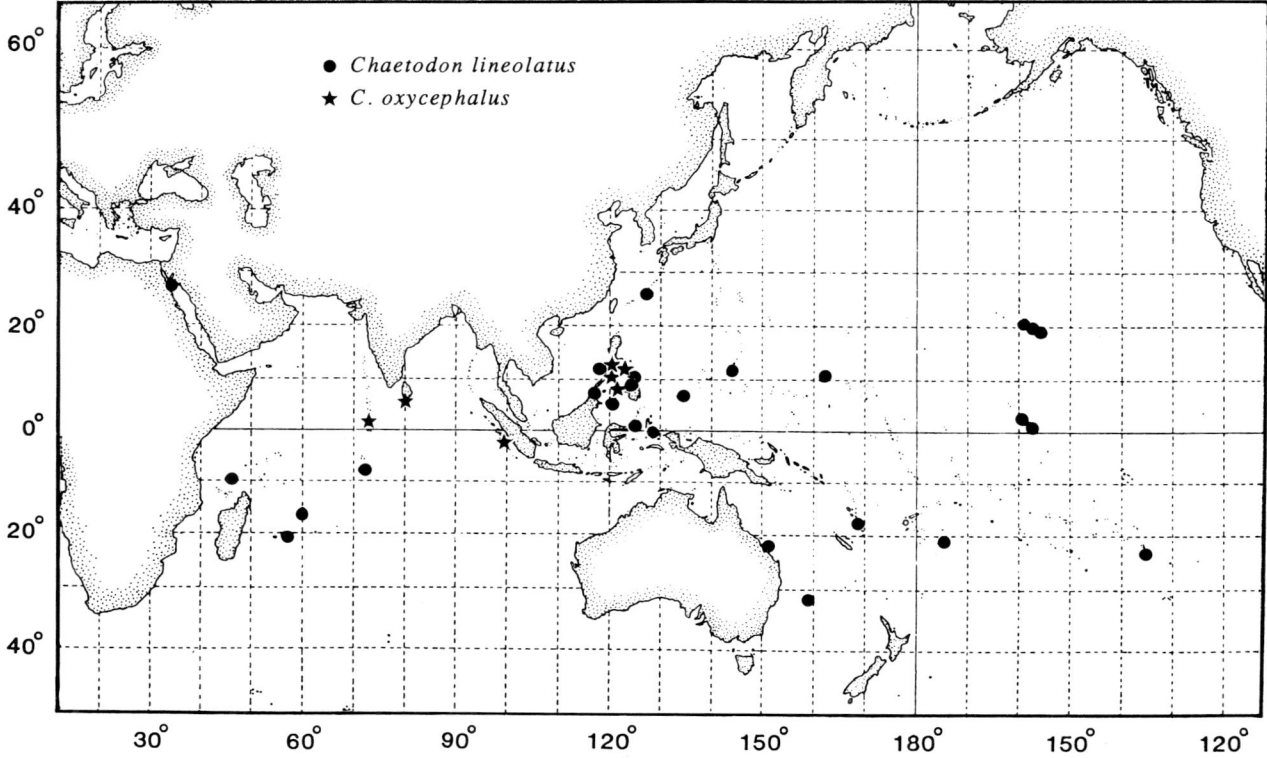

Fig. 9. Distributions of species in the *Chaetodon lineolatus* complex.

effort rather than true absence.) *C. oxycephalus* is more narrowly distributed, from the Laccadive Islands to the Philippines.

Chaetodon melannotus complex: *C. melannotus* Schneider 1801, and *C. ocellicaudus* Cuvier 1831 Burgess (1978) classified these two species as members of the subgenus *Chaetodon*. However, the relatively more derived morphology of the first dorsal pterygiophore infers that this placement is incorrect, and that they are instead more closely related to members of the subgenera *Rabdophorus* and *Chaetodontops*. These two species are characterized by a white background color, 14 to 18 oblique black lines that break up into dots peripherally, and a black ocellus on the caudal peduncle. (In *C. melannotus*, the ocellus is present only in juveniles, and becomes a ventro-laterally interrupted bar in adults.) These species are so similar that, before the work of Burgess (1978), *C. ocellicaudus* was more frequently misidentified as *C. melannotus*. All localities shown in Figure 10 represent specimens whose identifications were confirmed by the author.

Chaetodon melannotus is widely distributed throughout the Indo-Pacific (including the Red Sea, but absent in Hawaii), and appears to be completely sympatric with the more narowly distributed *C. ocellicaudus*. The range of *C. ocellicaudus* is centered in the Sulu, Moluccan, and Banda Seas, but extends eastward to New Ireland and the Solomons. Burgess (1978) reported that *C. ocellicaudus* also extends westward to Sri Lanka, Zanzibar, and possibly the east coast of Africa, but as noted by Allen (1980), records from the Indian Ocean have not been confirmed.

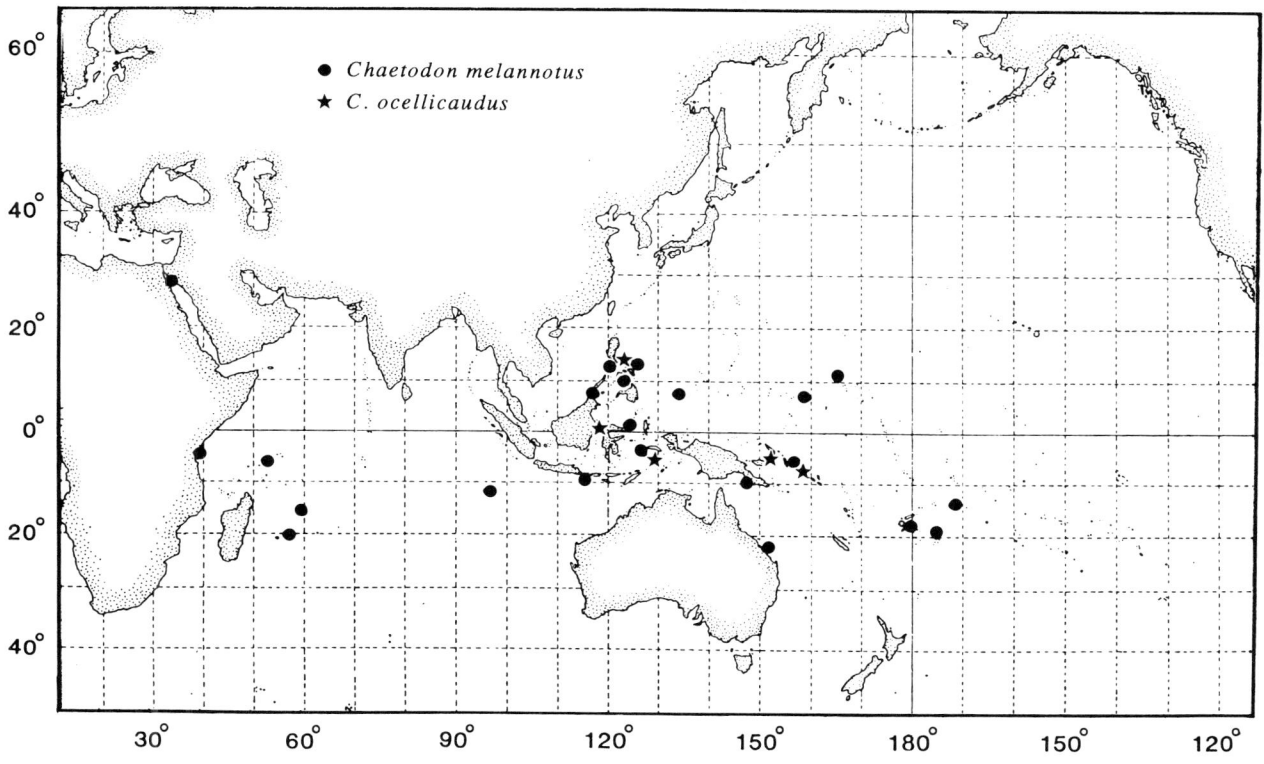

Fig. 10. Distributions of species in the *Chaetodon melannotus* complex.

Discussion

The barriers inferred by these species complexes are illustrated in Figure 11. The species distributions that determine each barrier are summarized in Table 2.

Winterbottom (1985a) noted two patterns of endemism among Red Sea endemics, one that includes only the northern region of the Red Sea (north of 20°N), and another that includes the entire Red Sea and adjacent Gulf of Aden. Among the evidence for the more limited track, he cited the distribution of *C. austriacus*. All museum records of this species in the present survey are from the Gulf of Aqaba. The more southern locality in Figure 3 was derived from the photograph of a juvenile published in Allen (1980). Absence of *C. austriacus* from the southern Red Sea is not due to a lack of collecting effort. Smithsonian expeditions collected extensively with piscicides along the Eretrean coast, but failed to obtain specimens of either *C. austriacus* or its sister species, *C. melapterus* (Springer in litt.). The presence of *C. melapterus* in the Gulf of Aden infers that the barrier responsible for the differentiation of these two species lies within the Red Sea proper, between Bab-el-Mandeb (the most narrow point of the south) and the southern most record of *C. austriacus* (22°N).

As noted above, color pattern indicates that *C. austriacus* and *C. melapterus* are more closely related to each other than either is to *C. trifasciatus*, the third species of *Corallochaetodon*. This relationship then infers that barrier C isolated the Red Sea after barrier D isolated the northwestern Indian Ocean (including the Red Sea and Persian Gulf) from the remainder of the Indian Ocean (Fig. 11). The only other species complex in this study which contains a Red Sea endemic (*Rhombochaetodon*) does not test this hypothesis. Only two species are present in this region, and the records of the Red

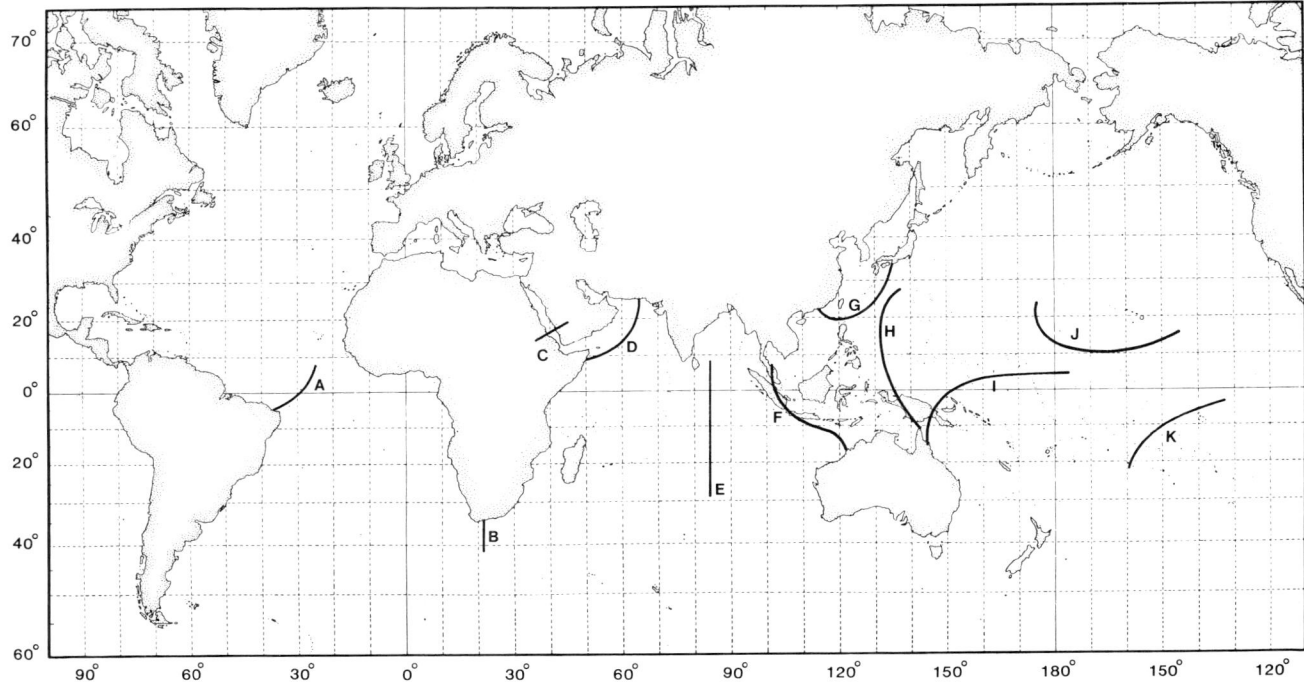

Fig. 11. Barriers inferred from the distributions of species within complexes. The species adjacent to each barrier are listed in Table 2.

Sea form, *C. paucifasciatus*, and the Indian Ocean form, *C. madagascariensis*, are so widely separated (Fig. 7) that the inferred distributions do not distinguish between barriers C and D. I am unaware of any other monophyletic group of three species with distributions that correspond exactly to those within *Corallochaetodon* (cf. Smith-Vaniz, 1976, Winterbottom 1985a, b, Carpenter 1987). However, few cladistic studies of Indo-Pacific marine organisms have been made, and thus the lack of corroboration may simply reflect the current lack of information.

The Malay archipelago has frequently been identified as a boundary between marine provinces (e.g. Schilder 1961, Klausewitz 1976). Although the Malay archipelago has never constituted a complete land barrier between the Indian and Pacific oceans, several recent authors have proposed that eustatic fluctuations in sea level (Pliocene to Quaternary) may have produced compounding environmental changes that enhanced its effectiveness as a barrier. Its permeability was apparently decreased sufficiently to allow specific or subspecific differentiation in many groups of marine fishes (e.g. *Myripristis:* Greenfield 1968, *Amphiprion:* Allen 1972, and the Siganidae: Woodland 1983, 1986).

Among the chaetodontids reviewed here, a barrier congruent with the Malay archipelago (F, Fig. 11) is inferred from the distributions of *C. madagascariensis* and *C. xanthurus*, *C. guttatissimus* and *C. punctatofasciatus*, and *C. assarius* and *C. guntheri*. It should be noted however, that the presence of *C. guntheri* in the southeastern end of the archipelago has been confirmed, and Allen (1980, Table 1 here) recorded *C. punctatofasciatus* in western Australia. Thus, the barrier may not have run the entire length of the archipelago, but may have diverged southward below Timor. Moreover, the Indian ocean members of these pairs have not been definitely recorded from the southwestern shores of Sumatra and Java. Thus, the precise nature of this barrier cannot be specified.

If the Malay archipelago is used to explain the

distributions of *C. falcula* and *C. ulietensis*, then their differentiation must have been followed by both dispersal and extinction. The Indian ocean form, *C. falcula*, is absent from Cocos Keeling and Christmas Islands. Instead, the Pacific species, *C. ulietensis* is found there. A more direct interpretation of these distributions is that their differentiation was caused by a barrier other that the Malay archipelago, one that lies to the west of Cocos Keeling and east of the Chagos-Laccadive ridge (E, Fig. 11).

Smith-Vaniz (1976), and Allen & Steene (1979) reported that the fish faunas of Cocos Keeling and Christmas Islands contain greater numbers of West-Pacific species than Indian Ocean species. Smith-Vaniz (1976) suggested that Christmas and Cocos Keeling Islands may have acted as stepping stones for dispersal into the Indian Ocean. However, only the sympatry of sister taxa on these islands requires the acceptance of dispersal after speciation. Any Pacific species recorded from Cocos Keeling that has an allopatric sister taxon in the western Indian Ocean actually adds support to the existence of a barrier to the west of Cocos Keeling. By documenting the Pacific character of these faunas, these authors may have discovered the generality of one or more barriers approximated by the position of E, rather than a path of dispersal.

Hocutt (1987; see also Springer 1988) proposed that the northward migration of the Indian plate through the proto-Indian Ocean may account for present-day endemism in the western Indian Ocean. However, this disruptive event ended approximately 50 mya (Davies & Kidd 1977, Hocutt 1987), well before the closure of the Indo-Tethys seaway 18 to 13 mya. As the latter event was hypothetically correlated to node 9 (Fig. 1), the separation of Atlantic and Indo-Pacific members of *Chaetodon*, the two tectonic and cladistic correlations are not congruent. One or both of the correlations must be incorrect. *Chaetodon falcula* and *C. ulietensis* show only slight morphological differences, comparable to the species pairs that span the Malay archipelago (discussed above) and the isthmus of Panama (e.g. *Prognathodes falcifer* – *P. aya*, and others noted in Rosenblatt 1963; Pleistocene, 3–5 mya, Woodring 1966). If the evolutionary rates among these pairs are roughly comparable (within an order of magnitude), then the differentiation of *C. falcula* and *C. ulietensis* cannot be plausibly linked to the northward migration of India. (This does not imply, however, that Hocutt's hypothesis lacks merit for other sister groups.)

Barrier J (Fig. 11) is inferred by the members of the *Chaetodon punctatofasciatus* complex, *C. multicinctus* in the Hawaiian and Johnston Islands, and *C. punctatofasciatus* plus *C. pelewensis* in the western and southern Pacific. The other Hawaiian endemic in this survey, *C. miliaris*, was recently photographed from a submarine at Johnston Island (94 m; Randall et al. 1985). These authors were inclined to believe that these individuals were waifs and did not necessarily constitute a range extention. However, even if its range is interpreted more broadly, *C. miliaris* remains so widely separated from its geographically closest relative, *C. guntheri* of the western Pacific, that a barrier at position J is

Table 2. Support for the inferred barriers. The adjacent species (assumed to be members of sister taxa) whose distributions indicate a barrier are listed. If more than one species in a complex approach a barrier from one side, they are listed in parentheses. If a pair of species is widely separated and their distributions are congruent with more than one barrier, the pair is listed in brackets for each barrier.

Barrier	Adjacent species
A	*Chaetodon sedentarius* - *C. sanctaehelenae*
B	*C. sanctaehelenae* – *C. dolosus*
C	*C. austriacus* – *C. melapterus*; [*C. pausifasciatus* – *C. madagascariensis*]
D	*C. melapterus* – *C. trifasciatus*; [*C. paucifasciatus* – *C. madagascariensis*]
E	*C. falcula* – *C. ulietensis*
F	*C. guttatissimus* – *C. punctatofasciatus*, *C. madagascariensis* – *C. xanthurus*, *C. assarius* – *C. guntheri*
G	*C. argentatus* – (*C. xanthurus* + *C. mertensii*), [*C. guntheri* – *C. miliaris*]
H	*C. mertensii* - (*C. xanthurus* + *C. argentatus*)
I	*C. punctatofasciatus* – *C. pelewensis*
J	*C. multicinctus* – *C. punctatofasciatus*, [*C. miliaris* – *C. guntheri*]
K	*C. kleinii* – *C. trichrous*
L	*C. oxycephalus* – *C. lineolatus*; *C. ocellicaudus* – *C. melannotus*

only weakly implicated as the cause of their differentiation.

Most species in the *C. miliaris* complex are widely separated. If their ancestral species was in fact distributed circumtropically, the large gaps between modern species imply that extinctions have occurred in adjacent island groups and along large tracts of continental shelf. The only two species of the *C. miliaris* complex not widely separated are *C. guntheri* and *C. assarius*. *Chaetodon guntheri* was recently collected in the southeastern region of the Malay archipelago (Gloerfelt-Tarp & Kailola 1984). The barrier between these two species is therefore inferred to lie along the northwestern coast of Australia.

Barrier K separates the Society Islands from the rest of the Pacific, and is inferred from the distributions of *C. kleinii* and *C. trichrous*. It is not corroborated by any other chaetodontid species pair or complex reviewed here, but several chaetodontid species are endemic to other islands of the southeastern Indo-Pacific (Randall 1975, Burgess 1978, Allen 1980). The sister taxa of these species cannot always be identified, but in each case, the cumulative range of any reasonable sister taxon (a closely related species or group of species) is limited to more northwestern areas of the Indo-Pacific. *Hemitaurichthys multispinosus* is endemic to Pitcairn Island. All other species of *Hemitaurichthys* are found further north and west in the Indo-Pacific. *Chaetodon declivis* is endemic to the Marquesas, and a member of the *tinkeri* complex. All other species of this complex are found more centrally in the Indo-Pacific. *Chaetodon litus* from Easter Island and *Chaetodon smithi* from Rapa are both members of *Exornator*. All other species of *Exornator* (except the two Atlantic species of the *Chaetodon miliaris* complex, *C. sedentarius* and *C. sanctaehelenae*) are found more centrally in the Indo-Pacific. These examples indicate that peripheral isolation and differentiation in the southeastern Indo-Pacific is a general phenomenon within the Chaetodontidae.

If the allopatric mode of speciation is assumed to be general in this family, then the distributions of the last two species pairs, *Chaetodon lineolatus – C. oxycephalus* and *C. melannotus – ocellicaudus*, infer that isolation and speciation has occurred more centrally within a widely distributed species. In both of these pairs, a widely distributed species is sympatric with its more narrowly and centrally distributed sister species. Sympatry infers that, minimally, dispersal has brought the more widely distributed species into the original range of the narrowly distributed species. The present ranges of the narrow species may or may not extend beyond their original limits.

The two narrowly distributed species in these pairs, *C. oxycephalus* and *C. ocellicaudus*, overlap in the Philippines, and perhaps the Sulawesi and Banda seas (collecting efforts in Indonesia are not as extensive as those in the Philippines). Thus, a single series of events may explain the differentiation of both species. The distribution of *C. xanthurus*, a member a third complex, *Rhombochaetodon* (Fig. 7), is also congruent with area of overlap between *C. oxycephalus* and *C. ocellicaudus*. The differentiation and distributions of *C. xanthurus* and its adjacent relatives are explained by the combined actions of barriers F, G, and H (Fig. 11). By the imposition of an additional requirement, barriers F and H can also account for the differentiation of *C. ocellicaudus* and *C. oxycephalus*. The additional requirement is that barriers F and H must have acted simultaneously, splitting the two wide-spread ancestral species into three populations each (Indian, Indonesian-Philippine, and oceanic Pacific). The central populations then differentiated (through drift or selection), whereas the Indian and Pacific ones did not. The lack of differentiation can be explained by the likelihood of similar stabilizing selection in the more oceanic habitats (i.e. those less affected by fluctuations in sea level). In the absence of differentiation, the Indian and Pacific populations would have been able to reintegrate after the barriers ceased to be effective, but the Indonesian-Philippine populations would have remained distinct. The crucial elements of this hypothesis are that barriers F and H acted simultaneously, and that the Indian and Pacific populations did not differentiate when they were isolated.

It is relevant to note that Allen (1980) and Randall (personal communication) have challenged the specific distinction between *C. mertensii* and *C. madagascariensis*, the species of *Rhombochaetodon* which flank *C. xanthurus* in the Pacific and Indian oceans, respectively. They have not, however, challenged the validity of *C. xanthurus*. Partial differentiation between the Indian and Pacific forms, and more extreme differentiation of the Indonesian-Philippine form are predicted by the hypothesis of synchronous flanking barriers. Moreover, the relevant 'species' within the *Rhombochaetodon* complex are still allopatric. Their distributions are congruent with barriers F and H, and therefore add significant support to the hypothesis of isolation by synchronous flanking barriers.

An alternative hypothesis is that a single barrier isolated the South China, Sulu, and Sulawesi basins (or a subset thereof) while a connection was maintained between the Indian and Pacific Oceans further south. These basins are individually and collectively known to be areas of endemism (e.g. *Plagiotremus celebesensis*, *P. iosodon*, and *P. spilistius*: Smith-Vaniz 1976, *Cirrhilabrus lubbocki*, *C. flavidorsalis*, and *C. rubripinnis*: Randall & Carpenter 1980, *Congrogadus hierichthys*: Winterbottom 1985b, and at least five species of *Ecsenius*: Springer 1988). It seems likely that the connection between the Indian and Pacific oceans would have been to the north of the Australian plate (through the Banda Sea) because the Torres Strait was closed before the Pleistocene (Winterbottom 1985b), and has been open only intermittently since then (Potts 1983). However, Springer (1988) characterized one species, *Ecsenius bandanus*, as having a distribution that extends across the Banda Sea, from the north coast of Java in the east, to Irian Jaya in the west. Additionally, Fleminger (1986) reported an endemic upwelling copepod, *Calanoides philippinensis*, in the Banda Basin, and presented a hypothesis of Pleistocene oceanographic conditions that accounts for its isolation from congeners. Thus, the Banda Basin also appears to have been isolated during the Pleistocene. If the Banda Basin is added to the list of more northern isolates, the resulting discontinuity between the Indian and Pacific oceans is equivalent to that required by the hypothesis of synchronous flanking barriers above.

Classical biogeographers have inferred the Indonesian-Philippine area to be the center of origin and differentiation for the Indo-Pacific region by virtue of its high diversity. Alternatively, Woodland (1983) hypothesized that much of the region's diversity may result from the overlap of faunas centered in the Indian and Pacific oceans. Donaldson (1986) supported Woodland's hypothesis with patterns of distribution within the Cirrhitidae. Among chaetodontids, species within the subgenera *Chaetodontops*, *Tetrachaetodon*, and *Discochaetodon* may also fit the model proposed by Woodland, as they are broadly sympatric within this region though certain species range broadly to one side or the other. Woodland's hypothesis does not, however, account for the distributions of species in the *Chaetodon lineolatus* and *C. melannotus* complexes, as the more narrowly distributed species are in fact centered in the center of origin. It should be noted that Woodland's hypothesis and the one advanced here are not mutually exclusive.

Directions for future research

The critical reader will have undoubtedly recognized two insufficiencies in this report. Data are not available to resolve the relationships among species within complexes, and the geographical limits of many chaetodontid species are poorly known. Further cladistic resolution will require the examination of rapidly evolving characters (e.g. mitochondrial DNA). However, it should be born in mind that speciation may not have occurred dichotomously in several of these complexes, and thus the differences among species may not be distributed hierarchically. There may be no cladistic structure to be inferred from any amount of data. Demonstrating the lack of structure, however, is a bit like 'proving' a null hypothesis; confidence in this result can only be obtained through an exhaustive inventory of species differences.

A more complete knowledge of species distributions will result from additional museum work; only three collections were reviewed here. However, the faunas of certain critical areas, such as the southern Red Sea, Gulf of Aden, southern Arabian peninsula, and the entire southwestern coast of the Malay archipelago, have been inadequately collected. A more precise understanding of the biogeography and evolution of marine organisms is contingent upon our knowledge of these faunas.

The collection of systematic data by more sensitive techniques should also further our understanding of differentiation and gene flow (dispersal) within species. Population genetic models predict that wide-spread species may be undifferentiated with respect to their peripherally isolated descendants. If sensitive techniques are applied to widespread species, their peripherally isolated sister taxa (e.g. *Chaetodon trifasciatus – melapterus – austriacus, C. kleinii – trichrous, Chaetodon lunula – fasciatus,* etc.), and their outgroups, we may find that many wide-spread species are paraphyletic.

Acknowledgements

Jeffrey Williams and Carl Ferraris provided countless pages of locality data from the catalogs of the USNM and ANSP, respectively. William Smith-Vaniz provided the excellent base maps, and confirmed the identifications of important specimens from Cocos Keeling Island. Gareth Nelson, Victor Springer and Richard Winterbottom provided timely and thorough reviews of the manuscript and additional information relevant to the distributions of chaetodontids and other marine fishes. Their agreement with my hypotheses and conclusions is not implied, and the responsibility for any remaining errors is entirely my own. I extend my sincerest thanks to all of these individuals for their considerable time and effort.

References cited

Ahl, E. 1923. Zur Kenntnis der Knochenfischfamilie Chaetodontidae, insbesondere der Unterfamilie Chaetodontinae. Arch. Naturg. A 89: 1–205.

Allen, G.R. 1972. The anemonefishes. T.F.H. Publications, Neptune City. 288 pp.

Allen, G.R. 1980. Butterfly and angelfishes of the world, Vol. 2. Wiley-Interscience, New York. pp. 145–352.

Allen, G.R. & R.H. Kuiter. 1978. *Heniochus diphreutes* Jordan, a valid species of butterflyfish (Chaetodontidae) from the Indo-West Pacific. J. Roy. Soc. West. Aust. 61: 11–18.

Allen, G.R. & R. Steene. 1979. The fishes of Christmas Island, Indian Ocean. Australian National Parks Service, Special Publication 2: 1–81.

Amesbury, S.S. & R.F. Meyers. 1982. Guide to the coastal resources of Guam: Vol 1., the fishes. University of Guam Press, Agana, 138 pp.

Berggren, W.A. & C.D. Hollister. 1974. Paleogeography, paleobiogeography and the history of circulation in the Atlantic Ocean. pp. 126–186. *In:* W.W. Hay (ed.) Studies in Palaeoceanography, Soc. Econ. Paleontol. Mineralog. Spec. Publ. 20.

Bizon, G. 1985. Mediterranean foraminiferal changes as related to paleoceanography and paleoclimatology. pp. 453–470. *In:* D.J. Stanley & F.C. Wezel (ed.) Geological Evolution of the Mediterranean Basin, Springer Verlag, New York.

Blum, S.D. 1988. Osteology and phylogeny of the Chaetodontidae (Pisces: Perciformes). Ph.D. Dissertation, University of Hawaii, Honolulu. 365 pp.

Briggs, J.C. 1974. Marine zoogeography. McGraw-Hill Book Company, New York. 475 pp.

Burgess, W.E. 1978. Butterflyfishes of the world. T.F.H. Publications, Neptune City. 832 pp.

Brothers, E.B., D.McB. Williams & P.F. Sale. 1983. Length of larval life in twelve families of fishes at 'One Tree Lagoon', Great Barrier Reef, Australia. Mar. Biol. 76: 319–324.

Carpenter, K.E. 1987. Revision of the Indo-Pacific fish family Caesionidae (Lutjanoidea), with descriptions of five new species. Indo-Pacific Fishes No. 15. 56 pp.

Cracraft, J. & R.O. Prum. 1988. Patterns and processes of diversification: speciation and historical congruence in some neotropical birds. Evolution 42: 603–620.

Crane, J. 1975. Fiddler crabs of the world (Ocypodidae: genus *Uca*). Princeton University Press, Princeton. 737 pp.

Donaldson, T.J. 1986. Distribution and species richness patterns of the Indo-West Pacific Cirrhitidae: support for Woodland's hypothesis. pp. 623–628. *In:* T. Uyeno, R. Arai, T. Taniuchi & K. Matsuura (ed.) Indo-Pacific Fish Biology, Proc. Second Internat. Conf. Indo-Pacific Fishes, Ichthyological Society of Japan, Tokyo.

Ekman, S. 1953. Zoogeography of the sea. Sidgwick & Jackson, London. 417 pp.

Fleminger, A. 1986. The Pleistocene equatorial barrier between the Indian and Pacific oceans and a likely cause for Wallace's Line. pp. 84–97. *In:* A.C. Pierrot-Bults, S. van der Spoel, B.J. Zahuranec & R.K. Johnson (ed.) Pelagic Biogeography, UNESCO Tech. Pap. Mar. Sci. 49.

Futuyma, D.J. & G.C. Meyer. 1980. Non-allopatric speciation in animals. Syst. Zool. 29: 254–271.

Gloerfelt-Tarp, T. & P.J. Kailola. 1984. Trawled fishes of southern Indonesia and Northwestern Australia. DGF Indonesia, GTZ Federal Republic of Germany, and Australian Development Assistance Bureau, Canberra. 406 pp.

Greenfield, D.W. 1968. The zoogeography of *Myripristis* (Pisces: Holocentridae). Syst. Zool 17: 76–87.

Hocutt, C.H. 1987. Evolution of the Indian Ocean and the drift of India: a vicariant event. Hydrobiol. 150: 203–223.

Kennett, J.P. 1978. The development of planktonic biogeography in the Southern Ocean during the Cenozoic. Mar. Micropaleontol. 3: 301–345.

Klausewitz, W. 1976. The zoogeography of the littoral fishes of the Indian Ocean based on the distribution of the Chaetodontidae. Rev. Trav. Inst. Peches Marit. 40: 636–640.

Klausewitz, W. & H.W. Fricke. 1985. On the occurrence of *Chaetodon jayakari* Norman in the deep water of the Gulf of Aqaba, Red Sea. Senckenbergiana Marit. 17: 1–13.

Kuiter, R.H. 1986. A new species of butterflyfish, *Chelmonops curiosus*, from Australia's south coast. Rev. fr. Aquariol. 13: 73–78.

Lande, R. 1980. Genetic variation and phenotypic evolution during allopatric speciation. Amer. Nat. 116: 463–479.

Lubbock, R. & A.J. Edwards. 1980. A new butterflyfish (Teleostei: Chaetodontidae) of the genus *Chaetodon* from St. Paul's Rocks. Rev. fr. Aquariol. 7: 13–16.

Maddison, W.P., M.J. Donoghue & D.R. Maddison. 1984. Outgroup analysis and parsimony. Syst. Zool. 33: 83–103.

Masuda, H. 1984. Marine fishes: an illustrated field guide. Tokai University Press, Tokyo. 228 pp. (In Japanese).

Maugé, A. & R. Bauchot. 1984. Les genres et sous-genres de Chaetodontides etudies par une methode d'analyse numerique. Bull. Mus. natn. Hist. nat., Paris, 4e ser., 6, section A, 2: 453–485.

Mayr, E. 1942. Systematics and the origin of species. Columbia University Press, New York. 334 pp.

Mayr, E. 1954. Geographic speciation in tropical echinoids. Evolution 8: 1–18.

Mayr, E. 1970. Populations, species and evolution. Harvard University Press, Cambridge. 453 pp.

Meyers, R.F. 1980. *Chaetodon flavocoronatus*, a new species of butterflyfish (Chaetodontidae) from Guam. Micronesia 16: 297–303.

Motta, P.J. 1985. Functional morphology of the head of Hawaiian and mid-Pacific butterflyfishes (Chaetodontidae, Perciformes). Env. Biol. Fish. 13: 253–276.

Motta, P.J. 1988. Functional morphology of the feeding apparatus of ten species of Pacific butterflyfishes (Perciformes, Chaetodontidae): an ecomorphological approach. Env. Biol. Fish. 22: 39–67.

Müller, C. 1985. Late Miocene to Recent Mediterranean biostratigraphy and paleoenvironments based on calcareous nannoplankton. pp. 471–486. *In:* D.J. Stanley & F.C. Wezel (ed.) Geological Evolution of the Mediterranean Basin, Springer Verlag, New York.

Nelson, G.J. & N.I. Platnick. 1981. Systematics and biogeography: cladistics and vicariance. Columbia University Press, New York. 567 pp.

Okamura, O., K. Amaoka, C. Araga, T. Uyeno & T. Yoshino (ed.). 1984. Fishes of the Kyushu-Palau Ridge and Tosa Bay. Japan Fisheries Resource Conservation Association, Tokyo. 435 pp.

Randall, J.E. 1975. Three new butterflyfishes (Chaetodontidae) from southeast Oceania. Uo 25: 12–22.

Randall, J.E. & K.E. Carpenter. 1980. Three new labrid fishes of the genus *Cirrhilabrus* from the Philippines. Rev. fr. Aquariol. 7: 17–26.

Randall, J.E. & G.H.P. de Bruin. 1988. The butterflyfish *Prognathodes guyotensis* from the Maldive islands, a first record for the Indian Ocean. Cybium 12: 145–149.

Randall, J.E., P.S. Lobel & E.H. Chave. 1985. Annotated checklist of the fishes of Johnston Island. Pac. Sci. 39: 24–80.

Rosen, D.E. 1975. A vicariance model of Caribbean biogeography. Syst. Zool. 24: 431–464.

Rosenblatt, R.H. 1963. Some aspects of speciation in marine shore fishes. pp. 171–180. *In:* J.P. Harding & N. Trebble (ed.) Speciation in The Sea, Syst. Assoc. Publ. No. 5.

Russell, B.C. 1983. Annotated checklist of the coral reef fishes from the Capricorn-Bunker Group, Great Barrier Reef, Australia. Great Barrier Reef Marine Park Authority. Hatfields Printers, Mackay. 184 pp.

Schilder, F.A. 1961. The geographical distribution of cowries (Mollusca: Gastropoda). Veliger 7: 171–183.

Smith-Vaniz, W.F. 1976. The saber-toothed blennies, tribe Nemophini (Pisces: Blenniidae). Acad. Nat. Sci. Phil. Monogr. 19. 196 pp.

Springer, V.G. 1988. The Indo-Pacific fish genus *Ecsenius*. Smith. Contr. Zool. 465. 134 pp.

Steene, R.C. 1978. Butterfly and angelfishes of the world, Vol. 1. Wiley-Interscience, New York. 144 pp.

Steininger, F.F. & F. Rögl. 1979. The paratethys history. A contribution towards the Neogene geodynamics of the alpine orogene. Ann. Géol. Pays Hell. 3: 1153–1165.

Winterbottom, R. 1985a. A revision of the congrogadid fish genus *Haliophis* (Pisces: Perciformes), with the description of a new species from Indonesia, and comments on the endemic fish fauna of the northern Red Sea. Can J. Zool. 63: 209–217.

Winterbottom, R. 1985b. Revision and vicariance biogeography of the subfamily Congrogadinae (Pisces: Perciformes: Pseudochromidae). Indo-Pacific Fishes No. 9. 34 pp.

Woodland, D.J. 1983. Zoogeography of the Siganidae (Pisces): an interpretation of distribution and richness patterns. Bull. Mar. Sci. 33: 713–717.

Woodland, D.J. 1986. Wallace's Line and the distribution of marine inshore fishes. pp. 453–460. *In:* T. Uyeno, R. Arai, T. Taniuchi & K. Matsuura (ed.) Indo-Pacific Fish Biology, Proc. Second Internat. Conf. Indo-Pacific Fishes, Ichthyological Society of Japan, Tokyo.

Woodring, W.P. 1966. The Panama land bridge as a sea barrier. Proc. Amer. Phil. Soc. 110: 425–433.

Chaetodon trifasciatus on the reef of Enewetak Atoll, Marshall Islands. Photo by E.S. Reese.

Circumtropical patterns in butterflyfish communities

James S. Findley[1] & Muriel T. Findley[2]
[1] *Department of Biology, University of new Mexico, Albuquerque, NM 87131, U.S.A.* [2] *POB 44, Corrales, NM 87048, U.S.A.*

Received 19.1.1988   Accepted 1.6.1988

Key words: Chaetodontidae, Community patterns, Diversity, Species richness gradient, Density compensation, Niche breadth

Synopsis

Butterflyfish species richness increases along a longitudinal circumtropical gradient from lows of 3–5 species in the tropical Atlantic and Eastern Pacific to highs of 40 or more in the Indo-Pacific region. Biomass of the fishes increases as species richness increases, and single-site (alpha) diversity increases as does between-site (beta) diversity. There is no evidence of density compensation in richer communities, but at the level of islands and regions, habitat breadth diminishes as species richness increases. Morphologically, species are added to communities both at the boundaries and in the middle of morphospace.

Introduction

Butterflyfish species richness (the number of species coexisting at a given site or in a given area or region) is greatest in the central Indo-West Pacific region. At least 40 species are known from the Great Barrier Reef (Steene 1977), and perhaps higher numbers may occur somewhat further to the north and west. Proceeding westward species richness declines. Fourteen species are recorded from the Red Sea (Randall 1983). Eastward across the equatorial Pacific numbers drop gradually to 28 in the Society Islands and 14 in the Marquesas (Randall 1985). In the Tropical Eastern Pacific only 3–4 species have been reported (Burgess 1978, Thomson et al. 1979). In the Caribbean 5–6 species are known (Randall 1968), but even here an east-west gradient is noticeable in number of species seen at individual stations, with median number of species seen per station by us rising from 1.5 in Martinique, Antigua, and St. Martin to 3 in Belize (Fig. 1).

The presence of this striking gradient suggests a number of questions concerning changes in the butterflyfish communities as the number of species rises in this dramatic way. These questions arise if one assumes that the fish communities are structured in some way, as for example by resource limitation and competition, or conversely, are random assemblages of species, each responding chiefly to environmental parameters other than the presence of closely related species of fish.

1. Is increased richness accomodated by increased number of species in single habitats at given sites (alpha diversity), or do different sites or habitats support different sets of species, resulting in a high regional total (increased beta diversity)? This is the chief question addressed by Gladfelter et al. (1978) in trying to account for the greater species richness of fishes in the Tropical West Pacific Ocean compared with the Tropical West Atlantic.

2. When species richness increases does the number if individual fish remain the same (or even decrease), suggesting resource limitation and den-

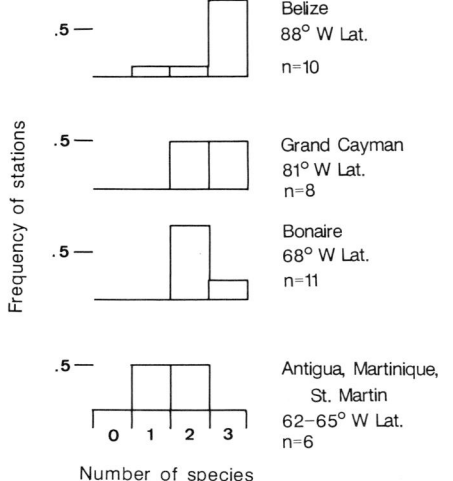

Fig. 1. The trans-Caribbean species richness gradient in *Chaetodon*. Frequency of 10-transect stations with various number of species expressed as a percent.

sity compensation, or does the number of species and of individuals rise together?

3. What changes in the use of space by the fish take place as species richness increases? If the fish are living in resource-limited, competition-based communities, one would expect that as species are added niche breadths would decrease. Is there evidence for such changes?

4. What changes take place in the morphological structure of the assemblage as more species are added? Does morphological packing increase, or does the morphospace occupied by the community simply expand? If the latter, how does this expansion take place? Here the assumption is that morphological similarity between two species is a measure of ecological similarity. In such a case, if a community was fully utilizing the available resources, a new species could be added only if it were sufficiently different ecologically, and hence morphologically, to utilize resources not already preempted. Such an addition would be at the periphery, rather than in the middle, of the multidimensional space defined be the morphological attributes of the fishes.

Our objectives are to address the above questions, and also to consider other implications for butterflyfish biology emanating from a set of censuses conducted along the circumtropical species richness gradient described above.

Methods

Within each geographic region we assessed species richness and individual abundance by conducting ten 10-minute censuses in each of several available habitats. Each set of 10 was carried out in a restricted locality usually by no more than several hundred meters in diameter and within one habitat type. We refer to each site where we conducted 10 such censuses as a station. Each of us ran 5 of the 10 censuses. Analyses of replicate census routes surveyed repeatedly by each of us revealed that there was no significant difference between observers. Each 10-minute census extended over approximately 100 meters. We made our observations while snorkeling at the surface and recorded all those fish which we could see within approximately 15 meters on either side of the transect. A variety of uncontrollable factors affected counts made in this way, but our methods were exactly the same at every site, and we believe that the results are as comparable at it was possible for us to make them.

In order to increase our understanding of the variance added to our results by using two observers, and also to assess the amount of variability which would be added by repeating each transect run, we conducted the following tests at Ambergris Cay, northern Belize. We marked out 5 consecutive 100 meter transects just inside of and parallel to the barrier reef. Each of us ran 10-minute censuses on these marked transects a minimum of 8 times, recording number of species and number of individuals. We subjected the resulting data to a 2-way analysis of variance designed to reveal the contributions to total variance of observer, transect, and transect-observer interaction. For both species number and number of individuals the effect of transect was highly significant, but neither observer nor observer-transect interactions contributed significantly to the variance. We thus concluded that real differences or similarities between transects would not be obsured because of the differences in counts by two observers. Furthermore we compared the variances obtained from ten replicate runs of each Ambergris Cay transect with variances from runs of ten different transects at each of several other stations near Ambergris Cay, and determined that the two sets of variances (one from repeated runs of the same transects, and one from runs of different transects at the same station) were not significantly different. Thus our method of running ten 10-minute transects at each station provided estimates as accurate as we would have obtained with more time consuming replicate censuses, and our confidence that our method gives a reasonably accurate picture of the number of species and individuals at each station was correspondingly increased.

We censused butterflyfishes in the Caribbean, in western Mexico, in French Polynesia, on the Great Barrier Reef, and in the Red Sea, for a total of 97 sites. A summary of the dataset is preseted in Table 1. We also censused surgeonfishes (acanthurids), angelfishes (pomacanthids), and triggerfishes (balistids) at each site. Analyses of the data for these other fishes is not complete at this writing.

We used the number of species recorded at each station as a straightforward measure of alpha diversity. Beta diversity was indexed by computing the coefficient of variation of each species' abundance across all stations in each region, following the suggestion of Schulman (1983). We then compared this population of coefficients from each region with that from the other regions. High coefficients indicate high species turnover from place to place, and hence high beta diversity.

We computed place niche breadth using the following formula (Simpson 1949, Fox 1981):

$$\left(\sum_k P^2{ik}\right)^{-1}/k,$$

where p is the proportion of the total number of individuals of species i counted on all transects at the station occurring on the kth transect at the given station. In an earlier study (Findley & Findley 1985) we calculated this statistic across the 10 transects of a given station for each species. Thus, for each station, we had a set of niche breadths, one for each species recorded at the station. Each breadth indexed the evenness with which the individuals of that species were distributed across the space covered by the transects. We averaged these breaths for each station and compared station averages with the number of species recorded at each station to look for evidence that increasing species richness resulting in decreasing use of space by the species coexisting at the station. Our findings in that case were that species richness and niche breadth were unrelated. Because we became concerned that the small scale of our stations (each a small area of similar habitat containing 10 hundred-meter transects) might fail to reveal a pattern of reduced space use in more speciose assemblages, we calculated niche breadth also at two larger scales: across stations at a given island (or limited stretch of coastline), and across islands in a given geographic region. In the former case we used the total count for a given species at each of the several stations at an island to compute the breadth of that species across the stations, and in the latter case we used the mean number of individuals of the given species seen at each island in a region (since the total number at each island would be a function of the number of stations at the island). We compared the niche breadths calculated at these three scales of measurement with the number of individuals and the number of species at stations, islands, and in regions.

Table 1. A summary of the dataset for chaetodontid fishes upon which this report is based.

Region	stations	transects	species	individuals	ind. per sta.
Caribbean	41	410	3	1852	45
Polynesia	25	250	21	7929	317
Australia	21	210	28	4216	201
Nayarit	3	30	2	54	18
Red Sea	7	70	7	1422	203

Clarke (1977) used the Shannon-Wiener Index, H', to express breadth of habitat use by chaetodontids and pomacanthids in the Bahamas, and weighted the indices following the method of Colwell and Futuyma (1971). Clarke's goal was to express habitat breadth and overlap. Ours is to study the affect upon space use of increased number of species, and for this reason we have followed Fox (1981) in using the unweighted Simpson Index.

We extracted morphological data for each species from Burgess (1978) as described by Findley & Findley (1985). We projected the coexisting species for each region into a morphospace defined by the set of morphological variables determined from Burgess, and recorded and analyzed the set of intertaxon distances for each region. We were interested in learning if intertaxon space increased, decreased, or stayed the same as species were added to the communities, or, in other words, in learning if the addition of species results in greater packing, or in increasing the morphological space occupied by the community, alternatives possibly resulting from resource limitation, or lack thereof, for the community as discussed above.

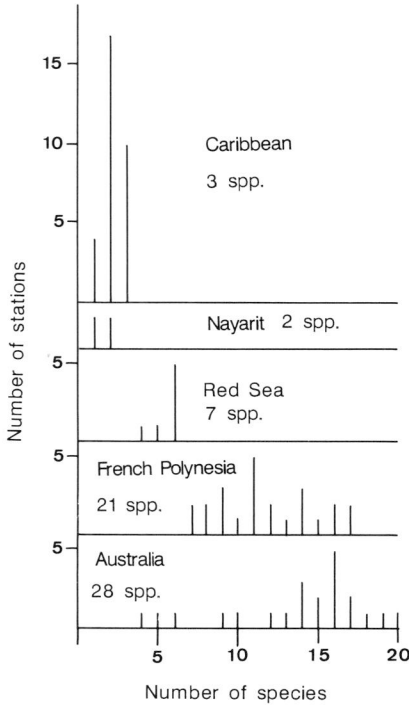

Fig. 2. The distribution of 10-transect stations with various numbers of species of chaetodontid in each region.

Results

1. Alpha and beta diversity

In regions with high numbers of species we consistently recorded high numbers of species at individual stations. Numbers of species recorded at individual stations are compared by region in Figure 2. Clearly alpha diversity is higher in regions with high species richness than it is in low richness areas. Coefficients of variation for numbers of individuals of each species across stations is compared with number of species recorded by us in each region in Figure 3. The two values are highly correlated suggesting than beta diversity is important in determining total species richness. To compare the relationship of alpha and beta diversity with species richness, we first standardized the coefficients of variation and the numbers of species at individual stations, then used these two as independent variables in a multiple regression with number of species seen by us in the appropriate region as the dependent variable. The regression is highly significant (F = 283, p<0.0035). Both independent variables contribute equally to the regression equation: regression coefficient of alpha = 6.1, of beta = 5.7. Each independently explains a highly significant amount of the variation in species richness: R^2 for alpha = 0.94, p<0.01; for beta = 0.94, p<0.01. Both together explain more than either separately: R^2 = 1.00, p<0.004. Both alpha and beta diversity increase as species richness increases. More kinds of butterflyfishes are found in local habitats, and there is more species turnover from habitat to habitat.

2. Biomass and density compensation

Numbers of individual chaetodontids seen at various stations in each region are plotted against number of species at the various stations in Figure 4.

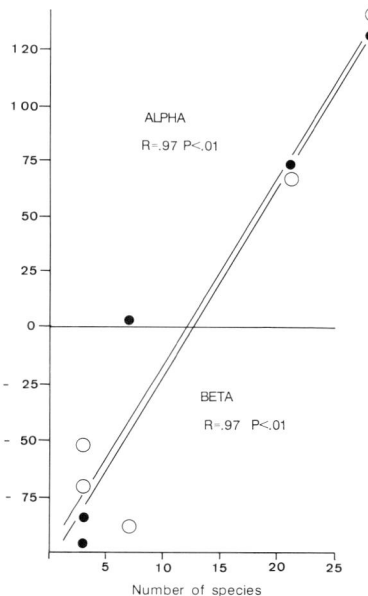

Fig. 3. The relationship of alpha and beta diversity to species richness. Dots indicate standardized mean number of species per station in each region (alpha diversity) plotted against total number of species seen by us in the region. Circles indicate standardized mean coefficient of variation of number of individuals in each species across stations in each region (beta diversity) plotted against number of species in the region. Both alpha and beta diversity are higher in regions with more species. Standardization involves setting the mean of the variables equal to 0 and the standard deviation equal to 1, then expressing each value in standard deviation units.

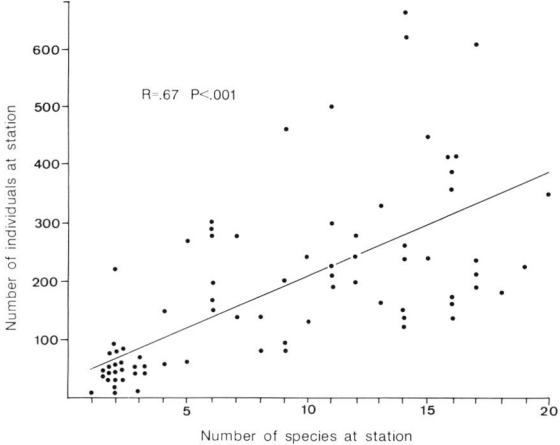

Fig. 4. Number of individual chaetodontids at a 10–transect station plotted against number of species at the station.

Stations and regions with more species have more individuals per station as well. There are highly significant differences between regions with respect to mean number of individuals per station ($F = 8.7$, $p<0.02$) except in the case of the Caribbean and Nayarit which do not differ significantly in this respect. In general, stations in the Indo-West Pacific region differ as a group from those in the Eastern Tropical Pacific or the Caribbean. When number of individuals per species is plotted against number of species at the station (Fig. 5) no significant relationship emerges: there is no reduction in number of individuals in each species as more species are added to the community. Neither does a trend appear when the data are viewed at the scale of islands or regions. Our data make no suggestion that number of individuals of species is systematically reduced as species richness increases: there is no evidence of density compensation.

3. Niche breadth

When mean niche breadth across transects of all the species at each station is plotted against the number of species recorded at the station no significant correlation emerges (Fig. 6). There is clearly no niche contraction as species are added to the local community. However, place niche breadth as measured by the Simpson Index within each station, as was done by Fox (1981), is mostly a function of the number of individuals of the species for which the measure is being calculated. As more individuals of a species enter a small area, the chance of finding some on each transect increases, and the evenness of their distribution increases. But across stations and across islands this effect diminishes, and instead an inverse relationship between niche breadth and species richness becomes evident (Fig. 6). At the scale of islands (kilometers or 10's of kilometers) and of regions (100's of kilometers) the reduction in the use of space as species richness increases is clear.

The results of three multiple regression analyses of the relationship between number of species and number of individuals as independent variables

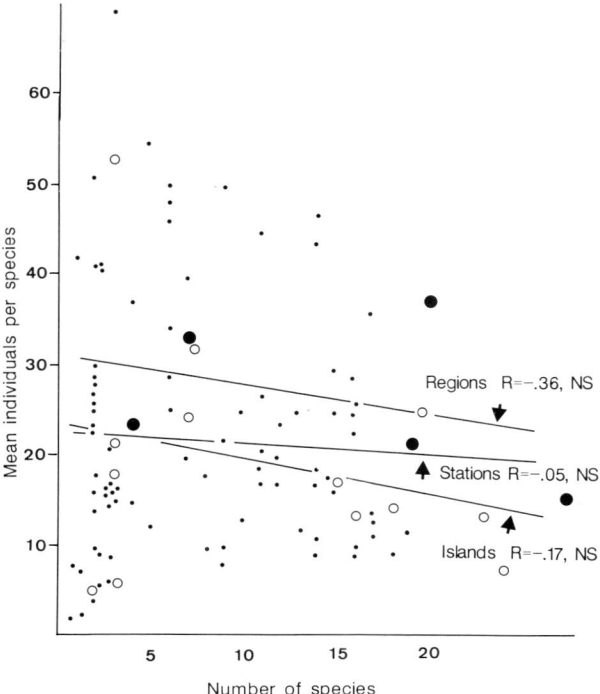

Fig. 5. Mean number of individuals per species as a function of number of species present at three scales. Dots indicate number of individual fish per species at each of 90 stations; circles, at each of 13 islands; large dots in each of 5 regions. Regression lines for the 3 relationships are shown. No relationship reaches significance at the 0.05 level. Density compensation is not seen.

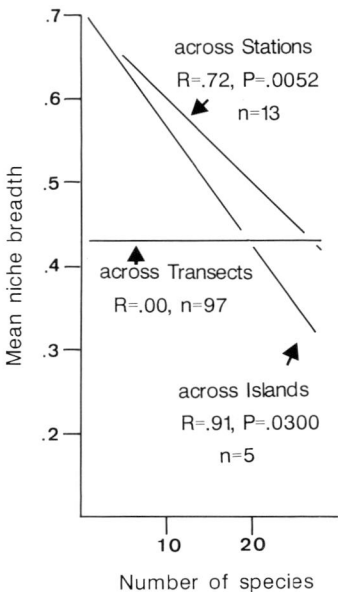

Fig. 6. Mean place niche breadth as a function of number of species across transects within stations, stations within islands, and islands within regions. Niche breadth is a significant negative function of species richness at the two larger scales.

and mean niche breadth as a dependent variable show that above the level of individual stations, (at the level of islands and regions) number of species shows a significant negative relationship with niche breadth: the more species at an island or in a region the lower is mean niche breadth. At these higher levels the number of individuals has an insignificant effect on niche breadth, but at the station level it is all-important: at individual stations the more individuals of a species the greater is its niche breadth.

4. Morphospace and species richness

The array of intertaxon morphological distances for a selection of communities of different sizes is depicted in Figure 7. More speciose communities display more large values as more distinctive species are added to them. Increase in the volume of community size is accompanied by an increase in community morphospace. But distances are added all across the spectrum: species are added interstitially as well as peripherally, and there seems to be no limit to the number of close morpho-relatives that can coexist so far as our data reveal. These results are consonant with those of Findley (1976) involving temperate and tropical bat faunas of various sizes.

5. The 1982–83 El Niño

At the time of our first censuses in French Polynesia predation by *Acanthaster* starfishes on the corals at Moorea had been noted by Bouchon-Navaro et al. (1985) on that island. In November-December, 1981, when we conducted our studies, we did not notice extensive predation, and at most stations the condition of the reefs was good, with a large amount of living hard coral. According to the cited

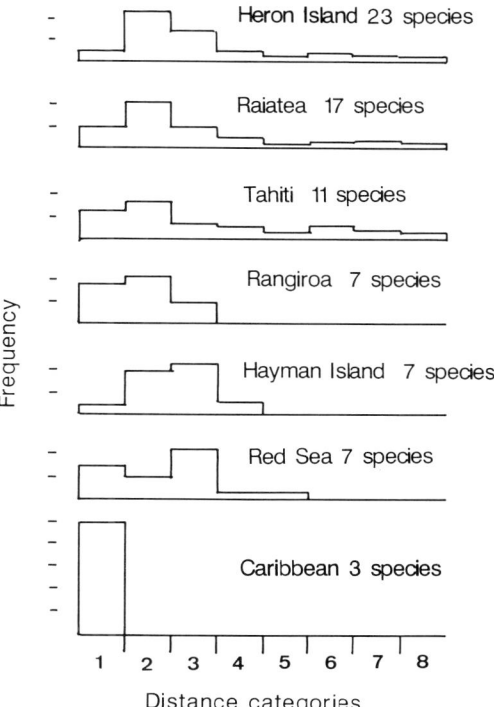

Fig. 7. Frequency distribution of intertaxon Euclidean distances, based upon 29 morphological characters, between all the species at each of the indicated places. Low distance values obtain between species which resemble one another closely.

seemed likely to us that coral specialists would have suffered most from these changes. To test this idea we assembled feeding data on French Polynesian chaetodontids from the work of Harmelin-Vivien & Bouchon-Navaro (1983) and conducted a multiple regression analysis with percent of numerical change in each species as the dependent variable and percent by volume of algae, scleractinian coral, alcyonarian coral, and all other food items, as the independent variables. No independent variable by itself or in combination with others explained a significant amount of variation in pre- and post-El Niño numbers of individuals. Surprisingly, scleractinian specialists fared as well as 9 generalists during this stressful episode (Fig. 8). Our results are in striking agreement with those of Bouchon-Navaro et al. (1985) who reported a 47 percent numerical decrease in chaetodontids between 1979 and 1982, but differ from theirs in that we did not note differential reduction in scleractinian specialists. Our

authors, the effects of predation were noticeable in 1982. In 1982–1983 an El Niño event of unusual severity struck French Polynesia. Chief among the manifestations were torrential rains which freshened the waters in the lagoons of high islands, and increased sea surface temperatures. Both phenomena seemingly had an adverse affect on corals, and the resultant mortality was followed by a considerable increase in benthic algae. When we returned to our study sites in December, 1986, most places supported considerably less living hard coral, and many areas of scleractinian skeleton were covered with brown algae.

Following El Niño we recensused a few of our old stations, specifically one on Moorea, two on Bora Bora, and two on Rangiroa. The results of this comparison are shown in Table 2. The number of individual fish counted dropped by about 40 percent, and the number of species from 20 to 18. It

Table 2. Comparisons of mean number of individuals of each of 20 species of chaetodontid seen at the same set of stations in 1981 and 1986. Arranged in order of decreasing drop in numbers.

Species	No. 1981	No. 1986	Change	Percent
Chaetodon mertensi	0.6	0.00	−0.60	−1.00
C. semeion	0.1	0.00	−0.10	−1.00
C. trifascialis	5.8	0.60	−5.20	−0.90
C. trichrous	0.7	0.10	−0.60	−0.86
C. pelewensis	1.0	0.20	−0.80	−0.80
H. chrysostomus	3.3	0.90	−2.40	−0.73
Heniochus acuminatus	0.1	0.03	−0.07	−0.70
C. bennetti	0.1	0.04	−0.06	−0.60
C. lunula	3.3	1.60	−1.70	−0.52
Forcipiger spp.	0.8	0.40	−0.40	−0.50
C. trifasciatus	14.0	7.80	−6.20	−0.44
C. citrinellus	2.5	1.60	−0.90	−0.36
H. monoceras	0.1	0.07	−0.03	−0.30
C. vagabundus	0.8	0.60	−0.20	−0.25
C. ulietensis	6.3	4.90	−1.40	−0.22
C. ornatissimus	0.7	0.60	−0.10	−0.14
C. ephippium	1.8	1.70	−0.10	−0.06
C. unimaculatus	2.7	3.00	0.30	0.11
C. auriga	2.1	2.90	0.80	0.30
C. reticulatus	0.4	0.80	0.40	1.00
Mean	2.36	1.39	−0.97	−0.39

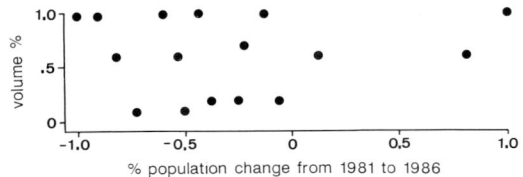

Fig. 8. Volume of scleractinian coral in stomach contents of 16 species reported by Harmelin-Vivien et al. (1983) plotted against percent of population change of those species between 1981 and 1986.

results do show a strong reduction in *Chaetodon trifascialis* as reported by those authors.

6. Structure of the chaetodontid communites

The Caribbean

The numerical relationship of the three shallow-water species in the Caribbean is depicted in Table 3. The dominance of *C. capistratus* and the rarity of *C. ocellatus* are absolutely consistent across the region. There is an increase in the frequency of stations with 3 rather than 2 (or 1) species as one procedes from east to west (Fig. 1). Across this cline the relative abundance of *C. ocellatus* in the local communities increases slightly chiefly because of the abundance of the species at Southwater and Curlew Cayes on the southern Belize Barrier Reef.

Chaetodon capistratus feeds chiefly upon scleractinians, anemones, and zoantharians, being an active generalist within this group of prey organisms, and also ingests polychaetes, crustaceans, eggs, and algae in small amounts (Birkeland & Neudecker 1981). Under some conditions it may feed preferentially on gorgonians (Lasker 1985). Comparable information is not available for the other two species. Our impression is that *C. capistratus* is usually seen picking at scleractinians, and that it does not cover very large areas in its foraging. *C. striatus* and *C. ocellatus*, on the other hand, were less often seen feeding, and appeared to move more steadily and to cover greater areas in their foraging. Our guess is that both are more generalist feeders among invertebrates than is *C. capistratus*. *C. ocellatus* is considered the most common butterfly fish in the northern Gulf of Mexico (Hoese & Moore 1977) which may be related to its increased frequency of occurrence in the western Caribbean.

Nayarit, Mexico

Our study sites at Punta de Mita and Isla Jaltemba, Nayarit, lie near the northernmost limit of coral reef formation on the Pacific coast of North America (Greenfield et al. 1970). The two species of chaetodontid were observed on a small percentage of the 39 transects (0.31 for *C. humeralis* and 0.28 for *Johnrandallia nigrirostris*). Both were seen several times in large aggregations near the ocean side of the coral formations. *Chaetodon humeralis* was also seen moving steadily in shallower areas. Its appearance and foraging behavior are reminiscent of those of *C. striatus*. The numerical relationship of the two Pacific species is approximately that of *C. capistratus* and *C. striatus* in the Caribbean. Our impression was that the abundance and diversity of corals in this area were not adequate to support a corallivore such as *C. capistratus*. That *C. humeralis* is not dependent upon reefs is attested to by its presence further north on the Pacific coast of Mex-

Table 3. Mean number of individuals of three species of *Chaetodon* recorded per station at each of four islands in the Caribbean. Number of stations in each place shown in parentheses.

	Antigua (4)		Bonaire (9)		Cayman (8)		Belize (10)	
	ind per sta	%	ind per sta	%	ind per sta	%	ind per sta	%
C. capistratus	8	0.82	48	0.81	51	0.79	38	0.75
C. striatus	2	0.18	11	0.18	12	0.19	8	0.17
C. ocellatus	0	0.00	0.4	0.01	0.3	0.02	4	0.08

ico, as in the San Carlos Bay area near Guaymas, Sonora, where we observed a few foraging over rocky reefs.

Society and Tuamotu archipelagos

In French Polynesia *C. trifasciatus* is the most abundant shallow water chaetodontid, occurring at all stations in substantial numbers, overall 2 to 2½ times as abundant as the next most common species, and comprising 26–29 percent of the total community. They are usually observed in pairs picking at scleractinians. Each pair appears to forage within a limited area. Seven pairs we followed for five-minute periods moved a median distance of 8.4 m, with extremes of 1 and 28 m. The general behavior of this species reminds us of the Caribbean *C. capistratus*. Ehrlich et al. (1977) estimated the home range of *C. trifasciatus* at 10–100 m^2, and observed movements of only 1 m in 2 individuals over 50 and 75 minute periods on Lizard Island, Queensland. Harmelin-Vivien & Bouchon-Navarro (1983) found only scleractinians in 20 specimens examined on Moorea. Despite the reduction in scleractinian cover from 1981 to 1986, we still observed *C. trifasciatus* to be the most common species in 1986. Half of the Polynesian fauna was made up of three species in 1981 and 1986. Two of the species in each year were *C. trifasciatus* and *C. ulietensis* (10–12 percent). In 1981 *C. trifascialis* was among the top three (12 percent), but in 1986 it had dropped to 11th place (3 percent) and was replaced as number 2 by *C. citrinellus* (13 percent), a generalist and herbivore (Harmelin-Vivien & Bouchon-Navaro 1983). The reduction in *C. trifascialis*, a scleractinian feeder, may have been a result specifically of the reduction in tabular *Acropora* species with which it is associated in a territorial manner, but clearly not all corallivores suffered a similar fate, e.g., *C. trifasciatus*. In 1986 we observed *C. trifascialis* defending territories around branching *Acropora* species in the absence of tabular forms.

In both years 6 species each comprised 5 percent or more of the total observations: *C. trifasciatus*, *C. ulietensis*, *C. citrinellus*, *C. vagabundus*, *C. lunula*, and *C. auriga*. *Chaetodon trifasciatus* and *C. auriga* were seen at every station each year. *Chaetodon auriga* is most commonly seen in pairs, foraging over sandy areas and travelling in sandy channels between coral heads. We have rarely observed this species picking at anything, but Harmelin-Vivien et al. (1983) record a lot of scleractinians and a diversity of other invertebrates in the stomachs of 7 individuals from Moorea. *Chaetodon auriga* travels widely. Ehrlich et al. (1977) record two movements of 500 m, and estimate home ranges of 100–500 m^2. The relationship between *C. trifasciatus* and *C. auriga* is thus reminiscent of that between *C. capistratus* and *C. striatus* in the Caribbean: one a corallivore foraging in a limited home range, the other an omnivore foraging over a wider area of lower coral cover. *Chaetodon citrinellus* and *C. vagabundus*, two more of the 'big 6', are both commonly seen foraging over sand and around the edges of coral heads. Both are seen picking at algae, and both are recorded as containing algae by Harmelin-Vivien et al. (1983). Both pick at items in the sand, perhaps polychaetes, which are also common in their stomach contents. *Chaetodon citrinellus* is frequently attacked by territorial damsel fishes. Both *C. citrinellus* and *C. vagabundus* occupy limited home ranges. Median distances moved in 5 minutes are 6.5 m for *C. citrinellus* and 10 m for *C. vagabundus*. Ehrlich et al. (1977) estimated home ranges of 10–100 m^2 and 100–500 m^2 respectively. We have observed territorial interactions between pairs of *C. vagabundus* and have seen a pair each of *C. citrinellus*, *C. vagabundus*, and *C. auriga* foraging together without agonistic interactions in the same one m^2 area over sand among coral heads at Moorea. These three are thus especially characteristic of habitats with reduced scleractinian cover and with extensive sandy areas such as inner reef flats. *Chaetodon ulietensis* is often observed in aggregations in areas where currents coming over the barrier reef or through passes are strong. Thus it may not be present in some back reef areas where *C. citrinellus*, *C. vagabundus*, and *C. auriga* are common. *Chaetodon lunula* is most often seen in shady places under overhangs where it is often relatively immobile. Rarely have we observed *C. lunula* foraging during the day, and we have the suspicion that it may be at least partially a nocturnal feeder as reported by Hobson (1974).

Six species each comprise between 2 and 4 percent of the fauna: *Forcipiger* spp., *C. pelewensis*, *C. ephippium*, *C. ornatissimus*, *C. reticulatus*, and *Heniochus chrysostomus*. Although these taxa are relatively uncommon, they seem to have regular places in the Polynesian community, and not to be satellite species in the sense of Hanski (1982). The places occupied by these species are not clear to us in detail but, except for *C. ephippium* and *H. chysostomus*, the species all seem to be most common in high energy habitats with moderate to high scleractinian cover. Typical examples are situations exposed to currents through passes, and reef flats near the barrier reef crest where strong currents come over the barrier from the open ocean. *Heniochus chrysostomus* is most common where the faces of fringing reefs or large coral heads provide dark shelters. *Chaetodon ephippium* was recorded at all but one of our stations, and its pattern of occurrence does not indicate a clear habitat preference. It is a wide-ranging species the home range of which was estimated at greater than 500 m^2 by Ehrlich et al. (1977) at Lizard Island. Its diet includes sponges as well as algae according to Harmelin-Vivien et al. (1983). It appears to be a rare but regularly occurring and wide-ranging species, perhaps selecting food less commonly used by other chaetodontids. Of the remaining species, those kinds comprising 1 percent or less of the total counts, several seemed rare because our census routes did not regularly cross their habitats. In this category are *C. quadrimaculatus*, mostly in deeper water outside the barrier reefs, and *H. monoceras*, mostly in shelters at some depth. *Chaetodon trichrous*, *C. benneti*, *C. mertensi*, *C. semeion*, and *H. acuminatus* all seem properly classified as satellite species in the shallow-water reef community.

The Great Barrier Reef

Our stations on the Australian Great Barrier Reef span a greater latitudinal range than those in other regions, and our data suggest some latitudinal replacement of species. A group most common at higher latitudes contains *C. flavirostris*, *C. aureofasciatus*, *C. rainfordi* and *C. lineolatus*. A group more common in the north (lower latitudes) includes *C. baronessa*, *C. citrinellus*, *C. vagabundus*, *C. trifasciatus*, *C. ephippium*, and *C. auriga*. A further grouping is of species the habitats of which exhibit some combination of high exposure to open ocean conditions, distance from the mainland, and prevalence of scleractinian cover. Members are *C. ornatissimus*, *C. speculum*, *C. unimaculatus*, *C. trifascialis*, *C. pelewensis*, *C. ulietensis*, *C. lunula*, *C. kleinii*, and *C. melannotus*.

At Heron Island, where the southern group predominates, *C. rainfordi* comprised 42 percent of the community, and the 4 members of the southern group comprised 61 percent of all the individuals of the 23 species. *Chaetodon aureofasciatus* was the second most common species, making up 10 percent. At Hayman Island, the calm sheltered bays, rich in soft corals, contained only 7 species. *Chaetodon rainfordi* and *C. aureofasciatus* were again the most common, but here *C. aureofasciatus* was dominant, 73 to *C. rainfordi's* 17 percent. Both species occurred in approximately equal numbers in the Lizard Island region, but together comprised only 6 percent of the 26 species. The 6 members of the northern group together made up 68 percent of the fauna.

In the Lizard Island area we sampled in a number of habitats around Lizard Island, and also at Yonge Reef on the outer barrier. This spectrum of stations allowed us to see part of the shore-to-Coral Sea species replacement noted by Anderson et al. (1981) and Williams (1982). For example two species, *C. ornatissimus* and *C. pelewensis*, were seen only on the outer barrier reef, while *C. aureofasciatus* and *C. rainfordi* were essentially limited to the immediate vicinity of Lizard Island, corresponding to Williams' (1982) midshelf localities. A large set of species included members that occurred at most sites sampled. This included *C. baronessa*, *C. citrinellus*, *C. vagabundus*, *C. ephippium*, *C. trifasciatus*, *C. plebeius*, and *C. melannotus*. *Chaetodon aureofasciatus*, *C. rainfordi*, *C. auriga*, and *Chelmon rostratus* were most common in sheltered sites with fair amounts of soft coral. Finally *C. rafflesi*, *C. trifascialis*, and *C. ulietensis* were most common where hard corals were best developed.

The Red Sea

Our stations were mostly grouped along the front of the flourishing, rich, and diverse fringing reef several miles south of Eilat on the Israeli cost of the Gulf of Aqaba. The 7-species fauna was dominated by two species: *C. paucifasciatus* (according to Burgess 1978 a Red Sea relative of *C. mertensi*) 54 percent, and *C. austriacus* (a relative of *C. trifasciatus*) 31 percent, the two comprising 85 percent of the community. Our quantitative findings are very close to those of Bouchon-Navaro (1980) for the Jordanian coast of the gulf.

The third commonest species at our stations was *C. fasciatus* (a relative of *C. lunula*) 7 percent. Individuals of this species were fairly conspicuous and were regularly seen foraging over the reef in the daytime, in contrast to *C. lunula* in the Pacific, which, as noted above, we suspected of being nocturnal. Conceivably the greater diurnal activity of *C. fasciatus* is a result of lessened competetive pressure.

Discussion

The overall and most obvious pattern that our data reveal is that in areas which support more species of butterflyfishes more kinds coexist at local sites and population densities of the fishes are higher. The impression one receives is that resources for butterflyfishes must be greater in the areas of the greatest species richness. When this pattern is examined at coarser degrees of resolution, by looking across, rather than within habitat patches, two additional patterns become evident: (1) in species-rich regions there is more replacement of species from station to station. Thus part of the increment of species in these places is contributed by beta diversity. (2) As examined at across-station or across-island scales, the eveness of distribution of individual species, as measured by the Simpson Index, decreases. These are all patterns to be expected if resources were limiting to the fishes in the richest areas, and if a community process, such as competition, were acting upon the populations. These processes obviously cannot be inferred from the pattern, and in any case other observations suggest caution in proposing process-level explanations.

If butterflyfish communities were using their resources maximally, and additional species were added, this would result in a reduction in number of individuals of some of the species (density compensation), but this does not happen in a significant way (Fig. 5). Rather, mean number of individuals per species stays about the same even though the number of species increases (density stasis; Williamson 1981). Why this result which seems contradictory to the evenness and place niche results discussed above? If the species newly added to communities are different enough ecologically to utilize untapped resources, they could enter the community without displacing resident species, and density stasis could be maintained. Indeed our morphological results support this possibility, since some of the added species are peripheral morphologically and, presumably, ecologically. In short, these results are at least consonant with conventional community theory.

Of the areas in which we have worked, the fringing reefs in the northern Red Sea seemed to us to be among the richest in terms of suitable habitat for butterflyfishes. There the amount of healthy, living sceractinian coral, and of other reef invertebrates, was high, and other habitats and sources of food were readily available. Yet only 7 species of chaetodontids were observed, and only a few more are even possible for the area. However, the number of individuals was high, well within the range seen in more speciose areas such as French Polynesia and the Great Barrier Reef. Thus in the Red Sea our impression was that the large numbers of the relatively few species was perhaps a response to the absence of the additional species which would have been seen in more species-rich areas. *Chaetodon fasciatus*, the Red Sea vicariant of the Indo-Pacific *C. lunula*, was common and was regularly observed foraging during the daylight hours when we conducted our censuses. In Australia and Polynesia we gained the impression that *C. lunula* was at least partly nocturnal, since a majority of our sightings were of individuals or groups sheltering in dark places under overhangs. This situation is suggestive of ecological release in the more species-poor but resource-rich environment.

The reduced number of species in the Red Sea, compared with the Indian Ocean, may have a number of explanations. It seems likely to us that isolation of the Red Sea during Pleistocene low sea-level stands (Briggs 1974), with the resultant depressive effect of the reduced area of the sea, is a likely important cause.

Population sizes of the species with which we worked are notably labile. Pre- and post-El Niño population densities in French Polynesia differ by over 40 percent, and densities from island to island in the Caribbean differ even more. In most cases these differences seem attributable to fairly obvious differences in habitat quality. That densities of individual species are not closely keyed to species richness on a local scale is demonstrated by examination of pre- and post-El Niño censuses in French Polynesia. Here, despite a 40 percent drop in number of individuals per species, the number of species present held almost constant. In the Caribbean, where 3 species are almost always recorded, the least common species, *C. ocellatus,* dropped out at only one island (Antigua) with exceptionally low population densities of all chaetodontids, probably attributable to exceptionally poor habitat conditions at all stations.

The Caribbean-Eastern Pacific (CEP) and the Indo-West Pacific (including the Red Sea) (IWP) are very different in terms of butterflyfish diversity. In the former the number of species and the biomass of chaetodontids are low. In the latter, both are high. Species richness for most reef organisms follows the same pattern (Stehli & Wells 1971, Hall 1962, Schilder 1965, Abbott 1960, Stehli et al. 1967, Sale 1980), but data to make comparisons of biomass are not so easily come by. Is the IWP simply more productive of resources suitable for butterflyfishes, or is the region more speciose largely because of historical factors? Milliman (1973) suggested that more drastic Pleistocene climatic changes may have been experienced by the Caribbean than by the Indo-West Pacific resulting in more extinctions in the western Atlantic. More recent reconstructions of Pleistocene climates seem to negate this possibility (CLIMAP Committee Members 1976, Gates 1976). Adey & Burke (1977) noted that the ratio of coral species in the two areas (14 times as many in the Indo-West Pacific) is almost exactly the ratio of shallow marine area in the two regions. Butterflyfishes may be more speciose in the Indo-West Pacific because they have suffered fewer extinctions there, because of greater area effect with increased habitat diversity and opportunity for speciation, or because the diversity and density of resources available to them there is greater. If both regions were approximately the same in suitability for chaetodontids, it might be expected that the few species present in the CEP would experience massive ecological release, occupy a broad range of habitats, and be more common, species for species, than their relatives in the IWP. Such is clearly not the case. Total chaetodontid biomass is much less in the CEP. The species that are present exist at approximately the same population densities as IWP kinds. The 3 commonly observed species in the Caribbean seem to show about the same range of habitat use as species elsewhere. Indeed, three species not appearing on our Caribbean transects seem to have specialized in life at greater depths rather than having insinuated themselves ecologically into the shallow-water community. The average number of individual chaetodontids per transect in the Carribean is 4–5; in the IWP it is 20–30. It is not subjectively obvious to us that the best IWP habitats are 5–6 times better for chaetodontids than the best Caribbean ones. We therefore incline toward historical explanations of the differences between the two faunas.

Future studies

The results reported here give no indication that butterflyfish communities are responding to increases in species richness in any of the ways predicted by resource-limited competition-based community theory, except that (1) at spatial scales above the level of the local community, that is, at scales including whole islands or whole regions, spatial niche breadth of species declines as the number of species inhabiting the larger area increases, (2) reduction in live scleractinian coral in Franch Polynesia resulting from the 1982–1983 El Niño was correlated with the dramatic decline in the abundance, though not the species richness, of chaetodontids, and (3) the chaetodontid community of the northern Red Sea gave some evidence suggestive of ecological release as a result of the reduced number of species in that rich habitat. Thus the suspicion remains that chaetodontids are resource-limited, but that the census approach used here is not sufficiently sensitive to detect the mode of limitation. Some approaches which suggest themselves which may help to determine if resources are limiting or not, and if so in what ways include: (1) an extension of the census method used here to the putatively most species-rich area in the Philippines-Malaysia-New Guinea triangle, asking if an upper asymptote to the size of local communities is reached when the local species total reaches maximal levels, (2) detailed studies of food and space use in a set of the same species in areas of differing species richnesses of local communities, asking if there is evidence of ecological release (niche broadening) in the less speciose communities, and (3) intensive search for evidence of inter- and intra-specific territorial behavior in communities of differing species richnesses and individual densities, asking if territoriality indeed exists, and if it is intensified in response to increasing species richness and/or population density.

Summary

1. Species richness of chaetodontid communities is positively correlated with number of individual chaetodontids.

2. Increasing species richness is accompanied by an increased number of species at single sites (alpha diversity) as well as by increased replacement of species from habitat to habitat (beta diversity).

3. Place niche breadth of individual species of chaetodontids averages lower at islands or in regions that support more species, but at the level of single-habitat census sites there is no relationship between niche breadth and species richness.

4. Density compensation is not observed in chaetodontid communities. Rather, density stasis seems to prevail, possibly because increased species richness is made possible partly by the addition of ecologically distinctive species which do not displace other kinds.

5. More species rich communities occupy larger morphological spaces than less rich ones because the additional species are often added at the periphery of morphospace in keeping with their morphological (and presumed ecological) distinctiveness.

Acknowledgements

Financial support for work in Australia and Belize was provided in part by the University of New Mexico Research Allocations Committee. We acknowledge with pleasure the field assistance received from Joan Findley, Robert Goslow, Norman Scott, Terry and Nancy Yates, Wayne and Jolene Maes, Donald Thomson, and Donald Duszynski. The paper benefitted in numerous ways from the constructive criticism of Philip Motta and of two anonymous reviewers.

References cited

Abbott, R.T. 1960. The genus *Strombus* in the Indo-Pacific. Mollusca 1(2): 33–146.

Adey, W.H. & R.B. Burke. 1977. Holocene bioherms of the Lesser Antilles-geologic control of development. Amer. Assoc. Petrol. Studies in Geol. 4: 67–81.

Anderson, G.V.R., A.H. Ehrlich, P.R. Ehrlich, J.D. Roughgarden, B.C. Russell & F.H. Talbot. 1981. Community structure of coral reef fishes. Amer. Nat. 117: 475–495.

Birkeland, C. & S. Neudecker. 1981. Foraging behavior of two Caribbean chaetodontids: *Chaetodon capistratus* and *C. aculeatus*. Copeia 1981: 169–178.

Bouchon-Navaro, Y. (1980). Quantitative distribution of the Chaetodontidae on a fringing reef of the Jordanian coast (Gulf of Aqaba). Tethys 9: 247–251.

Bouchon-Navaro, Y., C. Bouchon & M.L. Harmelin-Vivien. 1985. Impact of coral degradation on a chaetodontid fish assemblage (Moorea, French Polynesia). Proc. 5th Int. Coral Reef Symp., Tahiti 5: 427–432.

Briggs, J.C. 1974. Marine zoogeography. McGraw-Hill Book Co., New York. 475 pp.

Burgess, W.E. 1978. Butterflyfishes of the world. T.F.H. Publications, New Jersey. 832 pp.

Clarke, R.D. 1977. Habitat distribution and species diversity of chaetodontid and pomacentrid fishes near Bimini, Bahamas. Marine Biol. 40: 277–289.

CLIMAP Project Members, 1976. The surface of the ice-age earth. Science 191: 1131–1137.

Ehrlich, P.R., F.H. Talbot, B.C. Russell & G.R.V. Anderson. 1977. The behaviour of chaetodontid fishes with special reference to Lorenz's 'poster colouration' hypothesis. J. Zool. London 183: 213–228.

Findley, J.S. 1976. The structure of bat communities. Amer. Nat. 110: 129–139.

Findley, J.S. & M.T. Findley. 1985. A search for pattern in butterfly fish communities. Amer. Nat. 126: 800–816.

Fox, B.J. 1981. Niche parameters and species richness. Ecology 64: 625–630.

Gates, W.L. 1976. Modelling the ice-age climate. Science 191: 1138–1144.

Gladfelter, W.B., J.C. Ogden & E.H. Gladfelter. 1978. Similarity and diversity among coral reef fish communities: a comparison between tropical western Atlantic (Virgin Islands) and tropical central Pacific (Marshall Islands). Ecology 61: 1156–1168.

Greenfield, D.W., D. Hensley, J.W. Wiley & S.T. Ross. 1970. The Isla Jaltemba coral formation and its zoogeographical significance. Copeia 1970: 180–181.

Hall, D.N.F. 1962. Observations on the taxonomy and biology of some Indo-West-Pacific Penaeidae (Crustacea, Decapoda). Colonial Office Fishery Pub. 17: 1–229.

Hanski, I. 1982. Dynamics of regional distribution: the core and satellite species hypothesis. Oikos 38: 210–221.

Harmelin-Vivien, M.L. & Y. Bouchon-Navaro. 1983. Feeding diets and significance of coral feeding among chaetodontid fishes in Moorea (French Polynesia). Coral Reefs 2: 119–127.

Hobson, E.S. 1974. Feeding relationships of teleostean fishes on coral reefs in Kona, Hawaii. U.S. Fish. Bull. 72: 915–1031.

Hoese, H.D. & R.H. Moore. 1977. Fishes of the Gulf of Mexico, Texas, Louisiana, and adjacent waters. Texas A & M Press, College Station. 327 pp.

Milliman, J.D. 1973. Caribbean coral reefs. pp. 1–50. *In:* O.A. Jones & R. Endean (ed.) Biology and Geology of Coral Reefs, I, Geology I, Academic Press, New York.

Randall, J.E. 1968. Caribbean reef fishes. T.F.H. Publications, New Jersey. 318 pp.

Randall, J.E. 1983. Red Sea reef fishes. IMMEL Publishers, London. 192 pp.

Randall, J.E. 1985. Fishes. pp. 462–480. *In:* B. Delesalle, R. Galzin, & B. Salvat (ed.) French Polynesian Coral Reefs, Proc. 5th Int. Coral Reef Symp. 1, Tahiti.

Schilder, F.A. 1965. The geographical distribution of Cowries (Mollusca: Gastropoda). The Veliger 3: 171–183.

Schulman, M.J. 1983. Species richness and community predictability in coral reef fish faunas. Ecology 64: 1308–1311.

Steene, R.C. 1977. Butterfly and angelfishes of the world. Wiley-Interscience, New York. 144 pp.

Stehli, F.G., A.L. McAlester & C.E. Helsley. 1967. Taxonomic diversity of Recent bivalves and some implications for geology. Bull. Geol. Soc. Amer. 78: 455–465.

Thomson, D.A., L.T. Findley & A.N. Kerstitch. 1979. Reef fishes of the Sea of Cortez. Wiley-Interscience, New York. 302 pp.

Williams, D.McB. 1982. Patterns in the distribution of fish communities across the central Great Barrier Reef. Coral Reefs 1: 35–43.

Williamson, M. 1981. Island populations. Oxford University Press, Oxford. 286 pp.

Correlations between chaetodontid fishes and coral communities of the Gulf of Aqaba (Red Sea)

Yolande Bouchon-Navaro & Claude Bouchon
Laboratoire de Biologie Animale, Université des Antilles et de la Guyane, B.P. 592, Pointe-à-Pitre, 97167 Guadeloupe (France)

Received 3.6.1988   Accepted 10.11.1988

Key words: Butterflyfishes, Coral distribution, Relationships, Red Sea

Synopsis

The relationships existing between the chaetodontid fishes and the surrounding coral communities were investigated in the Gulf of Aqaba. Quantitative data were analysed by a correspondence and a cluster analysis. The results demonstrated a similarity in the spatial distribution of both communities. Significant correlations were found between the density of chaetodontid fishes and the diversity of the coral community as well as the substratum coverage by the coral colonies. The density of exclusive coral browsers was also correlated to the abundance of branching colonies. Among the different genera of branching corals, correlations were significant only for the genus *Acropora*. These results suggested the existence of strong links between coral and chaetodontid fish assemblages.

Introduction

The distribution pattern of butterflyfishes is well documented for different reefs throughout the world in several studies on the zonation of coral reef fishes or in works dealing more particularly on the chaetodontid family. Such works were made on a quantitative basis in the Caribbean region (Clarke 1977, Birkeland & Neudecker 1981, Findley & Findley 1985), the South China Sea (Wood 1979), the Red Sea (Bouchon-Navaro 1980), the northern Great Barrier Reefs of Australia (Anderson et al. 1981) and French Polynesia (Bouchon-Navaro 1981, Bell et al. 1985, Bouchon-Navaro et al. 1985, Findley & Findley 1985).

Some relationships between the distribution of Chaetodontidae and the percentage of live coral cover were searched in different regions of the Pacific (Bouchon-Navaro et al. 1985, Bell et al. 1985, Findley & Findley 1985) and the Caribbean (Birkeland & Neudecker 1981, Findley & Findley 1985). However, no detailed quantitative data are available on both coral and coral fish species distribution, except for French Polynesia (Bouchon-Navaro et al. 1985).

Quantitative data were obtained on the Chaetodontidae of the Gulf of Aqaba (Bouchon-Navaro 1980) and simultaneously on the coral communities at the same stations (Bouchon 1980). For the present paper, these data were used to search for the correlations between the distribution patterns of the two communities. Do the distribution of chaetodontid species coincide with that of the coral assemblages? Are there significant correlations between the parameters of the two taxa? What is the importance of the coral growth forms for the Chaetodontidae?

Materials and methods

The study area

The Gulf of Aqaba represents the north prong of the Red Sea. The Jordanian coast extends for about 25 km at the North-East end of the gulf. Details on the morphology of the Jordanian reefs were given by Mergner & Schuhmacher (1974), Bouchon (1980) and Bouchon et al. (1981).

The Jordanian coast is fringed by a discontinuous belt of reefs along 13 km. These reef formations are principally developed around headlands and are separated by sedimentary embayments. All of them are of fringing type. The reef where the study was conducted is located 10 km south of Aqaba town. It is 600 m long and the reef front runs parallel to the coast at about 60 m from the shore (Fig. 1). From the shore seawards, there is a succession of a back reef channel and a reef flat. The back reef channel is a sedimentary depression with a maximum width of 40 m and 1 to 2 m deep. A gentle slope grades from the channel bottom to the back reef beach. The bottom is colonized by scattered coral heads and patches of seagrasses (*Halophila stipulacea* and *Halodule uninervis)*. The reef flat, about 20 m wide is a dead coral flagstone with an irregular back reef margin. It is colonized by a coral community composed of small-sized colonies. The reef flat is followed by the outer reef slope which starts with a steep drop off of some meters high, which characterizes the Jordanian coral reef front. The inclination of the outer slope is about 20°. At 50 m deep, it is interrupted by a vertical drop off about 50 m high. The outer slope is characterized by a high coral coverage except in sedimentary pools which are numerous between 5 to 15 m.

Methods

A stratified sampling strategy was adopted to study the distribution of Chaetodontidae and that of corals. Quantitative data on both communities were collected in the study area at 9 stations distributed along an axis perpendicular to the shore line (Fig. 1). The stations were respectively located in the boat channel (station 1), the reef flat (stations 2, 3) the upper part of the reef front (station 4) and on the outer slope at 5 m (station 5), 10 m (station 6), 20 m (station 7), 30 m (station 8), and 40 m (station 9).

The Chaetodontidae were studied by means of visual counts made by snorkeling on the reef flat and by SCUBA diving on the outer reef slope. Fishes were counted along a band 400 m long by 5 m wide, parallel to the shore-line. The data were obtained during day-time between 10:00 h and 14:00 h so as to avoid the diurnal variations of the fish community.

The distribution of corals was quantitatively studied at the same stations using a line transect technique (Loya 1972, Bouchon 1980, 1981). At each station, several 10 m long transects were established parallel to the shore line to avoid variations in ecological conditions down the slope. In station 1, 20 transects were studied and 5 in each of the other stations. Any coral colony underlying the line transect was recorded and its intercept length measured in order to calculate the substrate coverage. The growth forms of the colonies was also noted as encrusting, massive or branching (small-branched and long-branched colonies). Considering that the fishes should be more sensitive to the surface occupied by the corals than to the number of colonies, only the data dealing with the coverage of the substratum were used for this study.

From the data thus obtained, synthetic parameters such as the Shannon & Weaver index of diversity (H') and the Pielou evenness (J') were computed for both coral and fish communities. The raw data obtained, even transformed, did not fit with Gaussian distributions. So, non parametric statistical treatments were used for data processing.

The coral and the fish distribution data were treated with a correspondence analysis in which the data on fishes acted only as supplementary factors. Moreover, two matrices of distances were computed, using the χ^2 metric, for fishes and for coral communities. These matrices were used for a cluster analysis. Sorting was based on the Lance & Williams (1967) flexible fusion strategy (with $\alpha = 0.625$ and $\beta = -0.25$). The resulting groups were projected on the graph obtained from the factorial

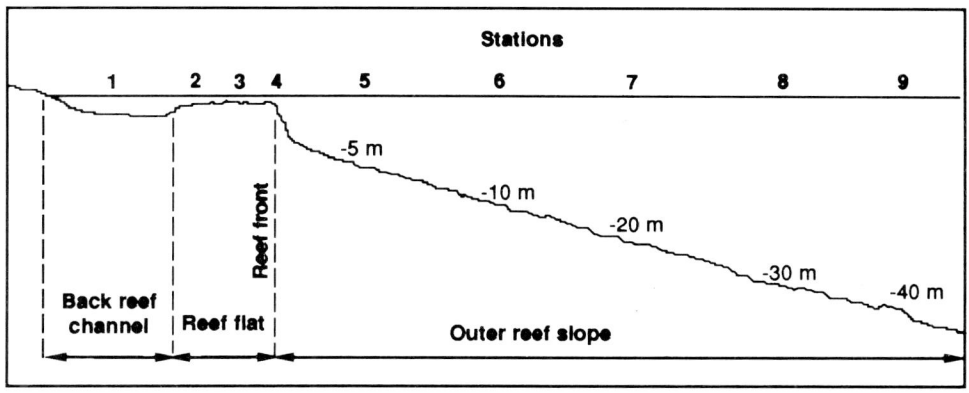

Fig. 1. Schematic cross section of the studied reef and location of the sampling stations.

analysis. The relationships between the parameters of the two communities were searched using the Spearman rank correlation coefficient.

Results

The chaetodontid community was composed of 7 species. Their distribution was detailed previously (Bouchon-Navaro 1980) and is quantitatively summarized in Table 1. A study on their diets revealed that three species had a strict diet based either on coral polyps (*Chaetodon austriacus* and *C. trifascialis*) or alcyonarians (*C. melannotus*) whereas the other species fed on a variety of organisms (Harmelin-Vivien & Bouchon-Navaro 1981). Thus, the density of exclusive coral browsers was calculated (Table 1).

The coral community of Aqaba includes 154 species (Bouchon, unpublished data) of which 115 species were observed on the study reef (Table 2). Species richness, percentage of coral cover, diversity indices and results concerning the distribution of coral growth forms (size of colonies) are available in Table 2.

Table 1. Quantitative distribution of the Chaetodontidae in the nine studied stations (number of individuals for 1000 m²).

Species	Stations								
	1	2	3	4	5	6	7	8	9
Chaetodon auriga	0	0.5	2.5	0	0	0	0	0	1.0
Chaetodon austriacus	0	3.5	14.5	22.5	35.5	27.0	21.5	8.0	5.5
Chaetodon fasciatus	0	1.5	4.0	1.5	2.0	0	0	1.0	0
Chaetodon melannotus	0	0	0.5	0.5	1.0	2.5	4.5	0	0
Chaetodon paucifasciatus	0.5	24.5	43.5	55.5	73.0	74.0	97.0	44.0	23.0
Chaetodon trifascialis	0	0.5	0.5	10.5	2.5	0.5	3.0	2.0	0
Heniochus intermedius	0	0.5	0.5	1.0	3.5	0	2.0	1.0	0
Total density	0.5	31.0	66.0	91.5	117.5	104.0	128.0	56.0	29.5
Density of coral browsers	0	4.0	15.0	33.0	38.0	27.5	24.5	10.0	5.5
Species richness	1	6	7	6	6	4	5	5	3
Shannon index	0	1.12	1.46	1.50	1.38	1.02	1.12	1.05	0.90
Pielou evenness	–	0.43	0.52	0.58	0.53	0.51	0.48	0.45	0.57

Table 2. Quantitative distribution of the coral species in the nine studied stations (coverage in cm for 50 m of transect).

Code	Species	Stations								
		1	2	3	4	5	6	7	8	9
1	*Psammocora explanulata*	0	0	0	0	0	5	5	0	5
2	*Psammocora haimeana*	0	0	0	0	0	5	10	0	13
3	*Psammocora profundacella*	0	0	0	0	10	0	5	15	29
4	*Stylophora pistillata*	130	424	308	70	135	135	50	0	15
5	*Stylophora sp.*	0	20	0	0	20	85	35	25	30
6	*Pocillopora damicornis*	0	0	0	10	0	20	0	0	0
7	*Pocillopora verrucosa*	0	0	10	45	0	0	0	0	0
8	*Seriatopora hystrix*	0	150	163	110	10	5	0	24	0
9	*Stylocoeniella guentheri*	0	0	0	0	10	11	30	28	36
10	*Acropora digitifera*	0	40	0	0	0	80	0	0	0
11	*Acropora cf erythraea*	0	0	0	0	10	0	0	0	0
12	*Acropora eurystoma*	0	0	0	25	0	0	0	0	0
13	*Acropora granulosa*	0	0	0	0	0	190	225	195	67
14	*Acropora hemprichii*	0	30	0	35	115	140	50	50	0
15	*Acropora humilis*	18	10	155	15	30	0	0	0	0
16	*Acropora hyacinthus*	3	50	70	20	0	0	0	0	0
17	*Acropora nasuta*	0	0	0	0	105	0	30	0	0
18	*Acropora cf squarrosa*	0	0	0	0	0	0	30	0	0
19	*Acropora valida*	0	0	0	105	0	60	25	115	15
20	*Astreopora myriophthalma*	0	0	0	0	0	45	65	10	0
21	*Montipora danae*	0	0	0	0	5	25	75	32	41
22	*Montipora cf erythraea*	0	0	10	210	220	275	255	185	50
23	*Montipora informis*	0	0	0	0	0	0	0	30	67
24	*Montipora monasteriata*	0	0	0	0	15	40	85	168	82
25	*Montipora spongiosa*	0	0	0	0	10	0	0	0	0
26	*Montopora cf spumosa*	0	0	0	0	55	55	25	40	45
27	*Montipora tuberculosa*	0	0	0	60	15	30	0	0	0
28	*Montipora venosa*	3	15	15	0	0	10	0	0	0
29	*Montipora verrucosa*	0	0	0	0	20	40	70	69	5
30	*Pavona cactus*	0	0	0	0	0	35	0	0	0
31	*Pavona decussata*	0	0	0	0	0	10	10	10	0
32	*Pavona divaricata*	0	5	0	15	20	20	0	0	0
33	*Pavona explanulata*	0	0	0	5	0	0	0	0	0
34	*Pavona maldivensis*	0	0	20	30	0	0	0	0	0
35	*Pavona varians*	0	0	10	40	26	70	10	7	17
36	*Leptoseris explanata*	0	0	0	10	0	0	0	0	15
37	*Leptoseris mycetoseroides*	0	0	0	0	0	0	0	0	20
38	*Leptoseris yabei*	0	0	0	0	0	0	10	0	30
39	*Coscinarea monile*	0	0	0	10	0	10	20	23	36
40	*Siderastrea savignyana*	0	0	0	0	0	0	0	5	0
41	*Cycloseris cf doderleini*	0	0	0	0	0	0	0	3	0
42	*Cycloseris patelliformis*	0	0	0	0	0	0	0	0	5
43	*Fungia fungites*	0	0	0	0	25	25	0	10	15
44	*Fungia granulosa*	0	0	0	10	10	10	12	0	0
45	*Fungia klunzingeri*	0	0	0	0	25	20	10	20	3
46	*Fungia mollucensis*	0	0	0	0	0	0	0	10	5
47	*Fungia paumotensis*	0	0	0	0	0	0	0	5	0
48	*Fungia scruposa*	0	0	0	0	15	15	0	0	5
49	*Fungia scutaria*	0	0	15	10	5	0	0	0	0
50	*Herpetoglossa simplex*	0	0	0	5	35	30	0	0	0

Table 2. (Continued).

Code	Species	Stations								
		1	2	3	4	5	6	7	8	9
51	*Herpolitha limax*	0	0	0	0	0	0	0	20	0
52	*Podabacia crustacea*	0	0	0	0	0	0	10	20	0
53	*Porites columnaris*	0	0	0	70	0	0	0	0	0
54	*Porites lobata*	0	0	0	0	0	20	35	106	48
55	*Porites lutea*	0	0	0	5	0	5	40	35	38
56	*Porites solida*	0	0	0	0	0	25	5	0	0
57	*Porites (Synarea) undulata*	0	0	0	0	0	0	30	0	0
58	*Goniopora djiboutiensis*	0	0	0	0	0	0	0	5	0
59	*Goniopora planulata*	0	0	0	0	20	5	0	0	0
60	*Goniopora savignyi*	0	0	0	0	10	20	50	5	0
61	*Goniopora somaliensis*	0	0	0	20	0	0	0	0	5
62	*Goniopora stutchburyi*	0	0	0	0	0	0	0	0	5
63	*Alveopora allingi*	0	0	0	0	0	0	0	60	75
64	*Alveopora fenestrata*	3	0	0	0	0	0	0	0	0
65	*Alveopora ocellata*	0	0	0	0	0	0	0	0	262
66	*Alveopora verrilliana*	0	0	0	0	5	10	50	120	550
67	*Barabattoia amicorum*	0	0	0	0	0	10	0	0	0
68	*Favia favus*	11	25	20	20	40	0	70	15	38
69	*Favia laxa*	0	0	0	0	15	20	50	15	0
70	*Favia pallida*	0	0	0	0	0	45	30	61	53
71	*Favia rotumana*	3	0	0	0	0	0	0	0	0
72	*Favia stelligera*	0	0	0	0	55	0	0	0	0
73	*Favites abdita*	0	0	0	0	0	10	72	10	5
74	*Favites chinensis*	29	65	95	0	0	0	0	0	0
75	*Favites complanata*	0	0	0	0	0	5	0	0	0
76	*Favites flexuosa*	0	0	0	0	35	5	20	0	0
77	*Favites melicerum*	1	0	0	0	0	0	0	0	0
78	*Favites pentagona*	0	0	0	0	0	10	15	35	27
79	*Favites peresi*	0	0	0	0	35	50	73	89	106
80	*Goniastrea aspera*	9	0	20	15	0	0	0	0	0
81	*Goniastrea edwardsi*	0	0	0	20	85	25	10	0	0
82	*Goniastrea pectinata*	0	0	0	0	85	195	160	34	35
83	*Goniastrea retiformis*	0	0	40	0	0	0	0	0	0
84	*Erythrastrea flabellata*	0	0	0	0	0	15	10	0	0
85	*Platygyra daedalea*	0	0	10	10	15	10	10	10	5
86	*Platygyra lamellina*	0	175	65	0	0	45	55	55	20
87	*Hydnophora exesa*	0	0	0	0	50	20	0	10	0
88	*Hydnophora microconos*	0	0	25	20	0	0	0	0	0
89	*Montastrea annuligera*	0	0	0	0	0	5	20	0	0
90	*Montastrea cf valenciennesi*	0	0	0	0	0	0	0	0	5
91	*Leptastrea bottae*	0	0	0	0	0	25	30	0	25
92	*Leptastrea inaequalis*	0	0	0	0	0	0	0	10	5
93	*Leptastrea purpurea*	1	0	20	5	10	38	35	15	32
94	*Cyhastrea chalcidicum*	0	0	0	0	0	11	10	0	10
95	*Cyphastrea microphthalma*	13	50	65	5	15	65	15	24	30
96	*Cyphastrea serailia*	6	10	5	0	5	0	0	0	0
97	*Echinopora fructiculosa*	0	0	0	0	160	235	68	0	30
98	*Echinopora gemmacea*	0	45	90	80	0	40	10	0	0
99	*Echinopora mammillosa*	0	0	0	0	100	125	90	50	25
100	*Galaxea fascicularis*	0	0	0	10	10	0	20	25	25

Table 2. (Continued).

Code	Species	Stations								
		1	2	3	4	5	6	7	8	9
101	*Acanthastrea echinata*	0	0	70	0	25	0	5	19	0
102	*Blastomussa merleti*	0	0	0	0	0	0	5	0	5
103	*Cynarina lacrymalis*	0	0	0	0	0	0	2	0	3
104	*Lobophyllia corymbosa*	0	0	20	30	250	20	20	0	0
105	*Lobophyllia hemprichii*	0	0	60	0	0	55	0	20	0
106	*Echinophyllia aspera*	0	0	0	10	10	10	45	28	5
107	*Mycedium elephantotus*	0	0	0	0	170	105	15	0	0
108	*Gyrosmilia interrupta*	0	0	0	0	0	0	0	5	25
109	*Dentrophyllia arbuscula*	0	0	0	10	0	0	0	0	0
110	*Tubastrea diaphana*	0	0	10	20	0	0	0	0	0
111	*Turbinaria mesenterina*	0	0	0	0	0	10	20	0	15
112	*Tubipora musica*	9	3	15	5	5	5	0	8	20
113	*Millepora dichotoma*	0	0	120	700	30	0	0	0	0
114	*Millepora cf exaesa*	18	20	0	0	10	8	28	51	35
115	*Millepora platyphylla*	0	0	130	415	10	0	0	0	0
	Small-branched colonies	151	729	973	1170	860	1038	576	515	682
	Long-branched colonies	0	0	60	485	40	60	0	30	0
	Massive colonies	71	323	330	320	845	1230	1367	1109	848
	Encrusting colonies	19	85	285	330	470	498	437	362	672
	Species richness	15	17	28	39	50	65	59	52	56
	% coral cover	5.1	22.7	33.1	46.2	44.1	56.2	47.5	40.8	44.4
	Shannon index	2.60	3.04	4.09	3.76	4.80	5.16	5.21	4.99	4.62
	Pielou evenness	0.70	0.74	0.85	0.71	0.85	0.86	0.89	0.88	0.80

Chaetodontid fishes and coral assemblages

The correspondence analysis results of the coral data are presented for the projection of the stations on the two principal axes in Figure 2 and for the species in Figure 3. The results of the cluster analysis for the coral community are presented for the stations in Figure 4 and for the species in Figure 5. The resulting groups were projected on the graphs obtained from the factorial analysis.

An examination of Figure 2 shows that the first axis of the correspondence analysis (41.3% of the inertia of the raw data matrix) separates the stations according to depth. The second axis (23.1% of the inertia of the data) have a low action on the coral communities of the outer reef slope but separates the stations from the back reef zone and the reef flat from those of the reef front. The cluster analysis completes these results (Fig. 4). Two groups of stations are separated from the data: one group gathering the stations of the outer reef slope and another one those of the reef flat and the back reef channel. The reef front occupies an intermediate position and is linked at a low level to the reef flat.

The projection of the species on the first plan of the factorial analysis shows similar results (Fig. 3). The species of the outer reef slope are distributed along the first axis in the upper quadrants of the graph. Those from the back reef channel, the reef flat and the reef front are mainly gathered in the lower quadrants and separated by the second axis. Examination of the cluster analysis (Fig. 5) shows the separation of two principal groups of coral species: one for the species of the reef slope (group A) and another for the species of the back reef zone and the reef flat (group B). To this last group, a small number of species of the reef front are linked

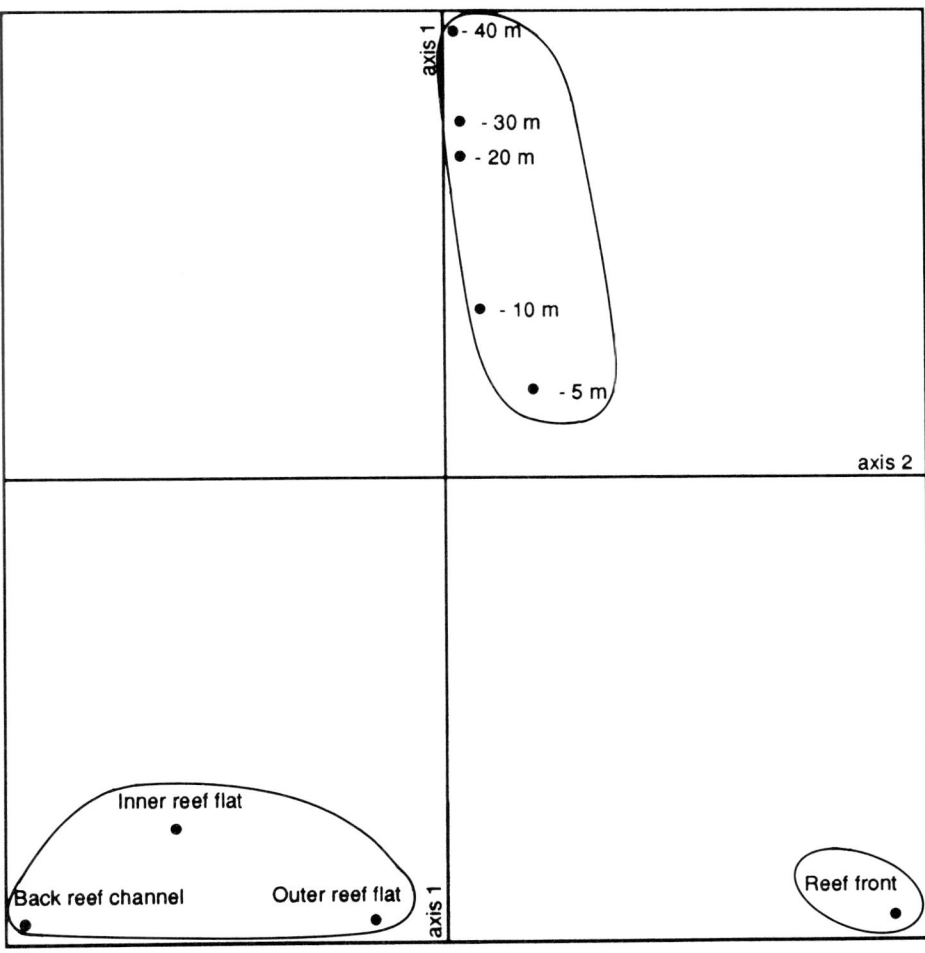

Fig. 2. Projection of the stations on the principal axis of the correspondence analysis. The groups obtained from the cluster analysis are circled.

at a lower level of similarity (group C).

So, two main coral assemblages composed the studied reef: one on the outer reef slope and another one on the back reef zone and the reef flat. They are separated by the reef front which is characterized by a particular coral assemblage of the reef flat.

The cluster analysis calculated on the distribution of the Chaetodontidae (Fig. 6) demonstrated the existence of two main groups: *Chaetodon auriga* and *C. fasciatus* on the one hand and *C. austriacus, C. melannotus, C. paucifasciatus* and *Heniochus intermedius* on the other hand. *C. trifascialis* is linked to this last group at a lower level of similarity.

In the correspondence analysis of the coral data, the Chaetodontidae were used as supplementary factors. The examination of the projection of the chaetodontid species on the first plan of the factorial analysis shows that the distribution of fishes is superimposed to that of corals (Fig. 3). So, it appeared that there is a correspondence between the distribution of chaetodontid and coral assemblages.

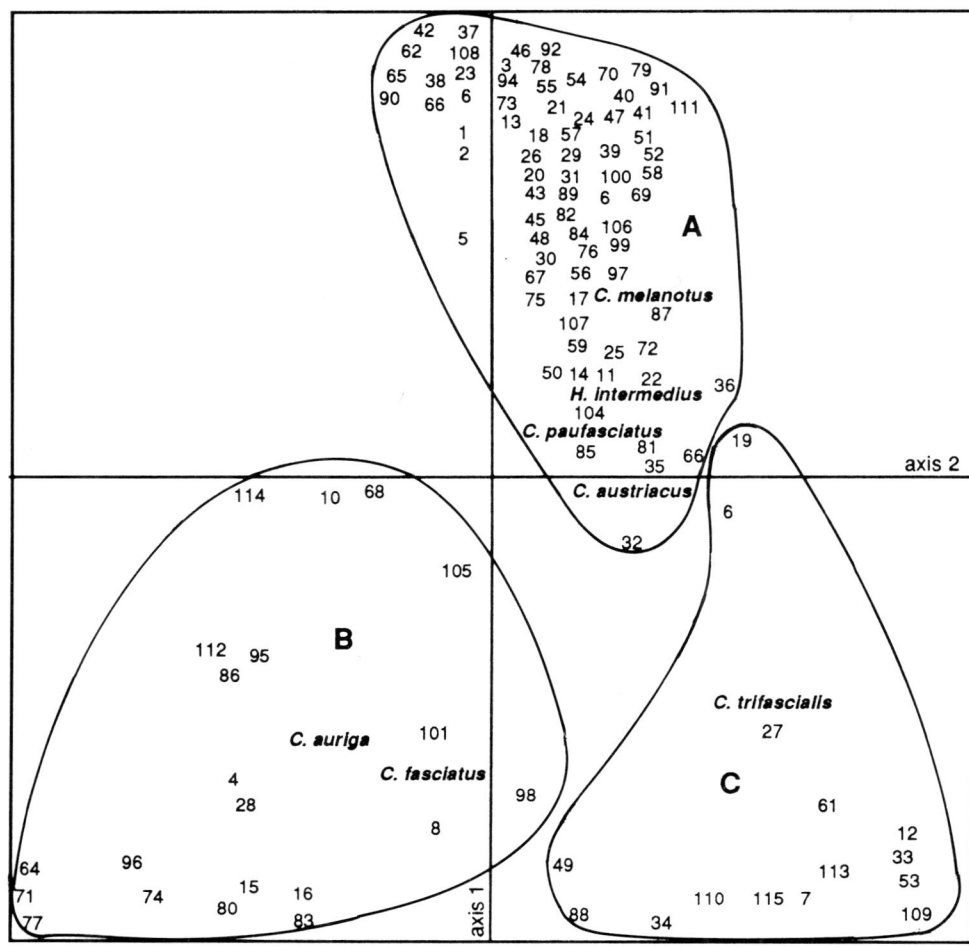

Fig. 3. Projection of the species on the principal axis of the correspondence analysis. The three groups (A, B, C) obtained from the cluster analysis are circled (the coral species code is available in Table 2).

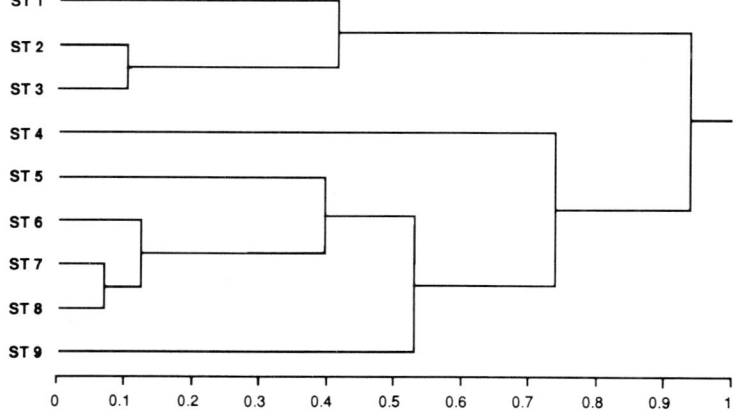

Fig. 4. Cluster analysis of the stations (ST1 to ST9) calculated from the quantitative data on the coral.

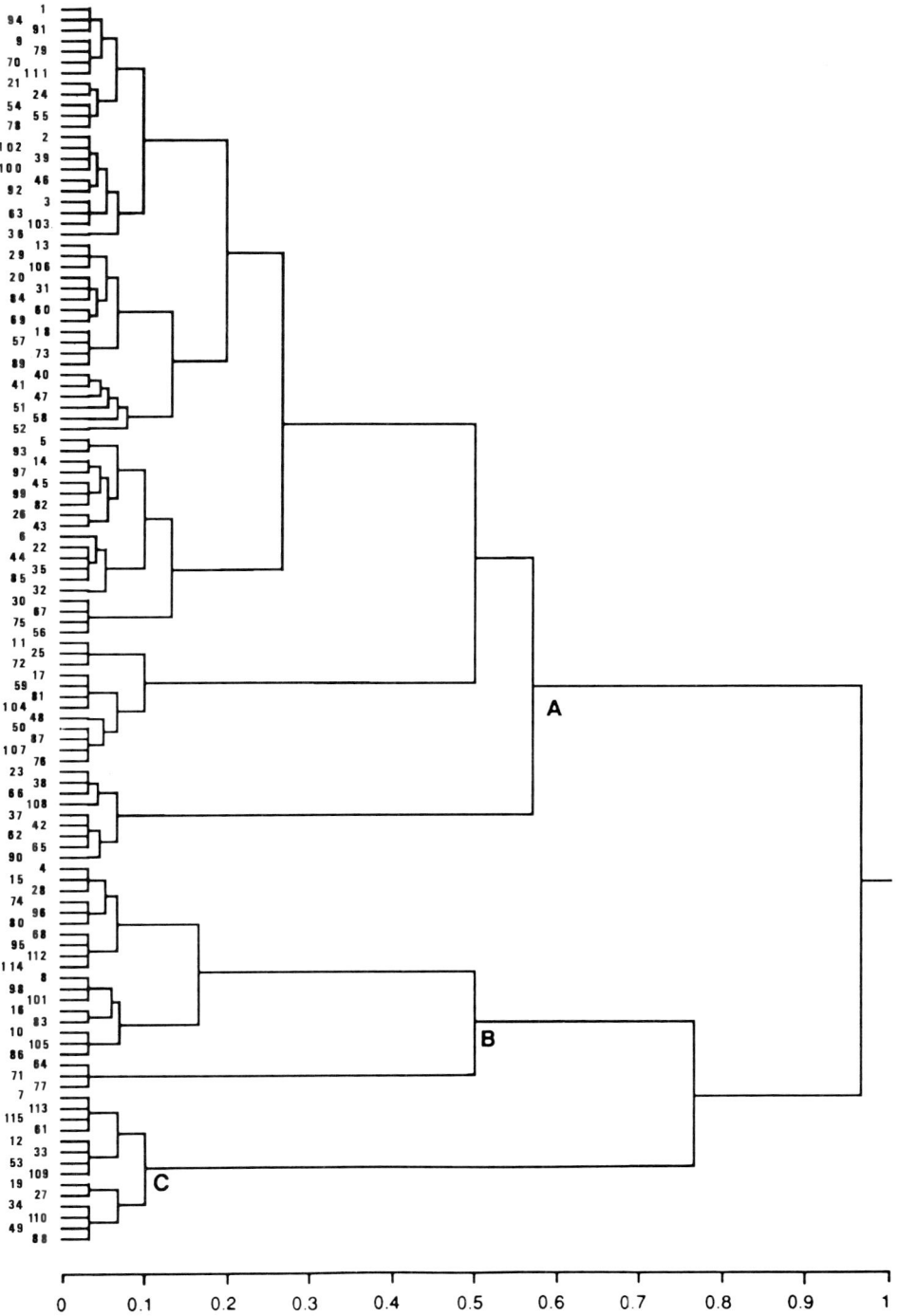

Fig. 5. Cluster analysis of the coral species (the coral species code is available in Table 2). The three main groups A, B, C are projected on the principal axis of the correspondence analysis (Fig. 3).

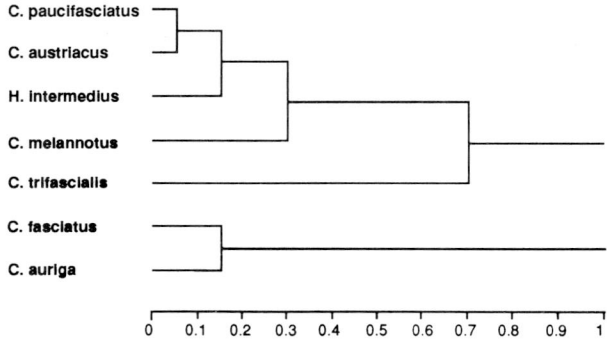

Fig. 6. Cluster analysis of the chaetodontid species. Two main groups of species are separated. *Chaetodon trifascialis* is isolated.

tions were found between the total density of chaetodontid fishes as well as the number of exclusive coral browsers and the percentage of live coral cover. The density of the fishes was also significantly correlated with the diversity index of the coral community. Significant correlations were found for the distribution of three chaetodontid species. The distribution of *Chaetodon paucifasciatus* was found to be correlated with the coral species richness, the substrate coverage by corals, and the Shannon index of corals (the diversity index integrates the species richness value and the last correlation may be an artefact). The distribution of *C. austriacus* and *C. melannotus* was significantly correlated with the percentage of the coral cover (Table 3).

Chaetodontid fishes and coral community parameters

The correlations between the parameters of the chaetodontid assemblage (species richness, density of the whole community, density of exclusive coral browsers, diversity indices, distribution of each chaetodontid species) and several parameters of the coral communities (the coral species richness, the percentage of coral cover, the diversity indices) were investigated (Table 3). Significant correla-

Chaetodontid fishes and coral growth forms

To check the importance of coral growth forms for the Chaetodontidae, Spearman rank correlation coefficients were calculated between the parameters of the fish community and the distribution of the main coral growth forms (Table 4). The density of exclusive coral browsers was significantly correlated with the distribution of branching corals (both small-branched and long-branched colonies). Significant correlations were also observed

Table 3. Spearman correlation coefficients between the chaetodontid community and the coral community parameters (*: correlations significant at 95%; **: 98%).

Chaetodontidae	Corals			
	Species richness	% coral cover	Shannon index	Pielou evenness
Species richness	−0.272	−0.102	−0.085	0.087
Total density	0.567	0.700*	0.717*	0.542
Density of coral browsers	0.467	0.683*	0.517	0.424
Shannon index	−0.151	0.134	−0.008	0.009
Pielou evenness	0.333	0.550	0.167	0.051
Chaetodon auriga	−0.248	−0.337	−0.317	−0.121
Chaetodon austriacus	0.550	0.717*	0.600	0.542
Chaetodon fasciatus	−0.463	−0.341	−0.280	−0.008
Chaetodon melannotus	0.568	0.726*	0.664	0.480
Chaetodon paucifasciatus	0.683*	0.783**	0.800**	0.627
Chaetodon trifascialis	0.238	0.468	0.391	0.182
Heniochus intermedius	0.094	0.171	0.359	0.183

between the Pielou evenness and the distribution of small-branched colonies. The distribution of each chaetodontid species, was significantly correlated with the branched corals for only *C. austriacus*. No other correlation was found between the Chaetodontidae and other forms of coral colonies (massive and encrusting).

We investigated the correlations between the Chaetodontidae and the main genera which form the branching colonies: *Stylophora*, *Pocillopora*, *Seriatopora* and *Acropora* (Table 5). Correlations were significant only with the genus *Acropora* (Table 5) and involved three species: *C. austriacus*, *C. melannotus*, *C. paucifasciatus*.

Discussion

By reviewing the relationships between reef fishes and coral reefs, Sutton (1983) underlined the fact that 'coral reefs and their fish faunas must be considered together, in terms of their influence on each other'. Various studies have revealed that the topographic complexity of coral reef habitats was correlated with the total fish species richness and diversity (Risk 1972, Talbot & Goldman 1972, Luckhurst & Luckhurst 1978, Gladfelter et al. 1980, Bradbury & Young 1981, Carpenter et al. 1981). The relationship between living coral cover and fish abundance and diversity was demonstrated by Carpenter et al. (1981) and Bell & Galzin (1984) but this relationship was not clear for Luckhurst & Luckhurst (1978). A simultaneous study

Table 4. Spearman correlation coefficients between the Chaetodontid community and the coral growth forms (*: correlations significant at 95%).

	Small branched	Long branched	Massive	Encrusting
Species richness	0.570	0.513	−0.170	−0.246
Total density	0.450	0.428	0.566	0.416
Density of coral browsers	0.683*	0.717*	0.317	0.483
Shannon index	0.652	0.623	−0.150	−0.150
Pielou evenness	0.683*	0.577	0.066	0.566
Chaetodon auriga	0.129	−0.130	−0.158	−0.069
Chaetodon austriacus	0.667	0.691*	0.417	0.550
Chaetodon fasciatus	0.472	0.509	−0.376	−0.314
Chaetodon melannotus	0.455	0.394	0.533	0.402
Chaetodon paucifasciatus	0.400	0.411	0.667	0.466
Chaetodon trifascialis	0.366	0.455	0.213	−0.085
Heniochus intermedius	0.077	0.179	0.205	0.051

Table 5. Spearman correlation coefficients between the Chaetodontidae and the branching coral genera (*: correlations significant at 95%; **: 98%; ***: 99%).

	Stylophora	*Pocillopora*	*Seriatopora*	*Acropora*
Chaetodon auriga	0.336	−0.023	0.382	−0.406
Chaetodon austriacus	0.336	−0.023	0.382	0.717*
Chaetodon fasciatus	0.050	0.485	0.101	−0.026
Chaetodon melannotus	0.209	0.374	−0.124	0.752*
Chaetodon paucifasciatus	0.000	0.306	−0.033	0.900***
Chaetodon trifascialis	−0.195	0.283	0.259	0.536
Heniochus intermedius	−0.145	−0.142	0.226	0.419

made by Mc Manus et al. (1981) on the corals and the coral reef fishes revealed few similarities between the two taxocenes. However, these authors suggested that the relationships might be stronger by restricting the study to a particular family. Reese (1981), Bouchon-Navaro et al. (1985) and Hourigan et al. (1988) have defended the idea that Chaetodontidae could be strongly linked to coral communities.

In the present study, the factorial analysis made on the Chaetodontidae and the coral communities supports that hypothesis. The cluster analyses showed similar assemblages for the two studied taxa. One assemblage of coral species was found on the reef flats of Aqaba. Two chaetodontid species (*C. auriga* and *C. fasciatus*) were preferentially distributed in these zones. Another main assemblage of corals was observed on the outer reef slope which corresponded to the preferential habitat for four chaetodontid species (*C. austriacus*, *C. melannotus*, *C. paucifasciatus* and *C. intermedius*). Another group of coral species characterized the reef front. *C. trifascialis* which corresponded to this coral assemblage was a characteristic species of the Aqaba reef front (Bouchon-Navaro 1980).

In Aqaba, the density of chaetodontid fishes seem to be favoured by a rich coral community (in terms of species diversity and percentage of live coral cover). However, no correlation was found between the species richness of the two communities whereas the correlation was significant in French Polynesia (Bouchon-Navaro et al. 1985).

Other studies have shown a relationships between hard-coral cover and chaetodontid fish abundance (Birkeland & Neudecker 1981, Bell & Galzin 1984, Bell et al. 1985, Findley & Findley 1985, Bouchon-Navaro et al. 1985) although some authors (Bell & Galzin 1984, Findley & Findley 1985) worked on a semi quantitative basis concerning the coral communities. The close relationships existing between chaetodontid fishes and coral communities, especially in terms of percentage of coral cover, was also demonstrated in studies made after the destruction of coral reefs. The Chaetodontidae, especially the exclusive coral browsers, were the most affected by the death of corals (Sano et al. 1984, Bouchon-Navaro et al. 1985, Williams 1986, Sano et al. 1987).

In the Caribbean, Birkeland & Neudecker (1981) found that the abundance of *Chaetodon capistratus*, which ingests mostly hexacorallians, was significantly correlated with total coral cover and with generic diversity of corals. On the reef of Aqaba, significant correlations were also found between the distribution of three chaetodontid species and the percentage of coral cover. Among the three species, *C. austriacus* is an exclusive coral browser and *C. paucifasciatus* ingest coral polyps (24% of the weight percentages of prey) (Harmelin-Vivien & Bouchon-Navaro 1981). However, the abundance of *C. melannotus*, which is a browser of soft corals, was also found to be correlated to the percentage of live coral cover. So, food cannot explain all the correlations found. Moreover, it seemed surprising that the distribution of *C. trifascialis*, which strictly feeds on corals was not found to be correlated with live coral cover. The explanation is that this species is known to settle around tabular colonies of *Acropora cytherea* in Aqaba (Bouchon-Navaro 1980) as well as in other Indo-Pacific areas (Reese 1981, Bouchon-Navaro et al. 1985). *Acropora cytherea* was rare on the studied reef and was not sampled on the coral transects.

The search for the correlations between the Chaetodontidae and the coral growth forms revealed that only the abundance of the main coral browsers was linked to the distribution of branching coral colonies and the correlation was only significant with *C. austriacus*. The distribution of that species as well as that of *C. paucifasciatus* and *C. melannotus* was found to be correlated to the distribution of the genus *Acropora*. Similar results were found in Moorea Island by Bouchon-Navaro et al. (1985) who observed that the abundance of Chaetodontidae was significantly correlated to the distribution of long branching coral colonies on the reef.

Thus, it seems that both the food available and the shelter offered by branching coral species among other factors determine the distribution of chaetodontids on a reef. That may explain why certain zones with a high coral cover can possess a poor chaetodontid community as was observed by Bell et al. (1985) in French Polynesia. If the percentage of coral cover is high in one place but if the coral community is mainly composed of massive or encrusting species, the abundance of Chaetodontidae may be low.

Further studies should focus on the obligate corallivores and include a closer analysis of their ties to the branching coral species and especially the coral species on which they feed. Such studies could be done through field observations. For example, by counting the bites of the fishes on the coral colonies, it may be possible to demonstrate the preferences of a chaetodontid for a particular species. Long term field observations would thus allow to detail the general trends and the relationships which have been enhanced in the present study.

Acknowledgements

The present work was made in the framework of the French-Jordanian scientific cooperation programme. Financial support was provided by the French Foreign Ministry and the logistic support by the Marine Science Station of Aqaba. Thanks are expressed to Philip Motta for his kind help.

References cited

Anderson, G.R.V., A.H. Ehrlich, P.R. Ehrlich, J.D. Roughgarden, B.C. Russell & F.H. Talbot. 1981. The community structure of coral reef fishes. Amer. Nat. 117: 476–495.

Bell, J.D. & R. Galzin. 1984. The influence of live coral cover on coral-reef fish communities. Mar. Ecol. Prog. Ser. 15: 265–274.

Bell, J.D., M.L. Harmelin-Vivien & R. Galzin. 1985. Large scale spatial variation in abundance of butterflyfishes (Chaetodontidae) on Polynesian reefs. Proc. Fifth Internat. Coral Reef Symp. 5: 421–426.

Birkeland, C. & S. Neudecker. 1981. Foraging behavior of two Caribbean chaetodontids: *Chaetodon capistratus* and *C. aculeatus*. Copeia 1: 169–178.

Bouchon, C. 1980. Quantitative study of the scleractinian coral communities of the Jordanian coast (Gulf of Aqaba, Red sea): preliminary results. Tethys 9: 243–246.

Bouchon, C. 1981. Quantitative study of the scleractinian coral communities of a fringing reef of Reunion island (Indian Ocean). Mar. Ecol. Prog. Ser. 4: 273–288.

Bouchon, C., J. Jaubert, L. Montaggioni & M. Pichon. 1981. Morphology and evolution of the coral reefs of the Jordanian coast (Gulf of Aqaba – Red Sea). Proc. Fourth Internat. Coral Reef Symp. 1: 559–565.

Bouchon-Navaro, Y., C. Bouchon & M.L. Harmelin-Vivien. 1985. Impact of coral degradation on a chaetodontid fish assemblage (Moorea, French Polynesia). Proc. Fifth Internat. Coral Reef Symp. 5: 427–432.

Bouchon-Navaro, Y. 1980. Quantitative distribution of the Chaetodontidae on a fringing reef of the Jordanian coast (Gulf of Aqaba – Red Sea). Tethys 9: 247–251.

Bouchon-Navaro, Y. 1981. Quantitative distribution of the Chaetodontidae on a reef of Moorea Island (French Polynesia). J. Exp. Mar. Biol. Ecol. 55: 145–157.

Bradbury, R.H. & P.C. Young. 1981. The effects of a major forcing function, wave energy, on a coral reef ecosystem. Mar. Ecol. Prog. Ser. 5: 229–241.

Carpenter, K.E., R.I. Miclat, V.D. Albaladejo & V.T. Corpuz. 1981. The influence of substrate structure on the local abundance and diversity of Philippine reef fishes. Proc. Fourth Internat. Coral Reef Symp. 2: 497–502.

Clarke, R.D. 1977. Habitat distribution and species diversity of chaetodontid and pomacentrid fishes near Bimini, Bahamas. Mar. Biol. 40: 277–289.

Findley, J.S. & M.T. Findley. 1985. A search for pattern in butterflyfish communities. Amer. Nat. 126: 800–816.

Gladfelter, W.B., J.C. Ogden & E.H. Gladfelter. 1980. Similarity and diversity among coral reef fish communities: a comparison between tropical western Atlantic (Virgin Islands) and tropical (Marshall Islands) patch reefs. Ecology 61: 1156–1168.

Hourigan, T.F., T.C. Tricas & E.S. Reese. 1988. Coral reef fishes as indicators of environmental stress in coral reefs. pp. 107–135. *In:* D.F. Soule & G.S. Kleppel (ed.) Marine Organisms as Indicators, Springer-Verlag, New York.

Harmelin-Vivien, M.L. & Y. Bouchon-Navaro. 1981. Trophic relationships among chaetodontid fishes in the Gulf of Aqaba (Red Sea). Proc. Fourth Internat. Coral Reef Symp. 2: 537–544.

Lance, G.N. & W.T. Williams. 1967. A general theory of classification sorting strategy. I – Hierarchical systems. Computer J. 9: 373–380.

Luckhurst, B.E. & K. Luckhurst. 1978. Analysis of the influence of substrate variables on coral reef fish communities. Mar. Biol. 49: 317–323.

Loya, Y. 1972. Community structure and species diversity of hermatypic corals at Eilat (Red Sea). Mar. Biol. 13: 100–123.

McManus, J.W., R.I. Miclat & V.P. Palaganas. 1981. Coral and fish community structure of Sombrero Island, Batangas, Philippines. Proc. Fourth Internat. Coral Reef Symp. 2: 272–280.

Mergner, H. & H. Schuhmacher. 1974. Morphologie, Okologie und zonierung von Korallenriffen bei Aqaba (Golf von Aqaba, Rotes Meer). Helgol. wiss. Meeresunters. 26: 238–358.

Reese, E.S. 1981. Predation on corals by fishes of the family Chaetodontidae: implications for conservation and management of coral reef ecosystems. Bull. Mar. Sci. 31: 594–604.

Risk, M.J. 1972. Fish diversity on a coral reef in the Virgin

Islands. Atoll Res. Bull. 153: 1–6.

Sano, M., M. Shimizu & Y. Nose. 1984. Changes in structure of coral reef fish communities by destruction of hermatypic corals: observational and experimental views. Pac. Sci. 38: 51–79.

Sano, M., M. Shimizu & Y. Nose. 1987. Long-term effects of destruction of hermatypic corals by *Acanthaster planci* infestation on reef fish communities at Iriomote Island, Japan. Mar. Ecol. Prog. Ser. 37: 191–199.

Sutton, M. 1983. Relationships between reef fishes and coral reefs. pp. 248–255. *In:* D.J. Barnes (ed.) Perspectives on Coral Reefs, Australian Institute of Marine Science, contrib. 200.

Talbot, F.H. & B. Goldman. 1972. A preliminary report on the diversity and feeding relationships of the reef fishes of One Tree Island, Great Barrier Reef System. Proc. Symp. Corals and Coral Reefs 1969. J. Mar. Biol. Assoc. India: 425–444.

Williams, D. McB. 1986. Temporal variation in the structure of reef slope fish communities (central Great Barrier Reef): short-term effects of *Acanthaster planci* infestation. Mar. Ecol. Prog. Ser. 28: 157–164.

Wood, C.R. 1979. The occurrences and distribution of butterflyfishes (Pisces, Chaetodontidae) in Sabah. The Malayan Nature Journal 33: 87–103.

Environmental determinants of butterflyfish social systems

Thomas F. Hourigan
Department of Zoology, University of Hawaii, Honolulu, HI 96822, U.S.A.
Present address: Laboratory of Reproductive Biology, National Institute for Basic Biology, 38 Nishigonaka, Myodaijicho, Okazaki, 444 Japan

Received 4.4.1988   Accepted 9.11.1988

Key words: Behavior, Chaetodontidae, Cooperation, Coral reef fish, Food resources, Mating system, Monogamy, Plankton, Reproduction

Synopsis

Butterflyfishes (Chaetodontidae) display a variety of social systems, including monogamous pair-bonds, harems, and schooling with group spawning. The range of reproductive options available to butterflyfishes is shaped by their general life history characteristics, such as broadcast spawning with widely dispersed pelagic larvae, large body size and low adult mortality. The distribution and quality of food resources are major determinants of group size and mobility, thereby influencing the relative costs and benefits of available options, and determining specific social systems. Planktivorous and corallivorous butterflyfishes exemplify the relationship between food resources and social systems. Pelagic plankton is a patchy, but temporally and spatially unpredictable food resource which is efficiently exploited by fish in mobile schools. Neither sex is able to monopolize food resources necessary for the other sex, and plantivorous butterflyfishes appear constrained to spawn in groups. In contrast, corals are stable and predictable in space and time, favoring residence in one area and territorial defense of that space by coral-feeding butterflyfishes. Females defend food resources from other females, and males defend territories containing a female from other males. Males attempt to defend areas containing more than one female, but are unsuccessful. A monogamous social system results. This system favors the evolution of cooperative behavior between mates to increase female fecundity, as long as the male has an opportunity of sharing in that reproduction. Mate removal experiments conducted on two monogamous coral-feeding species, *Chaetodon multicinctus* and *Chaetodon quadrimaculatus* reveal a division of labor between male and female pair-mates. Paired males assume most of the territorial defense activities, allowing their mates to feed more.

Introduction

The spatial and temporal associations of males and females, and how they mate, define the social systems of animals. These systems result from the sum of the interactions of individual males and females. Male and female behavior may be considered adaptations to the environment as individuals compete with members of their own sex for genetic representation in the following generations (Wilson 1975, Warner 1980, Wrangham & Rubenstein 1986). The observed social system is a compromise of optimal strategies of each sex, given environmental and phylogenetic constraints.

Butterflyfishes of the family Chaetodontidae are conspicuous inhabitants of tropical coral reefs (Burgess 1978, Allen 1979). They have a diversity of social systems which are often correlated with different feeding guilds (Reese 1975, Fricke 1986). Butterflyfishes may therefore serve as excellent models to study the effects of food resources on the evolution of social systems. Butterflyfishes are exceptional, because in many species individuals form apparently monogamous pair-bonds of long duration (Fricke 1973, 1986, Reese 1973, 1975, 1981, Steen 1978, Allen 1979, Neudecker & Lobel 1982, Tricas 1985, 1986, Hourigan 1987, Hourigan et al. 1988). Monogamy is uncommon among freshwater fishes and terrestrial vertebrates except birds, and is usually associated with biparental care (Williams 1966, Lack 1968, Wilson 1975, Kleiman 1977, Wittenberger & Tilson 1980, Wickler & Seibt 1983, Barlow 1984, 1986). Butterflyfishes spawn pelagic eggs, precluding all parental care.

In this paper I briefly summarize general environmental factors which determine reef fish social systems, and then examine specific life history characteristics of butterflyfishes that determine which of these environmental factors are most important. Examples drawn from planktivorous and corallivorous butterflyfishes show how the distribution, abundance and nutritional value of food resources interact with the life history characteristics of a species to determine the social system.

Environmental factors affecting male and female associations of reef fishes

The major environmental influences on the social systems of animals are those which exert selective pressures on the residence and movements of reproductive males and females (Emlen & Oring 1977). Primary among these are sources of mortality, and the density, variability and predictability of limiting resources (such as food, shelter and mates) in both time and space. Environmental or social factors which affect males and females differently are of special importance to the understanding of social systems. The following list is based in part on an analysis of the interactions of life history characteristics by Warner (1980).

Source of mortality and their effects

Most reef fish mortality occurs during the planktonic dispersal period (Sale 1980), but the factors involved are poorly understood. Predation is probably the major source of mortality for fishes after they recruit to reefs (Hobson 1978, Sale 1980).

Mortality may influence social and mating systems in several ways (Warner 1980):
(a) Low and unpredictable larval or juvenile survival, and high or consant adult survival should select for low reproductive effort and iteroparity (Murphy 1968, Charnov & Schaffer 1973).
(b) Predation pressure may limit the movement of animals to certain sheltered refuges. In this case, the effects of predation on social systems will be determined by the distribution of shelters (see below).
(c) Predation may also affect group size. Fishes which feed away from cover may form aggregations or schools to reduce the chances of predation on individuals (Hamilton 1971, Hobson 1978, Pitcher 1986). If groups are small, polygny may occur. As group size and/or mobility increase, individual males may have less control over mating by females, and group spawning is likely (Ralston 1981).
(d) Predation may affect males and females differently if one sex is more conspicuous than the other. Predation rates may be higher for males courting females or defending demersal eggs.
(e) Predation on eggs may cause the aggregation of adults around nesting or spawning areas which minimize this predation. If such areas are limited and defendable, this may favor male defense of these areas and polygynous mating. In this case the distribution of preferred nesting and spawning sites becomes an important determinant of social systems.

If predation on eggs varies predictably over time, spawning synchrony may be favored to avoid the time when predators are feeding, or to swamp predators; or asynchrony may be favored to reduce the chances of predators aggregating. Other sources of mortality on eggs or larvae (e.g. tidal rhythms, and seasonality in currents or food available for larvae) may also favor a degree of spawning synchrony. Spawning sychrony will reduce the operational sex ratio and thereby the environmental potential for polygamy for fish which spawn in pairs at dispersed spawning sites. If spawning sites are limited and defendable (as in lek-like spawning), or if group spawning occurs, synchrony may have the opposite effect.

(f) Finally, high levels of mortality of either adults (Talbot et al. 1978) or of larvae and juveniles (Doherty 1983) may reduce population densities below the level where other resources, such as food or shelters, are limiting.

Limiting resources

For polygyny to evolve, multiple females, or resources necessary for multiple females must be economically defendable (Emlen & Oring 1977). The distribution of limiting resources in space will determine the distribution and movements of females. The density, stability, and predictability of these resources in space and time will determine whether individual males are able to defend resources necessary for more than one female. Among reef fishes, shelters, spawning sites, nesting sites, and food are potential limiting resources. Since butterflyfishes are broadcast spawners, nesting sites will not be addressed.

(a) Shelters are often considered limiting resources for reef fishes (reviewed by Sale 1980, Walsh 1984). Shelters are discrete, and are stable and predictable in space and time, making them defendable. Competition for shelters and consistent use of shelters have been demonstrated by several authors (Robertson & Sheldon 1979, Walsh 1984, Shulman 1985), but most experimental manipulations have failed to show that shelters are a limiting resource for adult reef fishes (Robertson & Sheldon 1979, Sale 1980).

Small and spacially separated shelters will favor monogamy (e.g. gobies: Lassig 1977) and protandry (e.g. anemonefishes: Moyer & Nakazono 1978). In contrast, sheltered areas which are large enough for more than two fish (i.e. spatially aggregated shelters), will increase the environmental potential for polygyny (sensu Emlen & Oring 1977). The effect of shelter size on the social system is shown by the small damselfish, *Dascyllus marginatus* (Fricke 1980). Small coral heads contain one male and one female. In medium size coral heads, one male is able to defend shelters necessary for more than one female, resulting in haremic polygyny. In still larger coral heads, a single male is unable to exclude other males, and groups of multiple males and females occur.

(b) For many broadcast spawning fishes, spawning sites occur along the leeward edge of reefs, where eggs may be carried away from the reef (Johannes 1978). These sites are discrete, they are generally stable and predictable in space, and if limited, may be defended by large males. Since females are not restricted to these sites, except during the act of spawning, a lek-like mating system results. Such mating systems are common among many fishes, such as wrasses (Loiselle & Barlow 1978, Warner & Robertson 1978, Moyer & Yogo 1982). The restriction of males to a spawning territory will reduce the potential for males to defend or sequester mates, and may allow females a greater latitude of mate choice and opportunities to mate with more than one male.

(c) Food resources are often important determinants of the distribution of females, and therefore also of mating systems among terrestrial vertebrates (Orians 1969, Jarman 1974, Wilson 1975, Emlen & Oring 1977, Oring 1982, Wrangham & Rubenstein 1986). In contrast, while studies of coral reef fishes have indicated the importance of limiting shelters and nesting or spawning sites as determinants of social systems (Thresher 1984), the distribution of food resources has seldom been implicated.

Food resources differ from shelter and spawning sites, in that they may vary not only in spatial distribution, but also in stability and predictability in time and space, as well as in nutritional value. These factors add complexity to the food resource – social system interaction. For many fishes, female reproduction is likely to be food limited (Luquet & Watanabe 1986), since increased food ration is correlated with increased fecundity (Bagenal 1966, Hirshfield 1980) or increased number of clutches within a breeding season (Wootton 1977, 1985). In such conditions, female distribution and group size, and therefore the ability of males to monopolize mates, may be determined by the distribution of limited food resources. Male sperm production may also be food limited, an important consideration for species in which sperm competition occurs when groups of males spawn with one or more females. Food limitation may affect males directly by limiting the time or energy available for defending females; or indirectly by limiting growth, and thereby their ability to compete successfully for females. The effects of food limitation on the reproductive success of male fishes has rarely been investigated (Luquet & Watanabe 1986).

The effects of food resources on social systems will depend on the costs and benefits of mating within the foraging area versus elsewhere. For example, large acanthurids leave their feeding territories to spawn at the edge of the reef in groups or in leks, yet are able to return and reclaim their feeding territories (Robertson et al. 1979, Robertson 1983). In general, although the distribution of food resources is sometimes invoked as a determinant of reef fish mating systems (e.g. wrasses: Robertson & Hoffman 1977, acanthurids: Robertson et al. 1979), there have been few tests of this assumption, and the resources in question have seldom been quantified. Several features of the biology of butterflyfishes make food resources likely environmental determinants of social systems.

Life history characteristics of butterflyfishes

A number of life history characteristics of butterflyfishes contribute to the basic patterns of chaetodontid social organization. Certain characteristics shared by all butterflyfishes, such as broadcast spawning, are probably primary characteristics of the family. Others may be secondary characteristics related independently to different mating systems. Primary family characteristics represent phylogenetic constraints to possible environmental adaptations. These shared characteristics may help explain why the family is unique, but cannot be used to explain differences in mating systems among butterflyfish species.

1. Relatively large body size of reproductive adults

Adult butterflyfishes have large body sizes which decrease the risk of predation, but increase food needs. Average body sizes range between 100 mm and 260 mm total length (Allen 1979). By the time of sexual maturity, at approximately one year of age, butterflyfishes have reached 70% to 75% of their maximum size (Ralston 1976, Tricas 1986). These fishes are relatively large compared to many other reef fishes whose mating systems have been studied (e.g. most gobies, anthiids, damselfishes, as well as smaller angelfishes and wrasses; Thresher 1984).

Post-maturation mortality due to predation appears to be low in butterflyfishes compared to smaller fishes (Ehrlich et al. 1977, Reese 1981, Norris 1985, Hourigan 1987). Since most piscivores consume fish whole, larger fishes may be expected to be less subject to predation than smaller fishes (Hobson 1978, Shulman 1985). Predation is further reduced by the compressed, high body shape and sharp dorsal and anal spines of butterflyfishes (Gosline 1965).

Low predation pressure frees adult butterflyfishes from the constraints of limited diurnal shelters. Butterflyfishes consistently use the same nocturnal shelters (Hourigan 1987), however there is evidence that these shelters are not limited. When shelters of six individuals (three *Chaetodon multicinctus* and three *Chaetodon quadrimaculatus*) were covered, all individuals rapidly found new shelters within their original territories, and no mortality was observed during the following year (Hourigan 1987). There is no evidence that predation limits the movements of reproductive individuals, however, shelters may be important to juveniles.

Large body size also entails greater metabolic needs, and large individuals may require a large foraging area. In conjunction with low predation pressure, this will lead to increased mobility. Although individual *Chaetodon trifascialis* may defend interspecific feeding territories as small as $1 m^2$ to $2 m^2$ (Reese 1975), most adult butterflyfishes inhabit large home ranges of $50 m^2$ to over $10 000 m^2$ in size (Reese 1975, 1981, Ehrlich et al.

1977, Sutton 1985, Tricas 1985, Hourigan 1987, Driscoll & Driscoll 1988). Home range size generally increases with body size (Tricas 1986).

2. Long adult life

Related to low adult mortality, is a relatively long adult life. Long adult life allows individuals to develop social relationships which last beyond one breeding season. Long term studies of butterflyfishes have been limited to four to ten years, however many individually recognizable adults have been observed consorting with the same mates for the entire study time (Reese 1981, Fricke 1986, Hourigan et al. 1988, L. Fishelson personal communication).

3. Broadcast spawning, with extended pelagic larvae

Butterflyfishes, like the majority of reef fishes, spawn pelagic eggs which are externally fertilized in the water column and dispersed by currents. Embryos hatch about 30 hours after spawning (Suzuki et al. 1980), and larvae subsequently spend an average of 40 days in the plankton before metamorphosing and settling on the reef (Hourigan & Reese 1987). The life history characteristic of pelagic eggs, embryos and larvae with high dispersal capabilities has major adaptive consequences. First, parental care, a major determinant of many animal social systems, is not possible. Second, eggs and larvae from a single spawn are likely to be dispersed, and recruits settling in a particular area are probably unrelated, minimizing the effects of kin selection on social systems. Thus, environmental constraints such as the distribution of limiting resources, rather than parent-offspring or kin interactions, become major determinants of social systems (Thresher 1977, Barlow 1984).

Because of the high degree of unpredictable mortality suffered by larvae, there is selection for high fecundity spread over time, to maximize the chance that some larvae will sucessfully return to a reef (Murphy 1968). This will be reflected in behavioral adaptations by adult females to increase egg production, behavioral adaptations by males to increase the number of females with which they mate, and adaptations by both sexes to insure fertilization and successful dispersal. Generalist traits will be favored over specializations to local conditions, since dispersal reduces the genetic isolation of populations (Hourigan & Reese 1987), and disrupts local coadapted gene complexes (Strathmann 1986). Intraspecific territoriality may be such a generalist trait. It is an effective strategy for defending a variety of resources against conspecifics, the most likely competitors. It can be modified to respond to different types or levels of resources or competitors (e.g. Tricas 1986) depending on the local conditions encountered by a fish.

4. Large clutches spawned at intervals over an extended breeding season

Female butterflyfishes typically spawn thousands, to hundreds of thousands of of eggs at one time. Spawning may occur as often as once every two days in smaller fish (Colin 1989) or once or twice a month in larger fish, during an extended breeding season of many months (Lobel 1978, Ralston 1981, Thresher 1984, Fricke 1986, Tricas 1986, Hourigan 1987). Examination of the ovaries of four species of butterflyfishes in the genus *Chaetodon* indicate that most vitellogenic oocytes (yolked eggs) mature synchronously and all (>10 000) are probably spawned at the same time (*C. miliaris*: Ralston 1981, *C. multicinctus*: Tricas 1986, Hourigan 1987, *C. quadrimaculatus* and *C. fremblii*: Hourigan 1987). Cursory examination of the ovaries of four additional species (*C. ornatissimus, C. trifasciatus, C. unimaculatus, Forcipiger flavissimus*) showed similar trends (Hourigan 1987). In contrast, fish which spawn on a daily basis, such as the closely related pomacanthids, produce fewer eggs per spawn (e.g. Bauer & Bauer 1981, Hourigan & Kelley 1985). This may explain why so few observations of spawning by butterflyfishes have been reported compared to angelfishes (Thresher 1984).

If butterflyfishes do not spawn every day, the advantages of staying with a single mate to insure male access to a female ready to spawn, and synchronization of spawning behavior, may outweigh the disadvantages of limited access to other mates (Hourigan 1984a). Pair-bonding might be such a strategy (Reese 1975, Gronell 1984). This does not explain why some of these fishes are paired (*C. multicinctus, C. ornatissimus, C. quadrimaculatus, C. trifasciatus*), while others are schooling (*C. miliaris*), or haremic (*C. fremblii*), or occur in small groups (*C. unimaculatus, F. flavissimus*, Appendix 1).

5. Dusk spawning

Spawning by butterflyfishes has only been observed during a brief period at dusk (Lobel 1978, 1989, Neudecker & Lobel 1982, Thresher 1984, Fricke 1986, Hourigan 1987, Colin 1989). Combined with a strong lunar periodicity to spawning (Lobel 1978, Tricas 1986), this may result in high spawning synchrony within the population, and a resultant short period during which spawning is possible. This situation may limit the number of mates with which one male can spawn (Emlen & Oring 1977, Knowlton 1979, Barlow 1984). Since most observations of spawning have been of paired species, it is not known to what extent temporal synchrony may be a constraint resulting in pair-bonding in butterflyfishes. Fricke (1986) reported, however, that different pairs of *Chaetodon chrysurus* (= *C. paucifasciatus*) were not reproductively synchronized. Colin (1989) also offered evidence of little synchrony in nearby pairs. In addition, dusk spawning and lunar periodicity in the butterflyfish *C. fremblii* (Lobel 1978) do not preclude harem formation (Hourigan 1986a), nor do they preclude harems in the closely related angelfishes (Lobel 1978, Neudecker & Lobel 1982). Reproductive considerations alone fail to explain why pairs should remain together beyond the breeding season (Fricke 1986).

Most observers have reported butterflyfishes spawning in pairs, generally within their diurnal home ranges rather than at a lek (Lobel 1978, Neudecker & Lobel 1982, Thresher 1984, Fricke 1986, Hourigan 1987, Colin 1989). In these cases, spawning sites would not appear to be limited. The pressure of planktivorous egg predators may favor spawning at dusk, but does not explain differences in mating systems within the family Chaetodontidae. Generalizations about butterflyfish spawning are based on limited observations and should be viewed with caution. Different species of broadcast spawning acanthurids have different mating systems and different diel spawning rhythms, while certain individuals of the same species may spawn within the feeding territory while others spawn elsewhere (Robertson 1983).

6. No evidence of functional hermaphroditism

Few histological studies of butterflyfish gonads have been conducted, but in all cases, butterflyfishes appear to be gonochoristic (Yamsonrat 1980, Ralston 1981, Tricas 1986, Hourigan 1987). Tricas (1986) completed the only extensive histological study of a butterflyfish, the monogamous *Chaetodon multicinctus*. There was no evidence of sex change after sexual maturity, however, immature males occasionally showed remnants of immature female gonadal tissue. If functional hermaphroditism has not evolved in this group, certain reproductive options are precluded.

7. Sexual monomorphism

All butterflyfishes studied have been sexually monochromatic (Burgess 1978, Allen 1979, Thresher 1984). Sexual monomorphism is usually related to low levels of sexual selection, and is commonly a secondary characteristic associated with monogamous and group spawning mating systems. Size distributions of males and females in a population are similar in group spawning species (Ralston 1981) and monogamous species, although males in pairs are generally slightly larger than their mates (Yamsonrat 1980, Reese 1981, Hourigan et al. 1988)

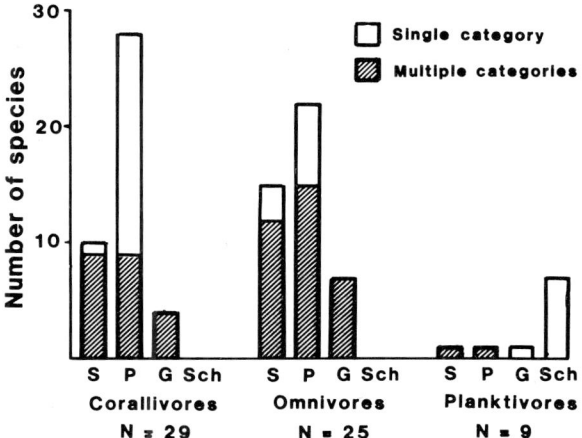

Fig. 1. Observed social grouping of corallivorous, omnivorous and planktivorous butterflyfish species. Social categories are: S = solitary, P = paired, G = small groups, Sch = schools. Social grouping may differ with age or locality resulting in certain species being counted in more than one social category (details in Appendix 1).

Food as a major determinant of butterflyfish social systems

Food resources appear to be the major determinant of different social and mating systems among butterflyfishes. Reese (1975) first noted the correlation between the feeding behavior of 20 butterflyfish species and their observed social grouping. In particular, corallivorous species occurred predominantly in pairs. Figure 1 shows a compilation of data from the literature for 62 butterflyfish species for which some data on feeding and social behavior are known. Many species have been observed in more than one social category. Data for some species are limited to casual observations of social groupings which may not reflect the social system. Monogamous pair-mates may be separated for up to 50% of the time observed, or may occasionally join groups of conspecifics for short periods (Hourigan 1987, Driscoll & Driscoll 1988). Social grouping may also vary, depending on sexual maturity (Fricke 1986, Hourigan 1987) or geographic area (Reese 1975, Colin 1989). Clear correlations occur, however, between planktivory and schooling, and between corallivory and pair-bonding.

Planktivory, schooling, and group spawning

Planktivorous butterflyfishes feed on pelagic plankton (Hobson 1974, Motta 1988) which drifts over the reef from offshore. Two aspects of the distribution of plankton determine the social system of these fishes. First, plankton is not evenly distributed in space, but often occurs in patches of varying density. This distribution favors the aggregation of individuals at food patches. Plankton is also unpredictably distributed in time and space, i.e. the time and location at which these plankton aggregations will cross the reef is not predictable. This makes it uneconomical to defend plankton resources. The mobility of small planktivores (e.g. many damselfishes) may be constrained by the need for shelter or nesting sites. These fish remain in one area, often grouped near the edge of the reef, and wait for plankton to come to them (Reese 1978).

Planktivorous butterflyfishes are generally found in aggregations or schools near the windward edges of coral reefs (Hobson 1974, Reese 1978, Ralston 1981, Fricke 1986, Hourigan 1987). Unlike smaller planktivores, schools of butterflyfishes move over large areas. Mobile schools of planktivores are most likely a response to the unpredictable occurrence of pelagic plankton aggregations which drift over reefs (Reese 1978, Pitcher 1986). In addition, grouping may reduce predation on fish which must forage far above the shelter of the reef. *Chaetodon sancthelenae* is reported to form pairs when foraging on the bottom, but groups when foraging on plankton in the water column (Allen 1979). Individual males cannot monopolize the ephemeral food resources necessary for females. In large, mobile heterosexual groups, males also cannot monopolize or sequester females directly. Group spawning is likely to result unless fish leave the group to spawn. Male reproductive success in a group will depend on the number of females in the group, the number of males in each spawning event, a male's relative sperm output, and access to each spawning female (related to spawning synchrony within the group, and perhaps to dominance relationships among males or to female choice). Females which spawn with groups of males have few opportunities for mate choice, but a high percent of their eggs should be fertilized. This may be an important consideration in a broadcast spawning species.

Spawning of planktivorous butterflyfishes has not been observed, however, inferences about their spawning behavior can be made from male gonadosomatic indices (GSI). These are much larger for mature, schooling plantivores (up to a maximum of 6.6% for one male *Chaetodon miliaris*: Ralston 1981), compared to those recorded from butterflyfishes which spawn in pairs (Table 1). Among other fishes, large male testes are an adaptation to sperm competition in mating systems where numerous males spawn simultaneously with one or more females (Warner & Robertson 1978). The effect of the distribution of plankton on social and mating systems is shown in Figure 2.

Corallivory and monogamous pair-bonds

Most studies of butterflyfish social behavior have concentrated on corallivorous species. The distribution and abundance of corals can be quantified, and the feeding of corallivores is easily observed (Reese 1975, Motta 1988). Most species of coral-feeding butterflyfishes form long-term pair-bonds (Reese 1975, Fig. 1). Heterosexual pair-bonds among butterflyfishes can last more than ten years (Reese 1981, Hourigan et al. 1988, L. Fishelson personal communication), with the male and female sharing the same feeding territory (Reese 1973, 1975, 1981, Sutton 1985, Tricas 1985, 1986, Fricke 1986, Hourigan 1986b, 1987, Hourigan et al. 1988). Spawning occurs between pair-mates (Neudecker & Lobel 1982, Fricke 1986, Hourigan 1987), indicating that these bonds represent monogamous social and mating systems (Wickler & Seibt 1983).

The stable social environment of corallivorous butterflyfishes is correlated with stable resources. Corals are long lived and self-regenerating, providing a temporally stable and predictable food resource. Temporal stability allows year-round residence and territorial defense of the area where the resource occurs (Brown 1964). Low predation pressure will also enhance the defendability of a territory (Geist 1974).

The abundance of corallivores on the reef is correlated with the spatial distribution of their coral food resources (reviewed by Hourigan et al. 1988, Bouchon-Navaro & Bouchon 1989), as one would expect if food resources were major determinants of fish distributions. In general, the distribution of

Table 1. Gonadosomatic indices (GSI) of male butterflyfishes with different social systems.

Species	Social system	Spawning behavior	Male GSI[1]	
			Mean	S.D.
Chaetodon milliaris[2]	Schooling	Group spawn promiscuity?	2.14% (N = 57)	1.80%
Hemitaurichthys polylepis[3]	Schooling	Group spawn promiscuity?	3.22% (N = 4)	0.71%
Chaetodon multicinctus[3]	Pair-bond monogamy	Pair spawn	0.27% (N = 12)	0.06%
Chaetodon quadrimaculatus[3]	Pair-bond monogamy	Pair spawn	0.22% (N = 12)	0.22%
Chaetodon fremblii[3]	Haremic polygyny	Pair spawn[4]	0.20% (N = 12)	0.11%

[1] Adult males during breeding season.
[2] Data calculated from Ralston (1981).
[3] Hourigan 1987.
[4] Lobel 1978.

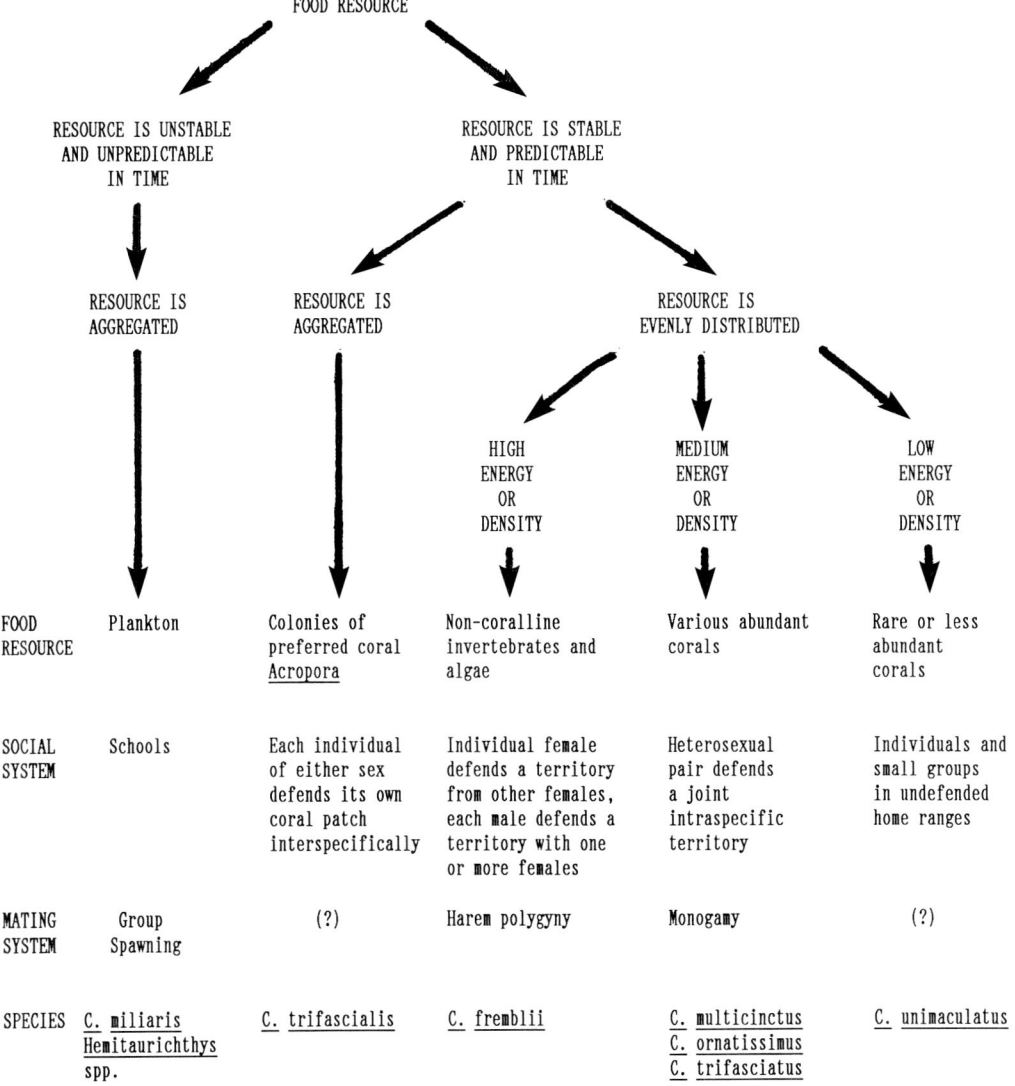

Fig. 2. The effects of the distribution of food resources in time and space and their relative energy value (or density) on social and mating systems of butterflyfishes. A question mark (?) denotes that the mating system of those species has not been determined.

fish is most closely related to preferred food resources, rather than to total coral cover (Hourigan et al. 1988). Feeding preferences of corallivores have seldom been taken into account, leading to poor correlations between total coral cover and butterflyfish abundances (Hourigan et al. 1988). Corallivores, such as *Chaetodon multicinctus*, *Chaetodon ornatissimus* and *Chaetodon trifasciatus*, are coral-feeding generalists, which feed on a variety of abundant coral species (Hourigan et al. 1988). A generalized diet will increase the apparent evenness of resource distribution. In contrast, a specialist on one coral species, such as *Chaetodon trifascialis* is more likely to encounter a patchily distributed resource.

Corals may be either aggregated or evenly distributed in space (Fig. 2). *Chaetodon trifascialis* is a specialized feeder on corals of the genus *Acropora* which form large plate-like tables (Reese 1975, 1981, Motta 1985, 1988, Hourigan et al. 1988, Irons 1989). These coral colonies are abundant and large enough to support a single butterflyfish, but represent a spatially aggregated resource (Fig. 2). An individual maintains exclusive access to one or more coral colonies by defending a small interspecific feeding territory against other corallivores (Reese 1975). The mating system of this species is not known, but probably depends on the distribution of males and females on adjacent patches, the ease of movement between patches, and whether mating occurs within the male or female's territory or at a spatially separated spawning site. The social system of *C. trifascialis* superficially resembles that of the dominant herbivorous acanthurid, *Acanthurus lineatus* (Robertson et al. 1979).

Most corals represent an evenly distributed resource relative to the high mobility of butterflyfishes. Temporal stability and even distribution may also characterize the food resources of some omnivorous butterflyfishes. Even distribution of food resources favors an even spacing of females feeding on those resources. Whether any single male is able to monopolize more than one female, will depend on the home range sizes of females relative to energetic or time constraints on his defense capabilities (Emlen & Oring 1977). These both depend on the density and energetic value of the food resource.

Corals are an energetically poor food resource. Coral tissue has a high water content and is relatively low in caloric value (Tricas 1986), leading to a low caloric intake per feeding bite (Hourigan 1987). Perhaps more important, however, is that coral-feeding butterflyfishes have very low absorption efficiencies: less than 25% of dietary energy intake is absorbed in the gut (Table 2). This value is low compared to the omnivorous butterflyfish, *Chaetodon fremblii* (42%, Hourigan 1987); and average values from the literature of 80%–85% absorption efficiency for carnivorous fishes, and 58%–60% absorption efficiency for herbivorous fishes (Brett & Groves 1979, Pandian & Vivekanadan 1985).

A low energy return per bite is equivalent to feeding on a low density food resource, and has several consequences. First, corallivorous butterflyfishes will require relatively large feeding ranges which may be difficult for one individual to defend. Second, since gamete production constitutes a significant energy investment by females, female reproduction is likely to be feeding-time limited. Female *Chaetodon multicinctus* of pairs in a coral-poor habitat in Hawaii, inhabit larger territories and have lower feeding rates than do females in a coral-rich habitat (Hourigan 1987). The former also have smaller gonadosomatic indices, fewer vitellogenic oocytes, and less stored fat than females from the coral-rich habitat. If female reproduction is food-limited, there is selection for behavior which increases female feeding rates. Foraging is the primary activity of female coral-feeding butterflyfishes (Tricas 1986, Hourigan 1987, Reese 1989). Females forage almost continuously during an active period of 12 to 13 h. Continuous foraging is compatible with the low amount of energy per bite ingested and absorbed by these fishes.

If food resource density is very low, resource defense is uneconomical. The butterflyfish *Chaetodon unimaculatus* is a specialist on corals of the genera *Montipora* and *Pocillopora*. These corals are not abundant on coral reefs off the island of

Table 2. Energy intake and absorption by two species of coral-feeding butterflyfishes (data from Hourigan 1987).

	Corallivore	
	Chaetodon multicinctus	*Chaetodon quadrimaculatus*[1]
Energy intake per bite: (in calories) Laboratory:		
Pocillopora meandrina	0.25	0.70
Porites lobata	0.19	–
Porites compressa	0.13	–
Field:	0.20	0.81
Absorption efficiency[2]:	22.4%	14.5%

[1] *C. quadrimaculatus* feeds primarily on the coral, *P. meandrina*, but supplements its diet with non-coralline invertebrates and algae.
[2] Absorption efficiency for field caught fishes calculated as:
$$100 \times 1 - \frac{(Cal\ mg^{-1}\ dry\ weight\ feces)}{(Cal\ mg^{-1}\ dry\ weight\ stomach\ contents)}.$$

Hawaii, and C. unimaculatus forage over large undefended areas as single individuals or small groups (Fig. 2, Hourigan 1987). The mating system of this species is unknown.

Most coral feeding butterflyfishes are generalists on several species of abundant corals, exploiting a relatively evenly distributed resource of moderate food value (Fig. 2). In these species, a heterosexual pair defends a joint feeding territory, primarily from conspecifics. Pair-bonds in fishes without parental care are generally associated with the defense of a feeding territory (Barlow 1986). Two fish can patrol a larger shared area than the total area that could be patrolled by two single fish defending separate territories (Robertson et al. 1979). Two fish together may also provide a more effective territorial advertisement than two fish separately (Fricke 1986). By itself, this energetic argument does not explain why there should be only two fish, or why pairs should be heterosexual.

Where individuals of known sex have been observed, males primarily chase males, and females primarily chase females (*Chaetodon chrysurus* = *C. paucifasciatus*: Fricke 1986, *C. multicinctus* and *C. quadrimaculatus*: Hourigan 1987). This suggests that intrasexual aggression may limit the number of females whose territories a male can encompass, and therefore with which he can mate.

In *C. quadrimaculatus* and *C. multicinctus*, females in monogamous pairs feed more than unmated females, while paired males feed less than their mates but contribute most to the defense of the territory. Based on these observations, I proposed a hypothesis to explain pair-bonding and monogamy in butterflyfishes and other coral reef fishes without parental care (Hourigan 1984b, 1987):

Pair-bonding and monogamy are of selective advantage to both sexes, because fecundity is food limited, and pairing: (a) increases time available to females for feeding, and (b) enables the paired male to share in his mate's increased fecundity.

This may occur by a division of labor between the sexes, whereby the male assumes an increased proportion of territorial defense activities, allowing the female to spend more time feeding. If males generally have exclusive access to their pair-mates for breeding, they will share in the female's increased fecundity. Part (a) of this hypothesis is similar to the explanation of Robertson et al. (1979) for pair formation in an acanthurid, *Acanthurus leucosternon*.

Tests of this hypothesis were conducted on two paired species *C. multicinctus* and *C. quadrimaculatus* in Hawaii (Hourigan 1987). When males were experimentally removed from their pairs, intrusions by conspecifics increased, and the feeding rates and territory sizes of the remaining females decreased as predicted by the hypothesis (Table 3). In contrast, males whose mates had been removed were able to defend their original territories with no decrease in feeding rates. In both cases, the remaining individuals courted members of the opposite sex and mate replacement occurred quickly, indicating an important advantage to being paired. Fricke (1986) conducted removals of one individual

Table 3. Results of mate removal experiments (Hourigan 1987). A single male or female was removed from its pair-mate, and changes in territory size, feeding rates and chase rates were measured.

Species		Remaining individual of pair			
		Territory size	Feeding rate	Chase rate	Mate replacement
C. multicinctus	Male removed (N = 12)	Decrease	Decrease	Increase	1–4 days
	Female removed (N = 6)	No change	Increase or no change	Increase	1–8 days
C. quadrimaculatus	Male removed (N = 6)	Decrease	Decrease	Increase	1–3 days
	Female removed (N = 5)	No change	Increase or no change	Increase	1 day

from each of numerous pairs of *Chaetodon chrysurus* with similar results. Reproduction may depend on forming a pair and defending a territory. Populations of both *C. multicinctus* and *C. quadrimaculatus* contain females which were excluded from the feeding territories of pairs. These non-territorial females had lower feeding rates than paired females and were not reproductive (Hourigan 1987). The generality of this system among other paired butterflyfishes remains to be determined.

Observations of *C. multicinctus* (Hourigan 1987) and *C. chrysurus* (Fricke 1986) have revealed spawning only between pair-mates, despite attempts at sneaking by neighboring males. Thus, the male will share the benefits of increased egg production gained by his mate which feeds more as a result of his assistance. In contrast, when males of the haremic butterflyfish *Chaetodon fremblii* were removed from their harems, female territory sizes, chase rates and feeding rates did not change (Hourigan 1986a), indicating that females received no feeding advantages from the presence of the male. Female *C. fremblii* accepted new mates without apparent courtship.

In monogamous species, a male may be limited to one mate by intrasexual aggression and a low energy food resource which allows him to assist only one female. Monogamous male *C. multicinctus* and *C. quadrimaculatus* attempted to defend territories containing their own mate and the neighboring female whose mate was removed. In no case was this polygynous situation stable for more than a few days (Hourigan 1987). The omnivorous *C. fremblii* exploits a higher energy food resource, and females are able to defend individual feeding territories without male assistance (Hourigan 1986a). Males defend larger territories containing up to four female territories (Fig. 2).

Long-term pair-bonds may allow a differentiation of sexual roles which preadapt fishes for the evolution of further social cooperation (Barlow 1974). Aside from the division of labor shown by *C. multicinctus* and *C. quadrimaculatus* in territorial defense, there is some evidence of additional intrapair cooperation. The feeding behavior of six pairs of *Chaetodon multicinctus* was observed in a shallow habitat dominated by two corals, *Porites lobata* and *Pocillopora meandrina*. The latter coral has a significantly higher energy content than the former (Tricas 1986, Table 2), and is the most preferred food-coral of both males and females in the field as well as in the laboratory (Hourigan et al. 1988). Although mates fed in close proximity to each other, females in all pairs took a significantly greater proportion of bites from the preferred food coral than did males (Fig. 3). Males directed more feeding toward algae on non-coralline hard sustrata, a less preferred food. It is tempting to speculate that this may represent a case of reciprocal altruism, in which a male refrains from feeding on preferred food resources, allowing the female to feed more and thereby produce more eggs which the male may later fertilize. If true, this shows a greater degree of cooperation in a non-reproductive setting than has been observed for any reef fish. Further studies on monogamous species may provide insights into the evolution of cooperative behavior between mates in long lived species.

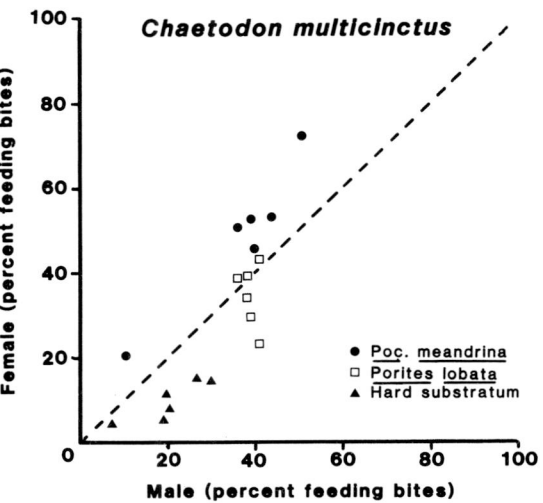

Fig. 3. Percent of feeding bites on different substrata by adult female and male *Chaetodon multicinctus* belonging to six pairs. Each point is the mean of 100 minute observations on each member of the pair. Points below the 45° line signify that the male of those pairs fed more on that substratum than his pair-mate, and points above the line signify that the female fed more on that substratum than her pair-mate.

Conclusions and directions for future research

I have tried to show how the life history charateristics of butterflyfishes act in a permissive fashion, to reduce the importance of parental care, kin selection, predation, and limited spawning and shelter sites on social systems, leaving food resources as a major determinant. The distribution, abundance and nutritional value of food resources shape social systems, sometimes through subtle effects on the behavior of males and females, as shown in studies of paired (Fricke 1986, Hourigan 1987) and haremic butterflyfishes (Hourigan 1986a, 1987). The study of reef fish social systems is just beginning, and several aspects of butterflyfish social behavior deserve special consideration:

1. Long term studies on recognizable individuals are essential to understanding the development and evolution of social behavior (Wrangham & Rubenstein 1986). Social systems are the sum total of individual male and female behavior, yet studies of the individual behavior patterns which contribute to social and mating systems of reef fishes has barely begun. Of special interest are the role of individual recognition and the extent of cooperative behavior in long term social relationships.

2. The social organization of juvenile and sub-adult butterflyfishes often differs from that of adults (Fricke 1986, Hourigan 1987). These differences may explain the variability in social organizations observed in several species, e.g. *Chaetodon capistratus* (Neudecker & Lobel 1982, Gore 1983, Colin 1989). The development of social behavior and the relationship of these patterns to adult mating systems deserves study.

3. Insights into the ultimate environmental factors which have led to the evolution of behavior and resulting social systems can be gained by observing the short-term, proximate responses to different ecological constraints (Warner 1980). Some species of butterflyfishes (e.g. *Chaetodon auriga*: Reese 1975, *Chaetodon capistratus*: Colin 1989) exhibit different social systems in different geographic areas, presumably in response to different environmental pressures. Experimental studies of changes in behavior in response to manipulations of food resources, competitors or mates have been used in studies of butterflyfish feeding (Tricas 1985, 1986, Hourigan 1986b), territoriality (Tricas 1985, 1986), and mating systems (Fricke 1986, Hourigan 1986a, 1987). From an evolutionary standpoint, such studies represent cases of reasoning by analogy, and there is no assurance that the proximate responses reflect the ultimate causes (Warner 1980). Nevertheless, such studies hold the greatest promise for understanding the ways in which food resources shape male and female behavior and thereby determine fish social systems.

4. Closely related species with different social systems are assumed to have diverged in response to different environmental pressures. Comparisons of the social behavior and related ecology of these species should therefore provided information on these pressures. The comparative method is particularly important when the behavior in question shows a degree of invariance within species. Many paired butterflyfishes (e.g. *Chaetodon trifasciatus*) appear to have similar social systems wherever they occur (Reese 1975, 1981, Sutton 1985).

In this paper, I have concentrated on the social systems of schooling planktivores and pair-bonded corallivores. Exceptions to the trends of paired corallivores (e.g. *Chaetodon trifascialis, Chaetodon rainfordi,* and *Chaetodon unimaculatus*) and exceptions to schooling planktivores (e.g. *Chaetodon kleinii*) may elucidate the selective pressures favoring certain mating systems. Much less is known of the mating systems of non-corallivorous omnivores. Many of these species also form pairs (e.g. *Chaetodon chrysurus*: Fricke 1986). Others may from harems, as does *Chaetodon fremblii* (Hourigan 1986a). Compared to studies of corallivores, studies of omnivorous butterflyfishes pose greater difficulties in measuring and manipulating the food resources.

5. Finally, the large number of species in the family Chaetodontidae may allow treatment of social behavior as a trait for systematic analysis. Together with studies of the feeding ecology and morphology (e.g. Motta 1985, 1988), such analyses many reveal the extent to which particular mating systems result from convergence to similar environmental factors, or from shared phylogenetic constraints.

Acknowledgements

I would like to thank Philip Motta, both for his organization of the conference on butterflyfishes, and for helping to introduce me to these beautiful animals. Many hours in the field with him and Timothy Tricas, as well as the long discussions which ensued have helped to develop the ideas presented here. Finally, I would like to thank Ernst Reese, whose work has shaped much of the field of butterflyfish research, and who has personally served as advisor, mentor and friend. This paper has benefited greatly from the comments of E. Reese, G. Barlow, P. Motta, T. Telecky, M. Lutnesky and an anonymous reviewer. The research on Hawaiian butterflyfishes was conducted with the help of numerous enthusiastic Earthwatch Volunteers under a grant from the Center for Field Research.

References cited

Allen, G.R. 1979. Butterfly and angelfishes of the world. Vol. 2. John Wiley, New York. 352 pp.

Bagenal, T.B. 1966. The ecological and geographical aspects of the fecundity of plaice. J. Mar. Biol. Assoc. U.K. 46: 743–751.

Barlow, G.W. 1974. Contrasts in social behavior between Central American cichlid fishes and coral-reef surgeon fishes. Amer. Zool. 14: 9–34.

Barlow, G.W. 1984. Patterns of monogamy among teleost fishes. Arch. FischWiss. 35: 75–123.

Barlow, G.W. 1986. A comparison of monogamy among freshwater and coral-reef fishes. pp. 767–775. In: T. Uyeno, R. Arai, T. Taniuchi & K. Matsuura (ed.) Indo-Pacific Fish Biology, Proc. 2nd Internat. Conf. Indo-Pacific Fishes, Ichthyological Soc. Japan, Tokyo.

Bauer, J.A., Jr. & S.E. Bauer. 1981. Reproductive biology of pigmy angelfishes of the genus *Centropyge* (Pomacanthidae). Bull. Mar. Sci. 31: 495–513.

Bouchon-Navaro, Y. & C. Bouchon. 1989. Correlations between chaetodontid fishes and coral communities of the Gulf of Aqaba (Red Sea). Env. Biol. Fish. 25: 47–60.

Brett, J.R. & D.D. Groves. 1979. Physiological energetics. pp. 279–352. In: W.S. Hoar, D.J. Randall & J.R. Brett (ed.) Fish Physiology, Volume 7, Academic Press, New York.

Brown, J.L. 1964. The evolution of diversity in avian territorial systems. Wilson Bull. 6: 160–169.

Burgess, W.E. 1978. Butterflyfishes of the world. T.F.H. Publications, Neptune City. 832 pp.

Charnov, E.L. & W.M. Schaffer. 1973. Life history consequences of natural selection: Cole's result revisited. Amer. Nat. 107: 791–793.

Colin, P.L. 1989. Aspects of the spawning of western Atlantic butterflyfishes (Pisces: Chaetodontidae). Env. Biol. Fish. 25: 131–141.

Doherty, P.J. 1983. Tropical territorial damselfishes: is density limited by aggression or recruitment? Ecology 64: 176–190.

Driscoll, J.W. & J.L. Driscoll. 1988. Pair behavior and spacing in butterflyfishes (Chaetodontidae). Env. Biol. Fish. 22: 29–37.

Ehrlich, P.R., F.H. Talbot, B.C. Russell & G.R.V. Anderson. 1977. The behavior of chaetodontid fishes with special reference to Lorenz's 'poster coloration' hypothesis. J. Zool. Lond. 183: 213–228.

Emlen, S.T. & L.W. Oring. 1977. Ecology, sexual selection and the evolution of mating systems. Science 197: 215–223.

Fricke, H.W. 1980. Control of different mating systems in a coral reef fish by one environmental factor. Anim. Behav. 28: 561–569.

Fricke, H.W. 1986. Pair swimming and mutual partner guarding in monogamous butterflyfish (Pisces, Chaetodontidae): a joint advertisement of territory. Ethology 73: 307–333.

Geist, V. 1974. On the relationship of social evolution and ecology in ungulates. Amer. Zool. 14: 205–220.

Gore, M.A. 1983. The effect of a flexible spacing system on the social organization of a coral reef fish, *Chaetodon capistratus*. Behaviour 85: 118–145.

Gosline, W.A. 1965. Thoughts on systematic works in outlying areas. Syst. Zool. 14: 59–61.

Gronnell, A.M. 1984. Courtship, spawning and social organization of the pipefish, *Corythoichthys intestinalis* (Pisces: Sygnathidae) with notes on two congeneric species. Z. Tierpsychol. 65: 1–24.

Hamilton, W.D. 1971. Geometry for the selfish herd. J. Theor. Biol. 31: 295–311.

Hirshfield, M.F. 1980. An experimental analysis of reproductive effort and cost in the Japanese medaka, *Oryzias latipes*. Ecology 61: 282–292.

Hobson, E.S. 1974. Feeding relationships of the teleostean fishes on coral reefs in Kona, Hawaii. U.S. Fish. Bull. 72: 915–1031.

Hobson, E.S. 1978. Aggregating as a defense against predation in aquatic and terrestrial environments. pp. 219–234. In: E.S. Reese & F.S. Lighter (ed.) Contrasts in Behavior, John Wiley, New York.

Hourigan, T.F. 1984a. The adaptive significance of pair bonding in reef fishes without parental care. Ethology Newsletters 18: 12–13.

Hourigan, T.F. 1984b. Pair bond formation and monogamy in two species of Hawaiian butterflyfishes (Fam. Chaetodontidae). Pac. Sci. 38: 363.

Hourigan, T.F. 1986a. A comparison of haremic social systems in two reef fishes. pp. 23–28. In: L.C. Drickamer (ed.) Behavioral Ecology and Population Biology, Privat, I.E.C., Touluse.

Hourigan, T.F. 1986b. An experimental removal of a territorial pomacentrid: effects on the occurrence and behavior of competitors. Env. Biol. Fish. 15: 161–169.

Hourigan, T.F. 1987. The behavioral ecology of three species of butterflyfishes, Ph.D. Dissertation, University of Hawaii, Honolulu. 496 pp.

Hourigan, T.F. & C.D. Kelley, 1985. Histology of the gonads and observations on the social behavior of the Caribbean angelfish, *Holacanthus tricolor*. Mar. Biol. 88: 311–322.

Hourigan, T.F. & E.S. Reese. 1987. Mid-ocean isolation and the evolution of Hawaiian reef fishes. Trends Ecol. Evol. 2: 187–191.

Hourigan, T.F., T.C. Tricas & E.S. Reese. 1988. Coral reef fishes as indicators of environmental stress in coral reefs. pp. 107–135. *In:* D.F. Soule & G. Kleppel (ed.) Marine Organisms as Indicators, Springer Verlag, Berlin.

Irons, D.K. 1989. Temporal and areal feeding behavior of the butterflyfish, *Chaetodon trifascialis*, at Johnston Atoll. Env. Biol. Fish. 25: 187–193.

Jarman, P.J. 1974. The social organization of antelope in relation to their ecology. Behaviour 58: 215–267.

Johannes, R.E. 1978. Reproductive strategies of coastal marine fishes in the tropics. Env. Biol. Fish. 3: 65–84.

Kleiman, D.G. 1977. Monogamy in mammals. Quart. Rev. Biol. 52: 39–69.

Knowlton, N. 1979. Reproductive synchrony, parental investment and the evolutionary dynamics of sexual selection. Anim. Behav. 27: 1023–1033.

Lack, D. 1968. Population Studies of Birds. Oxford Uiversity Press, Oxford. 341 pp.

Lassig, B.R. 1977. Socioecological strategies adopted by obligate coral-dwelling fishes. Proc. 3rd. Internat. Coral Reef Symp. 1: 565–570.

Lobel, P.S. 1978. Diel, lunar, and seasonal periodicity in the reproductive behavior of the pomacanthid fish, *Centropyge potteri*, and some other reef fishes in Hawaii. Pac. Sci. 32: 193–207.

Lobel, P.S. 1989. Spawning behavior of *Chaetodon multicinctus* (Chaetodontidae); pairs and intruders. Env. Biol. Fish. 25: 125–130.

Loiselle, P.V. & G.W. Barlow. 1978. Do fishes lek like birds? pp. 31–75. *In:* E.S. Reese & F.S. Lighter (ed.) Contrasts in Behavior, John Wiley, New York.

Ludwig, G.M. 1984. Contrasts in morphology and life history among Hawaiian populations of two longnose butterflyfishes, *Forcipiger longirostris* and *F. flavissimus:* a possible case of character displacement. Ph.D. Dissertation, University of Hawaii, Honolulu. 284 pp.

Luquet, P. & T. Watanabe. 1986. Interaction 'nutrition-reproduction' in fish. Fish Physiol. Biochem 2: 121–129.

Motta, P.J. 1985. Funtional morphology of the head of Hawaiian and Mid-Pacific butterflyfishes (Perciformes, Chaetodontidae). Env. Biol. Fish. 13: 253–276.

Motta, P.J. 1988. Functional morphology of the feeding apparatus of ten species of Pacific butterflyfishes (Perciformes, Chaetodontidae): an ecomorphological approach. Env. Biol. Fish. 22: 39–67.

Moyer, J.T. & A. Nakazono. 1978. Protandrous hermaphroditism in six species of the anemone fish genus *Amphiprion* in Japan. Japan. J. Ichthyol. 25: 25–39.

Moyer, J.T. & Y. Yogo. 1982. The lek-like mating system of *Halichoeres melanocher* (Pisces: Labridae) at Miyake-Jima Japan. Z. Tierpsychol. 60: 209–226.

Murphy, G.I. 1968. Pattern in life history and the environment. Amer. Nat. 102: 390–404.

Neudecker, S. & P.S. Lobel. 1982. Mating systems of chaetodontid and pomacanthid fishes at St. Croix. Z. Tierpsychol. 59: 299–318.

Norris, J.E. 1985. Trophic relationships of piscivorous coral reef fishes from the Northwest Hawaiian Islands, M.S. Thesis, University of Hawaii, Honolulu. 71 pp.

Orians G.H. 1969. On the evolution of mating systems in birds and mammals. Amer. Nat. 103: 589–603.

Oring, L.W. 1982. Avian mating systems. Avian Biol. 6: 1–92.

Pandian, T.J. & E. Vivekanadan. 1985. Energetics of feeding and digestion. pp. 99–124. *In:* P. Tytler & P. Calow (ed.) Fish Energetics: New Perspectives, John Hopkins University Press, Baltimore.

Pitcher, T.J. 1986. Functions of shoaling behaviour in teleosts. pp. 294–337. *In:* T.J. Pitcher (ed.) The Behaviour of Teleost Fishes, Croom Helm, London.

Ralston, S. 1976. Anomalous growth and reproductive patterns in populations of *Chaetodon miliaris* (Pisces: Chaetodontidae) from Kaneohe Bay, Oahu, Hawaiian Islands. Pac. Sci. 30; 395–503.

Ralston, S. 1981. Aspects of the reproductive biology and feeding ecology of *Chaetodon miliaris*, a Hawaiian endemic butterflyfish. Env. Biol. Fish. 6: 167–176.

Reese, E.S. 1973. Duration of residence of coral reef fishes on 'home' reefs. Copeia 1973: 145–149.

Reese, E.S. 1975. A comparative field study of the social behavior and related ecology of reef fishes of the family Chaetodontidae. Z. Tierpsychol. 37: 37–61.

Reese, E.S. 1978. The study of space related behavior in aquatic animals: special problems and selected examples. pp. 347–374. *In:* E.S. Reese & F.S. Lighter (ed.) Contrasts in Behavior, John Wiley, New York.

Reese, E.S. 1981. Predation on corals by fishes of the family Chaetodontidae: Implications for conservation and management of coral reef ecosystems. Bull. Mar. Sci. 31: 594–604.

Reese, E.S. 1989. Orientation behavior of butterflyfishes (family Chaetodontidae) on coral reefs: spatial learning of route specific landmarks and cognitive maps. Env. Biol. Fish. 25: 79–86.

Robertson, D.R. 1983. On the spawning behavior and spawning cycles of eight surgeonfishes (Acanthuridae) from the Indo-Pacific. Env. Biol. Fish. 9: 193–223.

Robertson, D.R. & S.G. Hoffman. 1977. The roles of female mate choice and predation in the mating systems of some tropical labroid fishes. Z. Tierpsychol. 45: 298–320.

Robertson, D.R., N.V.C. Polunin & K. Leighton. 1979. The behavioral ecology of three Indian Ocean surgeonfishes (*Acanthurus lineatus, A. leucosternon* and *Zebrasoma sco-*

pas): their feeding strategies and social and mating systems. Env. Biol. Fish. 4: 125–170.

Robertson, D.R. & J.M. Sheldon. 1979. Competition and the availability of sleeping sites for a diurnally active Caribbean reef fish. J. Exp. Mar. Biol. Ecol. 40: 285–298.

Sale, P.F. 1980. The ecology of fishes on coral reefs. Oceanogr. Mar. Biol. Ann. Rev. 18: 367–421.

Shulman, M.J. 1985. Coral reef fish assemblages: intra- and interspecific competition for shelter sites. Env. Biol. Fish. 13: 81–92.

Steen, R.C. 1978. Butterfly and angelfishes of the world. Volume 1. John Wiley, New York. 352 pp.

Strathmann, R.R. 1986. What controls the type of larval development? Bull. Mar. Sci. 39: 612–622.

Sutton, M. 1985. Patterns of spacing in a coral reef fish in two habitats on the Great Barrier Reef. Anim. Behav. 33: 1322–1337.

Suzuki, K., Y. Tanaka & S. Hioki. 1980. Spawning behavior, eggs, and larvae of the butterflyfish, *Chaetodon nippon*, in an aquarium. Jap. J. Ichthyol. 26: 334–341.

Talbot F.H., B.C. Russel & G.R.V. Anderson. 1978. Coral reef fish communities: unstable, high-diversity systems? Ecol. Monogr. 48: 425–440.

Thresher, R.E. 1977. Ecological determinants of the territorial behavior of reef fishes. Proc. 3rd. Internat. Coral Reef Symp. 1: 551–557.

Thresher, R.E. 1984. Reproduction in reef fishes. T.F.H. Publications, Neptune City. 399 pp.

Tricas, T.C. 1985. The economics of foraging in coral-feeding butterflyfishes of Hawaii. Proc. 5th. Internat. Coral Reef Symp. 5: 409–414.

Tricas, T.C. 1986. Life history, foraging ecology, and territorial behavior of the Hawaiian butterflyfish *Chaetodon multicinctus*. Ph.D. Dissertation, University of Hawaii, Honolulu. 248 pp.

Walsh, W.J. 1984. Aspects of nocturnal shelter, habitat space, and juvenile recruitment in Hawaiian coral reef fishes. Ph.D. Dissertation. University of Hawaii, Honolulu. 475 pp.

Warner, R.R. 1980. The coevolution of behavioral and life-history characteristics. pp. 151–188. *In:* G.W. Barlow & J. Silverberg (ed.) Sociobiology: Beyond Nature/Nurture, AAAS Selected Symposium 35, Westview Press, Boulder.

Warner, R. & D.R. Robertson. 1978. Sexual patterns in the labroid fishes of the western Caribbean, I. The wrasses (Labridae). Smithsonian Contrib. Zool. 254: 1–27.

Wickler W. & U. Seibt. 1983. Monogamy, an ambiguous concept. pp. 33–50. *In:* P. Bateson (ed.) Mate Choice, Cambridge University Press, Cambridge.

Williams, G.C. 1966. Adaptation and natural selection. Princeton University Press, Princeton. 304 pp.

Wilson, E.O. 1975. Sociobiology: the new synthesis. Belknap-Harvard Press, Cambridge. 697 pp.

Wittenberger, J.F. & R.L. Tilson. 1980. The evolution of monogamy: hypotheses and tests. Ann. Rev. Ecol. Syst. 11: 197–232.

Wootton, R.J. 1977. Effect of food limitation during the breeding season, on the size, body components and egg production of female sticklebacks (*Gasterosteus aculeatus*). J. Anim. Ecol. 46: 823–834.

Wootton, R.J. 1985. Energetics of reproduction. pp. 231–254. *In:* P. Tytler & P. Calow (ed.) Fish Energetics: New Perspectives, John Hopkins University Press, Baltimore.

Wrangham, R.W. & D.I. Rubenstein. 1986. Social evolution in birds and mammals. pp. 452–470. *In:* R.W. Wrangham & D.I. Rubenstein (ed.) Ecological Aspects of Social Evolution, Princeton University Press, Princeton.

Yamsonrat, S. 1980. Sex determination of butterflyfishes (Chaetodontidae). M.S. Thesis, University of Hawaii, Honolulu. 15 pp.

Appendix 1. Social grouping of butterflyfishes with reference to size and feeding guild. Social grouping may differ with age or locality. S = solitary, P = pairs, G = groups. When reasonable information is present concerning their mating system, this is presented as: (M) = monogamous, (H) = haremic, (GS) = group spawning.

Species	Adult[1] size (TL in mm)	Feeding guild[1, 2]		
		Corallivore	Omnivore	Planktivore
Chaetodon aculeatus	100		S[7, 15]	
C. capistratus	100	P[7, 9, 15] (M)[7]		
C. octofasciatus	100	P[1]		
C. punctatofasciatus	110	P[4, 15]		
C. larvatus	120	P[1]		
C. melapterus	120	P[1]		
C. multicinctus	120	P[2, 11, 14, 16] (M)[14]		
C. zanzibariensis	120	S[1], P[1], G[1]		
C. citrinellus	125	P[1, 2, 5]	P[1, 2, 5]	
C. assarius	130			School[1]
C. aureofasciatus	130	S[1, 2, 3], P[1, 15]		
C. austriacus	130	P[12]		
C. fremblii	130		S[2] (H)[13, 14]	
C. mertensii	130		S[1]	
C. nigropunctatus	130	S[1], P[1]		
C. pelewensis	130	P[4, 15]		
C. plebius	130	S[2, 3], P[2, 3]		
C. rainfordii	130	S[2, 3, 4], P[4, 9]		
C. kleinii	140			S[1, 14], P[14]
C. miliaris	140			School[7, 14], (GS)[7]
C. mitratus	140		S[1], P[1], G[1]	
C. paucifasciatus	140		P[12], (M)[12]	
C. xanthurus	140		S[1], P[1]	
C. baronessa	150	P[2, 3]		
C. melannotus	150		S[2, 3, 4, 12], P[3, 4, 12]	
C. meyeri	150	P[1]		
C. reticulatus	150	P[2, 5, 15]		
C. sanctaehelenae	150		P[1]	G[1]
C. speculum	150	S[2], P[3]		
C. triangulum	150	P[1]		
C. tricinctus	150	S[1], P[1], G[1]		
C. trifasciatus	150	P[2, 5, 10, 14], (M)[2, 14]		
Forcipiger flavissimus	150		S[3, 8, 14], P[3, 8, 14], G[3, 8, 14]	
Heniochus diphreutes	150			School[1]
C. collare	160	P[1]		
C. dichrous	160		P[1]	
C. quadrimaculatus	160	P[14, 16], (M)[14]	P[14, 16], (M)[14]	
C. striatus	160	P[1, 9, 15]	P[1, 9, 15]	
Hemitaurichthys zoster	160			School[4]
C. smithii	170			School[1]
C. humeralis	180		P[1], G[1]	
C. ornatissimus	180	P[2, 5, 15, 16], (M)[2, 15]		
C. trifascialis	180	S[2, 3, 12, 15]		
C. ulietensis	180		S[3, 4], P[4], G[4]	
C. weibeli	180		S[1], P[1]	
Hemitaurichthys polylepis	180			School[14] (GS)[14]
Hemitaurichthys thompsoni	180			School[14] (GS)[14]
Heniochus intermedius	180		P[1, 12], G[1]	

Appendix 1. (Continued).

Species	Adult[1] size (TL in mm)	Feeding guild[1, 2]		
		Corallivore	Omnivore	Planktivore
F. longirostris	190		P[8, 14]	
C. argentatus	200		P[1], G[1]	
C. auripes	200	P[15]		
C. ephippium	200		P[2, 3, 5, 15]	
C. falcula	200		P[1]	
C. flavirostris	200		P[1]	
C. unimaculatus	200	S[4, 14], P[2, 3, 4], G[4, 14]		
C. xanthocephalus	200		S[1], P[1]	
Chelmon rostratus	200		S[1, 15], P[1, 3]	
C. auriga	230		S[2, 3], P[2, 14]	
C. vagabundus	230		S[3], P[2, 3], G[4]	
C. fasciatus	250	S[12], P[12], G[12]	S[12], P[12], G[12]	
C. lunula	250		S[1], P[2, 3], G[14]	
C. lineolatus	260		S[2, 3], P[3, 12, 14], G[3, 15]	

Social grouping based on: R = observation of recognizable individuals; C = collection of observed individuals; Q = quantification of occurrence in social groupings; N = non-quantified observations. Data from 1. Allen 1979 (N), 2. Reese 1975 (R, Q), 1981 (R), 3. Ehrlich et al. 1977 (N), 4. Steen 1978 (N), 5. Yomsonrat 1980 (C), 6. Ralston 1981 (C, N), 7. Neudecker & Lobel 1982 (R), 8. Ludwig 1984 (R), 9. Thresher 1984 (N), 10. Sutton (R), 11. Tricas 1985, 1986 (R), 12. Fricke 1986 (R, Q), 13. Hourigan 1986a (R), 14. Hourigan 1987 (R, C), 15. Hourigan, pers. obs. (R), 16. Driscoll & Driscoll 1988 (R).

Orientation behavior of butterflyfishes (family Chaetodontidae) on coral reefs: spatial learning of route specific landmarks and cognitive maps

Ernst S. Reese
Department of Zoology and Hawaii Institute of Marine Biology, University of Hawaii, Honolulu, HI 96822, U.S.A.

Received 7.5.1988   Accepted 8.11.1988

Key words: Reef fish, Foraging, Cognition, Memory, Habit formation, Piloting

Synopsis

Foraging butterflyfishes follow predictable paths as they swim from one food patch to another within their territories and home ranges. The pattern is repeated throughout the day. The behavior is described in species belonging to the coral feeding guild. Habit formation and spatial learning are implicated. Foraging paths are based on learned locations of route specific landmarks. When a coral head is removed the fish look for it in its former location. If pairs of foraging fish are deflected from the path, they resume their routine pattern at the first landmark they encounter. Periodically, fish make excursions of 30 m or more to distant parts of the reef. Usually they follow different paths on the outbound and homeward legs of these excursions. The critical question is: Are the paths novel? If they are, it is evidence for the use of cognitive maps. Certainly fishes living in the highly structured coral reef environment are prime candidates to use cognitive maps in their orientation behavior.

Introduction

The purpose of this paper is threefold. First I ask two questions of fundamental importance to chaetodontid biology and of theoretical interest to comparative animal behavior research. Second, I present preliminary observational evidence which provides glimpses to the answers to these questions. Third, in keeping with the goal of the symposium, I point to a new direction of research on butterflyfishes.

The questions of interest are: How do butterflyfishes find their way on a coral reef? Do they use a system of learned landmarks specific to a route? Do they have the ability to develop and use a cognitive map of their habitat area during orientation? The term cognitive map refers to the ability of animals to develop a neural model or representation of relationships between stimulus objects or landmarks in their environment through spatial learning (Toates 1980, Gould 1986). Such a neural construct of spatial relationships would enable an animal to travel from one place to another along different and novel paths and thereby remove it from the constraints of following a sequence of learned route specific landmarks.

There is a subset of secondary questions. For example, are the orientational abilities equally developed in all species, or are they better developed in strongly site attached species, such as the corallivores, than in other species which are less closely tied to the surface of the reef such as planktivorous species? To date my observations are exclusively on the corallivores. Does early experience play a role and is its effect greatest during a critical or sensitive period? If early experience is important, is the mechanism of its action through imprinting or early perceptual experience or, since they are not mutually exclusive, both? To what extent are social transmission and tradition involved?

Fortunately, Dodson (1988) provides a current review of the role of learning in the orientation and migratory behavior of fishes. He correctly points out that orientation mechanisms based on spatial learning are expected intuitively to be of great importance to fishes such as those which live on coral reefs. Coral reef fishes are relatively site attached, but may make excursions over the reef of an order of magnitude of tens to perhaps a few hundred meters but not more. Furthermore, a coral reef provides much contour and therefore many potentially useful landmarks.

In contrast, one would suspect spatial learning to be less important in species which make annual or once-in-a-lifetime migrations of much greater magnitude. For these species, which must travel great distances through unfamiliar waters where landmarks may be few or absent, one must look to complex navigational mechanisms working in conjunction with passive transport systems such as wind-driven and tidal currents (Smith 1985, Herrnkind & Thistle 1987, Dodson 1988).

Spatial learning is demonstrated in a number of species of fishes and is reviewed by Dodson (1988). The now classic experiments by Aronson (1951, 1971) on the tidepool gobiid fish *Bathygobius soporator* are perhaps most relevant to the present study, because they indicate that *B. soporator* is able to learn the spatial relationships of one tidepool to another and then orient correctly on the basis of this information. It is noteworthy also that there is evidence for one trial learning in fishes (Beukema 1970, Riege & Cherkin 1971). Furthermore, it is becoming increasingly clear that foraging behavior of animals is based on learning and memory (Kamil & Roitblat 1985).

In this paper, I describe the foraging paths of three species of butterflyfishes in order to determine the fidelity of the path and the effect of experimentally altering the behavior. In addition, I report observations of voluntary long distance excursions of these strongly site attached species.

Materials and methods

Species studied

The observations described in this paper are derived from my own studies of the patterns of foraging behavior of *Chaetodon trifascialis* (formerly called *Megaprotodon strigangulus*) and *C. trifasciatus* at Enewetak Atoll, Marshall Islands, from 1972 to 1979, and of *C. ornatissimus* at Puako, Hawaii, in 1982. Similar observations are reported by Hourigan (1987) for *C. multicinctus* and *C. quadrimaculatus* at Puako, Hawaii, from 1980 to 1983. The five species are considered to be obligate coral feeders (Reese 1975, 1977, Hourigan 1987, Hourigan et al. 1988).

Study sites

The principal study reef at Enewetak Atoll was Medren pinnacle. It is a coral pinnacle which rises 20 m from the sandy lagoon floor almost to the surface. At its highest point it is only 3 m beneath the surface. The most luxuriant growth of corals and the greatest abundance of fishes occur at depths of around 10 m. This area of the reef is highly contoured with coral covered ridges transected by sand and rubble filled valleys. The vertical relief from valley floor to ridge crest is on the order of 3 to 5 m.

The territories of six pairs of *C. trifasciatus* and five solitary *C. trifascialis* were mapped and subsequent observations were focused on these individual fishes. *Chaetodon trifasciatus* is territorial and forms monogamous pairs (Reese 1975, Sutton 1985), while *C. trifascialis* characteristically establishes solitary but adjacent territories (Reese 1975, 1981, Irons 1986). *Chaetodon trifasciatus* shows little interspecific agonistic behavior and maintains its territorial boundaries by what appears to be visually mediated, mutual avoidance behavior, leading me (Reese 1975) originally to classify this species as a home-ranging one. In contrast, *C. trifascialis* is a typical territorial species displaying agonistic defensive behavior to intruding conspecifics as well as other species belonging to the corallivore guild.

The study site at Puako, Hawaii, is located at a depth of about 10 to 15 m in an area of rich coral cover. The area is described in detail by Tricas (1985, 1986) and Hourigan (1987). They observed the behavior of more than 50 pairs of *C. multicinctus*. I studied the behavior of eight pairs of *C. ornatissimus* at Puako. The carpet of coral cover is more complete at Puako than at Medren Pinnacle but the diversity of corals is much less. At Puako two species of corals dominate, *Porites compressa* and *P. lobata,* while at Medren Pinnacle corals of the genus *Acropora* are most conspicuous but at least a dozen other species occur in the study area. The vertical relief at Puako is less than at Medren Pinnacle. The massive mounds of *P. lobata* are on the order of 2 to 3 m high, while the more delicate growths of *P. compressa* rarely exceed 1 to 2 m in height. Nevertheless, landmarks of conspicuous coral heads are present in both habitats.

Behavioral observations

Underwater observations of foraging behavior were made by following focal pairs or individuals, in the case of *C. trifascialis*, at a discreet distance using SCUBA. An observer quickly learns what a 'discreet distance' is for each species. If one gets too close, the fish simply stops foraging and turns its attention to the diver. The fishes become tame, after a few, perhaps five to ten, exposures to divers. The habituation of the fishes to the benign presence of divers is extremely important if one wishes to observe normal patterns of behavior.

A map or sketch of the study area is drawn on underwater paper, territories are identified, and conspicuous landmarks within the territories are noted. The movements of the fishes are recorded on underwater paper often in conjunction with underwater audiotapes. Individual fishes are recognized on the basis of individual variation in their markings (Reese 1973) and by taking portrait photographs. Usually three 20 minute observation bouts are made on each of three different pairs of fishes per dive. Occasionally, a pair of fish is followed for the entire hour.

Results

Normal foraging paths

All of the species are consistently repetitive in their foraging behavior (Table 1). They follow a specific route through their territory moving from one coral head to another, feeding on certain heads while bypassing others of the same species and which appear to the observer to be equally attractive as a food source. It is possible to see whether the coral polyps are extended or retracted. When the polyps are extended, it is easier for a chaetodontid to 'pick' them without damaging the coral skeleton (Tricas 1986, Motta 1988). A typical, generalized feeding path is illustrated in Figure 1. The composite pair of *C. trifasciatus* swims through its territory in a pattern which is repeated three to four times

Table 1. Repetition and change of direction of foraging paths.

Species	Sample	Length of observation (min)	Number of observations each sample	Total observations	Number of repetitions[1]	Number of changes of direction		
						0	1	2
C. trifasciatus	6 pairs	20	3	18	3–4	14	3	1
C. trifasciatus	2 pairs	60	1	2	10	2	0	0
C. trifasciatus	1 pair	20	6[2]	6	No data	6	0	0
C. trifascialis	3 indiv.	20	3	9	4	No data		

[1] Analysis of audiotapes is incomplete.
[2] Three of these observations were made at about 10:00 h; three more at about 16:00 h, with paired observations on different days.

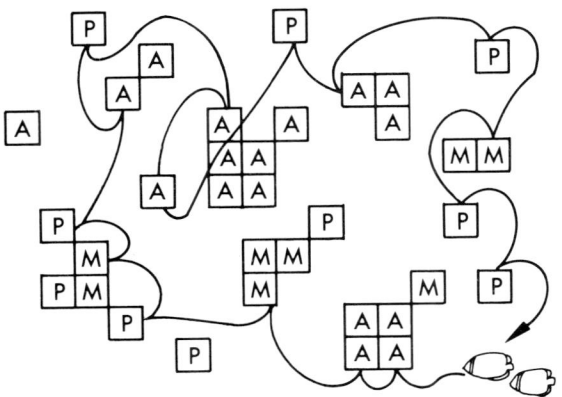

Fig. 1. The schematic diagram illustrates the composite foraging path of a typical pair of *Chaetodon trifasciatus* at Medren Pinnacle, Enewetak Atoll. The pair routinely visits, inspects, and feeds on certain coral heads while bypassing others of the same species. The foraging path is repeated habitually throughout the day with very little variation.

during a 20 minute observation period. The same coral heads are inspected and may be fed on or by passed, and then the pair swims to the next head of the same or different species of corals and so on.

Both *C. trifascialis* and *C. trifasciatus* rigidly follow a strict foraging path (Fig. 1). On two occasions at Medren Pinnacle I followed a pair for the entire 60 minutes of the dive. One pair made 11 circuits of its territory and fed on only three new coral heads, that is ones not fed on previously, during the entire period. The second pair made 9 circuits. I did not record its pattern of feeding because it was interrupted on the second and subsequent circuits by the presence of a suspected, potential predator the grouper, *Cephalopholis argus*. The pair continued in their foraging path but their normal foraging behavior seemed to be constrained by the presence of the grouper in the sense that they swam more slowly and closer together. They continued to inspect the coral heads but fed less. Interestingly, on a number of occasions at Enewetak Atoll when *C. argus* was present, I observed similar 'cautious' behavior by potential prey species but not the mobbing behavior observed by Motta (1984) in Hawaii when moray eels show themselves on the reef.

Another repetitious characteristic of foraging paths shown by all the species I observed is that once the direction of a foraging path is established, whether clockwise or counterclockwise, it persists throughout the observation period and even throughout the entire day (Table 1). Coral feeding butterflyfishes feed throughout the day. Time budget analyses by Tricas (1986) and Hourigan (1987) show that *C. multicinctus* spends in excess of 90% of its time feeding, and Irons (1986, personal communication) finds the *C. trifascialis* feeds throughout the day. I have observed all species reported on in this paper feeding from dawn to dusk. On three occasions I followed the same pair of *C. trifasciatus* for 20 minutes during the first dive in the morning and again for 20 min in the late afternoon. In all three cases, the pairs were following the same foraging path in the same direction. It seems, therefore, that a pair begins the day foraging in either a clockwise or counterclockwise direction and then persists in this direction throughout the day.

Data collected to date do not permit a definitive statement regarding the fidelity of the foraging path from one day to the next. However, it is clear that certain conspicuous landmark heads of coral are always visited day in and day out and even from one year to the next. At Medren Pinnacle, these are usually large heads of *Acropora hyacinthus*, often at the highest points of the territories of *C. trifascialis* and *C. trifasciatus*. I refer to these as 'crown coral heads' and they appear to be important territorial landmarks to the fish. Frequently, the pair returns to them and swims slowly and close to them following bouts of agonistic territorial defensive behavior. Similar behavior is observed in *C. ornatissimus* at Puako, Hawaii, where pairs consistently return to conspicuous heads of coral. In one instance, a pair foraged in two territories that were separated by about 10 m. When they swam from one territory to another, they always went immediately to the 'crown coral head', remained there for a minute or two not feeding, and then resumed their characteristic foraging path.

Female butterflyfishes spend more time feeding than do males who spend more time in territorial defense (Hourigan 1987). I suspect that females lead the foraging pair, and, therefore, decisions regarding the specific details of the foraging path – which direction to forage in, which coral heads to inspect and feed on – may be made by the females.

Experimentally altered foraging paths

I obstructed foraging behavior in two ways. First, 'crown coral heads' were simply removed at Medren Pinnacle, Enewetak. Second, pairs of *C. ornatissimus* were harrassed and deflected from their foraging path at Puako, Hawaii.

In the first case, *C. trifasciatus* and *C. trifascialis* search for missing coral heads in the place from which they were removed (Table 2). After the searching interruption the fishes resumed their pattern of foraging behavior. In two cases, the pair of *C. trifasciatus* under observation stopped on each circuit to look for the missing coral head. In one of these cases, this occurred on four sequential repetitions of the foraging path. Clearly, the fish appear to know the location of specific coral heads in their territories.

When *C. ornatissimus* pairs are deflected from their foraging path, they flee into the coral for cover following chasing. It is difficult to chase them out of their territory. If one succeeds, they swim a detour path away from the diver and back into their territory before seeking refuge in the coral. How long they remain in hiding is variable. Of the 17 trials, the time of day varied from 1000 to 1600 h and hence perhaps the hunger motivation for foraging, and the degree of diver harrassment necessary to chase them into hiding was variable. What is important to note, however, is that when foraging is resumed the pairs swim from coral head to coral head, inspecting and biting, in the regular pattern. Moreover, they seem to be able to reenter the path from wherever they sought refuge. It is possible, however, that they always seek refuge in known locations and thus know the relationship between the refuge and the coral heads in their foraging path. Individual *C. multicinctus* and *C. quadrimaculatus* return to the same shelters each night. They do not shelter as a pair. Individual *C. quadrimaculatus* visit their pair mate's shelter, making straight apparently goal directed swims (Hourigan, personal communication). Walsh (1983) describes similar behavior for the triggerfish *Melichthys niger*. In any case, the observations indicate that the fishes have a keen sense of spatial location of coral heads, routes, and refuges in their territories.

Voluntary long distance excursions

An interesting behavior is observed in pairs of *C. trifasciatus* and single *C. trifascialis*. The fishes leave their territories and swim directly to a distant location on the reef often 30 m or more from their own home area of the reef. The results of my observations of these excursions are given in Table 3. The excursions take them through territories of conspecifics as well as other species, and they are usually chased by the territory holders. Characteristically the fishes swim as fast as they can, often

Table 2. Experimentally altered foraging paths.

Species	Sample	Number of observations each sample	Total observations	Experimental manipulation	Behavioral response
C. trifasciatus	3 pairs	1	3	remove 'crown coral head'	all 3 pairs stopped and searched, and 1 pair did this on 4 consecutive repetitions
C. trifascialis	3 indiv.	1	3	same	all 3 individuals stopped and searched
C. ornatissimus	5 pairs	3, 3, 3, 3, 5	17	fish chased from path	all reentered path without change of direction

with their dorsoventral body axis parallel to the surface of the reef. They swim at least 1 to 2 m above the reef and in a straight line during these excursions. Usually they swim the entire distance without stopping. One *C. trifascialis* was followed on three occasions, and it swam to a different distant location on Medren Pinnacle each time. The return leg of the excursion characteristically is slow and meandering. The fishes stop and show hurried inspection and rapid feeding movements on coral heads along the return route. Similar long distance excursions of 25 to 30 m are made by *C. multicinctus* at Puako (Tricas, personal communication).

What is especially interesting about this behavior is that it appears to be strongly goal directed and yet there is no indication of how they recognize the distant goal. Furthermore the rapid outward bound leg of the excursions seems to preclude the use of route specific landmarks so much in evidence during normal foraging behavior. The slower, less direct return path is difficult to evaluate. The question of interest is whether or not these pathways are novel. At the present time we do not know. The direct, fast, apparently goal-directed outbound swim suggests that the excursions are not simply exploratory in nature. These voluntary, long distance excursions suggest the tantalizing possibility that butterflyfishes are indeed capable of developing and using cognitive maps for orientation on the reef.

Discussion

The routine pattern of movement followed by butterflyfishes as they forage within their territories appears to be an example of piloting or landmark orientation. Dodson (1988) reviews other examples, especially the classic experiments of Aronson (1951, 1971) in which he demonstrated spatial learning in the gobiid fish *Bathygobius soporator*. A fascinating combination of spatial learning based on both individual ontogenetic experience and social transmission of pathways occurs in Caribbean grunts of the family Haemulidae (Helfman et al. 1982, Helfman & Schultz 1984). These examples are explained by the fishes learning a sequence of route specific landmarks. Whether cognitive maps (Toates 1980) are involved is questionable, but Roitblat et al. (1982) demonstrate that the Siamese fighting fish, *Betta splendens*, has the ability to master a complex spatial learning problem not so much through the use of a high capacity, working memory system but rather by utilizing a system in which a preferred foraging strategy is adopted and memorized. Furthermore, Hourigan (1987) suggests that foraging along a path, which not only provides food but along which the fish does not encounter predators, may reinforce the rigidity of the foraging pathway. Thus a strategy of risk reduction may be involved in the formation of a rigid foraging habit.

The long excursions over the reef made by butterflyfishes in which they characteristically follow a different outward bound route to their distant goal than the subsequent return pathway to their home territory could be explained on the basis of other navigational systems (Schöne 1984), but both inertial and bicoordinate navigational systems seem overly complex and are extremely difficult to demonstrate. These systems are based on the organism having a psychic map of its movements, or, respec-

Table 3. Voluntary long distance excursions.

Species	Sample	Excursions observed	Behavioral observations of swimming	
			Outbound path	Return path
C. trifasciatus	2 pairs	2	2 direct and fast	2 indirect and slow
C. trifascialis	3 indiv.	5[1]	4 direct and fast, 1 indirect and slow	5 indirect and slow

[1] Two fish made one excursion each; a third individual was followed on three excursions.

tively, a bicoordinate map and compass system. Presumably in vector orientation, vector directions are followed rather than route specific landmark stimuli (Schöne 1984), but the maps must nevertheless be built on experience of movements of the organism in its environment. In this regard, the development of a cognitive map based on spatial learning as originally proposed by Tolman (1948, in Toates 1980) seems a simpler and more plausible hypothesis because clearly butterflyfishes are capable of spatial learning.

When a sequential order of landmarks is learned, then landmark orientation or piloting becomes possible and may be used in conjunction with sun compass navigation. Spatial memory is not necessarily involved. The stimulus controlling the response, the learned landmarks, is present at the time the response is made. The ability to navigate by sun compass is known for coral reef dwelling parrot fishes (Winn et al. 1964) as well as other species of fish (Hazler & Scholz 1983, Smith 1985, Dodson 1988). Here again, the system is based on a learning process in which the animal learns relationships between important goal or destination locations, for example a feeding ground and a refuge, which are spatially separated, and the position of the sun.

The simplest way to test the role of sun compass orientation is to determine whether or not fishes are able to make long distance excursions on overcast days when the sun's position cannot be determined. One can also catch fish and displace them. If they swim in a predictable parallel path and miss their destination, they may be using a sun compass. If they adjust their direction and reach their goal via a novel path, then something akin to a cognitive map may be involved. The latter seems to have occurred when fish were deflected from their paths.

It is exceedingly difficult to clearly establish the existence of a cognitive map. The idea is that an organism remembers the relationships between landmarks and stores them in a neural map using spatial memory. The animal can then orient to a goal via short cuts and alternate routes when the normal pathway is blocked, and furthermore it can do this without the landmarks being physically present.

For example, a displaced animal which relies on sequential landmark orientation, or piloting, must search randomly or in some systematic way for a familiar landmark, whereupon it proceeds from one familiar landmark stimulus to another in an ordered and known sequence until the goal is reached. In contrast, an animal using a cognitive map of the relationships between landmarks and the goal, should be able to proceed via short-cuts or alternative routes from the first landmark encountered directly to the goal.

Finally, it is postulated that the coral feeding species are more adept at orientation on the reef than those species which are less closely tied to the surface of the reef for food and shelter. Roitblat et al. (1982) correctly point out that each species' cognitive abilities evolve as adaptive specializations to the species' ecological niche. In this regard, the orientation ability of species which are less closely tied to the surface of the reef for food, such as planktivores, may not be as well developed as it is in coral feeding species, such as those reported on in this paper, which ecologically are more closely tied to the surface of the reef.

The recognition of a destination or goal, whether it is a distant location on the reef or the fish's home territory, presents a problem. Presumably recognition is based on specific stimuli learned during the fish's ontogenetic experience. For coral reef fishes such as butterflyfishes vision seems to be the sensory modality primarily involved, but olfactory clues must not be ruled out. In addition, directional clues provided by tidal currents over the reef deserve consideration.

Overall, the most plausible explanation and the hypothesis, which we intend to test, is that the butterflyfishes use a system of spatially learned route specific landmarks or underwater guideposts, as Hasler (1966) so aptly called them, possibly in conjunction with sun compass orientation, and possibly a cognitive map.

Acknowledgements

I thank Herbert L. Roitblat and R.J.F. Smith for their interest in this work and for commenting on portions of the manuscript. The research was supported by the Department of Zoology and the Hawaii Institute of Marine Biology of the University of Hawaii, and by the U.S. Department of Energy through its support of the Mid-Pacific Research Laboratory, Enewetak Atoll, Marshall Islands. Research at Puako, Hawaii, was made possible through a grant from the Earthwatch Organization. I thank Lori Yamamura for typing the manuscript and Susan Nakamura for preparing the figure. I am indebted to my friends and former students Tom Hourigan, Phil Motta, and Tim Tricas for many hours of stimulating discussion of butterflyfish behavior. It is with great pleasure and a feeling of deep gratitude that I dedicate this paper to Arthur D. Hasler. Many years ago he provided me with my first opportunity to experience the excitement of observing the orientation behavior of fishes underwater. Contribution No. 778, Hawaii Institute of Marine Biology.

References cited

Aronson, L.R. 1951. Orientation and jumping behavior in the gobiid fish, *Bathygobius soporator*. Amer. Mus. Novit. 1486: 1–22.

Aronson, L.R. 1971. Further studies on orientation and jumping behavior in the gobiid fish, *Bathygobius soporator*. Ann. N.Y. Acad. Sci. 188: 378–392.

Beukema, J.J. 1970. Angling experiments with carp (*Cyprinus carpio* L.). II. Decreasing catchability through one-trial learning. Neth. J. Zool. 20: 81–92.

Dodson, J.J. 1988. The nature and role of learning in the orientation and migratory behavior of fishes. Env. Biol. Fish. 23: 161–182.

Gould, J.L. 1986. The locale map of honey bees: do insects have cognitive maps? Science 232: 861–863.

Hasler, A.D. Underwater quideposts. University Wisconsin Press, Madison. 155 pp.

Hasler, A.D. & A.T. Scholz. 1983. Olfactory imprinting and homing in salmon (Zoophysiology, Vol. 14). Springer-Verlag, Berlin. 134 pp.

Helfman, G.S., J.L. Meyer & W.N. McFarland. 1982. The ontogeny of twilight migration patterns in grunts (Pisces: Haemulidae). Anim. Behav. 30: 317–326.

Helfman, G.S. & E.T. Schultz. 1984. Social transmission of behavioural traditions in a coral reef fish. Anim. Behav. 32: 379–384.

Herrnkind, W.F. & A.B. Thistle (ed.) 1987. Signposts in the sea. Florida State University, Tallahassee. 210 pp.

Hourigan, T.F. 1987. The behavioral ecology of three species of butterflyfishes (Family Chaetodontidae). Ph.D. Dissertation, University of Hawaii, Honolulu. 496 pp.

Hourigan, T.F., T.C. Tricas & E.S. Reese. 1988. Coral reef fishes as indicators of environmental stress in coral reefs. pp. 107–135. *In:* D.F. Soule & G.S. Kleppel (ed.) Marine Organisms as Indicators, Springer-Verlag, Berlin. 342 pp.

Irons, D. 1986. Feeding behavior of the butterflyfish, *Chaetodon trifascialis*, at Johnston Atoll. Pac. Sci. 40: 121.

Kamil, A.C. & H.L. Roitblat. 1985. The ecology of foraging behavior: implications for animal learning and memory. Ann. Rev. Psychol. 36: 141–169.

Motta, P.J. 1984. Response by potential prey to coral reef fish predators. Anim. Behav. 31: 1257–1259.

Motta, P.J. 1988. Functional morphology of the feeding apparatus of ten species of Pacific butterflyfishes (Perciformes, Chaetodontidae): an ecomorphological approach. Env. Biol. Fish. 22: 39–67.

Reese, E.S. 1973. Duration of residence of coral reef fishes on 'home' reefs. Copeia 1973: 145–149.

Reese, E.S. 1975. A comparative field study of the social behavior and related ecology of reef fishes of the family chaetodontidae. Z. Tierpsychol. 37: 37–61.

Reese, E.S. 1977. Coevolution of corals and coral feeding fishes of the family Chaetodontidae. Proc. Third Internat. Coral Reef Symp. 1: 267–274.

Reese, E.S. 1981. Predation on corals by fishes of the family Chaetodontidae: implications for conservation and management of coral reef ecosystems. Bull. Mar. Sci. 31: 594–604.

Riege, W.H. & A. Cherkin. 1971. One-trial learning and biphasic time course of performance in the goldfish. Science 172: 966–968.

Roitblat, H.L., W. Tham & L. Golub. 1982. Performance of *Betta splendens* in a radial arm maze. Anim. Learn. & Beh. 10: 108–114.

Schone, H. 1984. Spatial orientation. Princeton University Press, Princeton. 347 pp.

Smith, R.J.F. 1985. The control of fish migration (Zoophysiology, Vol. 17). Springer-Verlag, Berlin. 243 pp.

Sutton, M. 1985. Patterns of spacing in a coral reef fish in two habitats on the Great Barrier Reef. Anim. Behav. 33: 1322–1337.

Toates, F.M. 1980. Animal behaviour – a systems approach. John Wiley, New York. 299 pp.

Tolman, E.C. 1948. Cognitive maps in rats and men. Psych. Rev. 55: 189–208.

Tricas, T.C. 1985. The economics of foraging in coral-feeding butterflyfishes of Hawaii. Proc. Fifth Internat. Coral Reef Symp. 5: 409–414.

Tricas, T.C. 1986. Life history, foraging ecology, and territorial behavior of the Hawaiian butterflyfish, *Chaetodon multicinctus*. Ph.D. Dissertation, University of Hawaii, Honolulu. 248 pp.

Walsh, W.J. 1983. Stability of a coral reef fish community following a catastrophic storm. Coral Reefs 2: 49–63.

Winn, H.E., M. Salmon & N. Roberts. 1964. Sun-compass orientation by parrot fishes. Z. Tierpsychol. 21: 798–812.

Larval biology of butterflyfishes (Pisces, Chaetodontidae): what do we really know?

Jeffrey M. Leis
Division of Vertebrate Zoology, The Australian Museum, PO Box A285, Sydney, NSW 2000, Australia

Received 18.11.1987   Accepted 17.5.1988

Key words: Fish larvae, Fish eggs, Coral reef, Distribution, Dispersal, Feeding

Synopsis

Relatively little is known of the pelagic portion of the life history of butterflyfishes. Eggs are small (<1 mm), pelagic and hatch in less than 30 hours. Most species pass through a so-called tholichthys larval interval characterized by elaborate, distinctive head spinatation: *Coradion* larvae have different head spinatation. While older chaetodontid larvae can be identified by adult characters, young (preflexion) larvae generally cannot now be identified below family. In tropical plankton studies chaetodontid larvae averaged <0.1% of larvae captured, and occurred in 13% of samples. This rarity is a major hindrance to further work, but is not unexpected in view of adult abundance. Larvae of a few taxa are most abundant in shelf waters, but larvae of many chaetodontid taxa seem to be most abundant in oceanic waters. In either case, waters near reefs have the fewest chaetodontid larvae. Offshore maxima of larvae appear to exist a few kilometers seaward of Great Barrier Reef ribbon reefs. Chaetodontid larvae may prefer the upper portion of the water column. Both size and age at settlement vary widely within the family and the large genus *Chaetodon*, and the latter varies widely within species. Average size at settlement is less than 20 mm and age is less than 40 days. No correlation was found between size and age at settlement. Behaviour and feeding of chaetodontid larvae are essentially unstudied. Chaetodontid larvae seem to be least abundant in winter. The implications of these conclusions are discussed and some suggestions for further research are made. In all areas more work is needed.

Introduction

Chaetodontids are among the most conspicuous and intensively studied fishes on coral reefs as this volume attests, but little is known of the pelagic interval of chaetodontid life history. It is likely that the vast majority of mortality and dispersal in chaetodontids occurs between spawning and settlement, and that chaetodontid populations may be recruit-limited, so the importance of understanding the pelagic larval stage seems obvious. It has been known for over 100 years that chaetodontids have a specialized pelagic larva commonly called tholichthys (Burgess 1978), but early larval stages of chaetodontids were unknown until the mid 1970s (Leis & Miller 1976) and the complete development of eggs and larvae has not yet been described for any chaetodontid. There is no published study devoted to the ecology of chaetodontid larvae, and most studies of fish larvae in tropical waters mention chaetodontids in passing, if at all.

Surprisingly little information was available on biology of chaetodontid larvae in published sources and the data I could extract from my own work were limited. Although some interesting things have emerged from this review, and much of the information contained here from my own work is new, I view this paper more as an outline of what is not known about the pelagic phase of butterflyfishes than a compendium of our knowledge of it. If this paper serves as a prod to further study on chaetodontid larvae, then it will have served a useful purpose. Unfortunately, as noted below, there are some formidable obstacles to further understanding the biology of butterflyfish larvae.

Results

Taxonomy

Eggs
Although few chaetodontid eggs have been described, they are all apparently pelagic (6 species of *Chaetodon* and *Forcipiger*: Burgess 1978, Moe 1976 [cited by Thresher 1984], Suzuki et al. 1980a, b, Colin & Clavijo 1988) range in size from 0.6 to 0.75 mm (*C. nippon, C. ocellatus* and *C. aculeatus* – Moe 1976, Suzuki 1980a, Colin & Clavijo 1988) to 'about 1 mm' (*C. sanctaehelenae* Burgess 1978), and contain a single oil droplet, 0.18–0.19 mm (Suzuki et al. 1980a, Burgess 1978). Incubation time for the two species so far described was 28–30 hours at 22–26°C (Suzuki et al. 1980a, Colin & Clavijo 1988). Because chaetodontid eggs are similar to the vast majority of reef fish eggs it is unlikely they could be readily identified in plankton hauls.

Larvae
Although older chaetodontid larvae have been known for 120 years, young (i.e. preflexion) stages were first illustrated only relatively recently (Leis & Miller 1976), and only a few descriptions and illustrations of young chaetodontid larvae have appeared since (Suzuki et al. 1980a, b, Leis & Rennis 1983). Relatively few of these young larvae (which are of the size normally caught by plankton nets) can be identified with certainty to genus and almost none can be identified to species. This can be ascribed to the relative scarcity of both specimens (see below) and larval taxonomists. However, the result is that there is much taxonomic work to be done. The amount of ecological insight that can be gained on chaetodontid larvae without this taxonomic base is limited. Table 1 summarizes current taxonomic knowledge of chaetodontid larvae.

The so-called tholichthys interval of chaetodontids is characterized by a round, deep and compressed body, but more importantly by head spination. The head is covered by fused plates which extend posteriorly over the trunk in the form of broad, flat, more or less blunt, rugose plates originating from the post-temporal and supracleithrum dorsally and the preoperculum ventrally (Johnson 1984). This is the 'classic' tholichthys (Fig. 1). There are however, variations in head spination and shape of the head and body. In addition, some species have relatively elongate dorsal and pelvic fin spines which are strong and rugose (Fig. 1 and Leis & Rennis 1983), and some have a rounded supraoccipital crest. *Chelmon, Chaetodon, Hemitaurichthys, Heniochus, Parachaetodon* and *Johnrandallia* apparently have classic tholichthys larvae. *Chelmonops* and *Amphichaetodon* larvae are unknown. *Forcipiger* larvae have an elongate, armoured snout, reduced supracleithral and post-temporal plates, a large, ragged supraoccipital crest, and long, sharp preopercular spines. Young *Forcipiger* larvae are similar to other young chaetodontid larvae as described by Leis & Rennis (1983), which implies that *Forcipiger* is a derived genus relative to the other chaetodontids. Larvae of *Coradion* depart enough from the tholichthys pattern that use of this term for them seems inappropriate (Leis & Rennis 1983). *Coradion* larvae have more conventional percoid head spination including serrate spines along the preopercular border, a very large, sharp and serrate spine at the preopercular angle, serrate infraorbital, dentary, supraocular and supraoccipital ridges and reduced posttemporal and supracleithral spination (Fig. 1). None of these spines form broad, flat, blunt plates extending over the trunk posteriorly, and the rugosity is only moderately developed. Young *Coradion* larvae are very different from the young larvae of other chaetodontids. *Coradion* larvae are unique but the features of their head spination are found in various combinations among a wide variety of percoid fish larvae. This is not true of classic tholichthys head spination. Either *Coradion* larvae represent a highly specialized offshoot of the classic tholichthys larva, or a primitive larva. The former implies that *Coradion* is a highly specialized chaetodontid genus, while the latter implies that *Coradion* is the primitive sister group of the other chaetodontids. However, more detailed work on larvae of all genera is required before any firm conclusions can be reached. It seems obvious that a thorough study of chaetodontid larvae could shed considerable light on interrelationships, including those within the genus *Chaetodon*.

Chaetodontid larvae are highly distinctive and supply a number of characters which provide a good basis for defining the family. It is ironic that confusion over the identity of chaetodontid larvae resulted in a long controversy as to distinctiveness of the Chaetodontidae and the Pomacanthidae. The first description of a chaetodontid larva appeared in 1868, when Günther described *Tholichthys osseus* and allocated it to the scombroid fishes. Day (1870) suggested *T. osseus* was the young of a squamipinne fish (the group including pomacan-

Table 1. Chaetodontid taxa for which information is available on larval stages. Descriptions vary widely in detail, often being little more than illustrations. Burgess (1978) provides photographs, but no additional information, of late tholichthys or just transformed juveniles of 12 *Chaetodon* spp.: *auriga*, *auripes*, *blackburni*, *capistratus*, *ephippium*, *hemichrysurus*, *lineatus*, *litus*, *melannotus*, *plebeius*, *punctatofasciatus* and *sedentarius*. Burgess (1978) examined but did not describe tholichthys larvae of *Chelmon*, *Johnrandalla*, *Hemitaurichthys* and *Heniochus*, although his statement that 'all chaetodontids have a tholichthys larva' implies that these genera have classic tholichthys larvae.

Species	Stages				Reference
	Yolksac	Preflexion	Postflexion	Specialized pelagic stages	
Chaetodon					
C. auriga				×	Watanabe (1946)
C. aya				×	Hubbs (1963)
C. collare				×	Watanabe (1946)
C. citrinellus				×	Günther (1871)
C. excelsa				×	Burgess (1978)
C. lunula				×	Watanabe (1946), Günther (1873)
C. melannotus				×	Watanabe (1946)
C. meyeri				×	Burgess (1978)
C. modestus				×	Schmidt & Lindberg (1930)
C. nippon	×	×			Suzuki et al. (1980a)
C. ocellatus				×	Fowler (1945)
C. ornatissimus				×	Burgess (1978)
C. sedentarius?				×	Lütken (1880)
C. unimaculatus?		×	×	×	Leis & Rennis (1983)
C. vagabundus				×	Watanabe (1946)
Subgenus *Citharoedus*[2]				×	Günther (1871), Fraser-Brunner (1933), Burgess (1978), Fourmanoir (1976),[1] Bourret et al. (1979)[1]
Chaetodon. spp.		×	×	×	Günther (1868), Leis & Miller (1976), Lütken (1880), Burgess (1974, 1978)
Coradion sp.		×		×	Leis & Rennis (1983)
Forcipiger longirostris				×	Kendall & Goldsborough (1911), Randall (1961), Burgess (1978)
Parachaetodon ocellatus?[3]				×	Lütken (1880)

[1] mis-identified as *Heniochus* spp.
[2] identifications after Burgess 1978.
[3] larvae from Australia tentatively identified as *P. ocellatus* do not match Lütken's description.

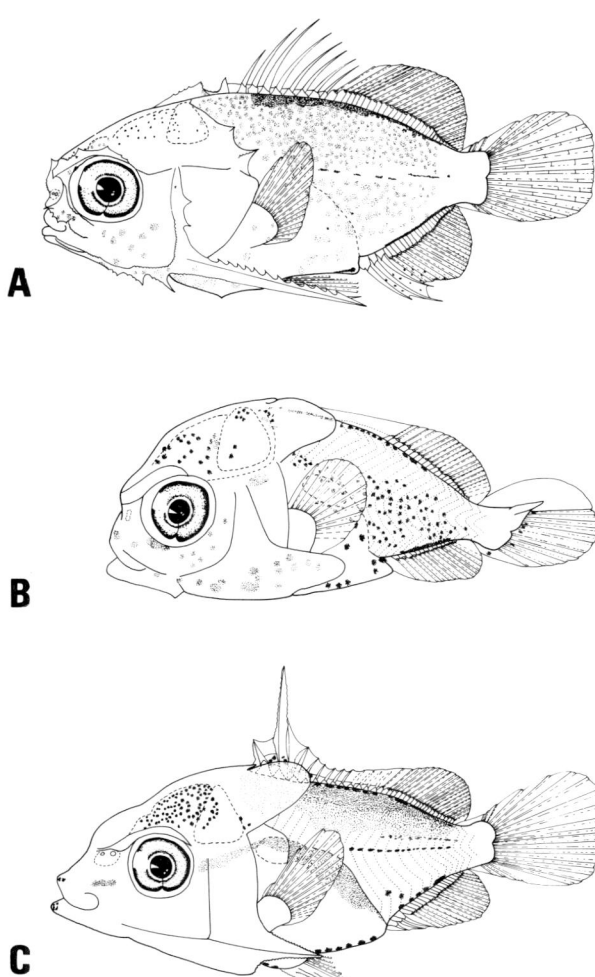

Fig. 1. Chaetodontid larvae (after Leis & Rennis 1983). A – *Coradion* sp., a larva which differs from tholichthys (6.5 mm); B – Unidentified species of classic tholichthys (4.0 mm); C– *Chaetodon* (*unimaculatus* ?) modified tholichthys (6.6 mm).

thids and chaetodontids). Günther (1871) agreed, describing what he now called *Tholichthys* larvae of *Chaetodon citrinellus* and another unidentified species. He noted he could not identify *T. osseus* further. However, he also stated (without evidence) that 'I have but little doubt that *Holacanthus* passes through a *Tholichthys*-stage'. Lütken (1880) described tholichthys larvae of two *Chaetodon* species and *Parachaetodon ocellatus*. He also described the non-tholichthys young of *Holacanthus* and *Pomacanthus* noting that 'neither of these genera, so far as we know, passes through the so-called "*Tholichthys*" phase' (Lütken 1881). This should have clarified the matter, but Fraser-Brunner (1933) illustrated a chaetodontid larva (of the *Chaetodon* sub-genus *Citharoedus*, according to Burgess 1978) as '*Pomacanthus* sp.' and noted how similar the illustrated larva was to 'the *Tholichthys* stage of the Chaetodontinae'. Carrying on in the same tradition, Okada & Watanabe (1947) described and illustrated the recently settled (15 mm) young of the pomacanthid, *Pomacanthus semicirculatus*, and referred to it as 'Tholichthys stage' in spite of the absence of any head plates or other tholichthys characteristics, except perhaps, body shape. It remained for Burgess (1974) to clearly show the difference between pomacanthid and chaetodontid larvae and prove Lütken correct after over 100 years of confusion. Leis & Rennis (1983) described and illustrated larvae of both families and showed they are distinct even at less than 2 mm. However, many workers continue to lump the two families, as does the Zoological Record. Tradition dies hard.

Abundance

Chaetodontid larvae are rare in plankton or even mid-water trawl samples. Some reports of larval fish studies in tropical areas do not mention chaetodontid larvae (Ahlstrom 1971, 1972, Nellen 1973, Houde et al. 1986), and it is unclear if they were omitted as too rare, because none were taken, or because of identification difficulties. Others mention capturing chaetodontid larvae but do not give quantitative data (Bourret et al. 1979, Belyanina 1975, Chen & Wei 1978, 1982). Data on relative abundance of chaetodontid larvae from a number of studies from the Atlantic and Indo-Pacific summarized in Table 2 reveal low relative abundance, ranging from 0 to 0.45% (mean = 0.09%) of all larvae captured. The proportion of samples containing chaetodontid larvae varied widely from 0 to 47.6% (mean = 13%). Even in samples which contained chaetodontid larvae, the mean number of larvae per sample was very low – ranging from 1 to

Table 2. Chaetodontid larvae (as percentage of total larvae captured) in ichthyoplankton studies of warm water areas which included coastal areas. pn, plankton net; nn, neuston net; mwt, mid water trawl; – information not supplied.

Area	% chaeto-dontids	% samples with chae-todontids	mean chaeto-dontids per positive sample	type of sampling gear	principal sampling environment	number of samples	total larvae sampled × 10^3	Reference
Eastern Gulf of Mexico	0.022	–	–	pn	shelf	756	143	Houde et al. (1979)
Gulf of Mexico and Caribbean	0.137*	9.2*	1.3*	pn	oceanic	109	9.5	Richards (1984)
	0.095*	5.6*	1.6*	nn	oceanic	89	8.4	Richards (1984)
Hawaii	0.060	2.0	2.8	mwt, pn	shelf/oceanic	562	51.9	Leis & Miller (1976)
Great Barrier Reef Lagoon/ Coral Sea 1979/80	0.052	22.2	1.8	pn	shelf/oceanic	36	23.2	Leis & Goldman (1984)
Great Barrier Reef Lagoon/ Coral Sea 1983	0.175	47.6	2.0	pn	shelf/oceanic	21	11.4	Leis, unpublished
	0	0	–	nn	shelf/oceanic	14	0.4	Leis, unpublished
Great Barrier Reef Lagoon/ Coral Sea 1984/85	0.054	15.6	1.9	pn	shelf/oceanic	116	57.6	Leis, unpublished
	0.445	10.0	1.0	nn	shelf/oceanic	20	0.4	Leis, unpublished
Great Barrier Reef Lagoon (open water) 1981/82	0.049	22.3	1.5	pn	shelf	112	76.0	Leis & Goldman (1987)
Great Barrier Reef Lagoon (open water) March 1983	0.019	14.8	1.2	pn	shelf	88	83.5	Leis, unpublished
Great Barrier Reef Lagoon (immediately around reefs)	0.011	4.4	1.3	pn	shelf	180	81.0	Leis (1982, 1986a) Leis & Goldman (1987)
Northwest Australian Continental Shelf	0.036	3.7	1.5	pn, mwt	shelf	355	53.4	Young et al. (1986)

* Overestimate because chaetodontids and pomacanthids were combined.

2.8 (mean = 1.7). In contrast, Bourret et al. (1979) reported capturing swarms ('essaims') of chaetodontid larvae in a few mid-water trawls.

In summary, overall abundance and relative abundance of chaetodontid larvae are low, the larvae generally occur in low numbers even in positive samples, and there is some indication that their distribution is patchy.

All this makes quantitative work on chaetodontid larvae very difficult and is a major limitation to the kinds of analyses which can be done with resulting data. When it is considered that there are 10 genera and 114 species of chaetodontids worldwide, and that the number of species in an area like Lizard Island on the Great Barrier Reef exceeds 36 (D.F. Hoese, personal communication), it is clear there are extremely limited possibilities for meaningful ecological work on larvae at the species level using towed nets. It seems unlikely that studies specifically directed at chaetodontid larvae using towed nets could be justified or would be very productive, but some information might emerge as spin-off from studies directed at more abundant larvae.

Are these low numbers of chaetodontid larvae what one would expect given adult abundance? There are relatively few estimates of adult chaetodontid abundance as a portion of the entire reef fish community. Brock et al. (1979) estimated that chaetodontids constituted 9.3–10.6% of all individuals and 9.9–13.3% of fish biomass on a patch reef in Kaneohe Bay, Hawaii. However, if one ignores *C. miliaris*, an extremely common, plankton-feeding, endemic species which Ralston (1981) showed did not mature in Kaneohe Bay, numbers reduce to 0.15–0.61% of individuals and 0.19–0.80% of biomass. Randall (1963) estimated that chaetodontids made up 0.04–0.34% of individuals and 0.01–0.60% of biomass on two coral reefs and one artificial reef in the Virgin Islands. Williams & Hatcher (1983) estimated that chaetodontids constituted 0.02–0.06% of the fish biomass in Great Barrier Reef explosive stations. Adult chaetodontids are really relatively rare, albeit highly conspicuous, reef fishes. Given these estimates of low adult chaetodontid abundance (except *C. miliaris* in Kaneohe Bay), and considering that plankton tows capture larvae from many habitats as well as coral reefs, the abundance values of chaetodontid larvae (Table 2) do not appear to be gross underestimates.

This conclusion is in general agreement with data on other fishes when appropriate sampling methods and areas are used. For example, Leis & Goldman (1987) concluded that only a few families such as haemulids, chaetodontids and acanthurids were unexpectedely rare in the Great Barrier Reef Lagoon. Subsequent work has shown that larvae of the latter two families are primarily distributed in the Coral Sea. Similarly, the larvae Leis & Goldman (1987) found to be most abundant grow into some of the most abundant adults on the reef. It is unlikely that larval abundance is directly representative of adult abundance – differences in batch fecundity, time to maturity, spawnings per unit time, and the like, in addition to differences in survival and dispersal of larvae will ensure that – but such a relationship does seem to hold as a first-order, albeit rough, approximation.

It is possible that chaetodontid larvae occupy habitats other than those sampled by conventional plankton nets or midwater trawls but there is little or no evidence for this. Older larvae may reside in the neuston, although the evidence for this is extremely limited. In one survey a relatively high percent of chaetodontids (0.45%) was found in 20 neuston hauls (Table 1), but the total catch constituted 2 chaetodontid larvae. I have observed numbers of chaetodontid larvae in a few nocturnal neuston tows in the Coral Sea. The only study of epibenthic fish larvae in a coral reef area – albeit a limited one – did not capture any chaetodontid larvae (Leis et al. 1989). On available evidence, chaetodontid larvae do appear to occupy habitats sampled by conventional plankton nets, but in low numbers.

Older chaetodontid larvae of at least some species are photopositive and readily collected under night lights (personal observations) or in automated light traps (P.J. Doherty & M. Milicich, personal communication). This could provide a viable method of studying certain aspects of the pelagic phase of chaetodontid larvae.

Horizontal distribution

There is no information on horizontal distribution based on studies specifically directed at chaetodontid larvae, so information must be gleaned from multi-species studies which are frequently very broadly-based surveys. The published information is very limited, so is supplemented here by my data.

On very broad scales, Richards (1984) characterized chaetodontid larvae (but including pomacanthids) as 'widely distributed' in a study of the Caribbean. In the eastern Gulf of Mexico, 75% of chaetodontid larvae were found at stations seaward of the 50 m isobath (Houde et al. 1979). In the Tuamotu Archipelago, chaetodontid larvae occurred in 3 of 14 midwater trawls taken within 20 km of atolls but in none of nine taken 37–95 km from these reefs (Bourret et al. 1979). Leis & Miller (1976) included chaetodontid larvae among those which increased in relative abundance from nearshore to offshore in the Hawaiian Islands. The data for chaetodontid larvae were not included in the paper, but are: 0.5–2 km, 0.06% (of all larvae from that station); 2–8 km, 0.18%; and 12 km, 0.75%, with none taken in Kaneohe Bay or 0.5 km from shore. Thus, published information indicates chaetodontid larvae do not stay close to reefs, but may be considered at least neritic, if not oceanic.

The following information is based on my work in Australia, and although aspects of these studies have been published, little or no information on chaetodontid larvae was included because chaetodontid larvae were so rare. This rarity combined with identification problems preclude general analysis below the family level. Usually data from different seasons or stations had to be lumped in order to do any analysis. Therefore, these results must be treated with caution.

No chaetodontid larvae were captured in the small lagoon at Lizard Island, Great BarrierReef nor were any taken in limited sampling in Osprey Atoll Lagoon in the Coral Sea (Leis 1982, 1986b). There was no indication of a windward/leeward effect in the distribution of five larvae caught around Lizard Island (Leis 1982, 1986a).

Within the outer portion of the Great Barrier Reef Lagoon (between Lizard Island and the outer ribbon reefs) 64 chaetodontid larvae have been captured. In two studies of horizontal distribution in this area, more chaetodontid larvae were found at mid-lagoon, 7 km from any reefs, than at locations closer to reefs (Leis & Goldman 1984 and unpublished). In the first study, 11 chaetodontid larvae were taken at mid-lagoon and only 3 larvae were taken among the patch reefs in the lee of Carter Reef (Leis & Goldman 1984). In April 1981, four stations were sampled (32 plankton samples) in this area, and 13 chaetodontid larvae were captured at mid-lagoon (11 elsewhere) (Leis & Goldman unpublished). A Chi Square test revealed the distribution of chaetodontid larvae among the four stations was significantly different from random ($p = 0.01$). Similar sampling in November 1981 took only two chaetodontid larvae, and in January/February 1982 the distribution of the 11 chaetodontid larvae captured did not appear to differ from random (Leis & Goldman unpublished), but could not be tested due to low numbers.

A preliminary study of distribution of fish larvae in the Great Barrier Reef Lagoon and Coral Sea near Lizard Island (Leis & Goldman 1984) captured 17 chaetodontid larvae in the Lagoon and none at two stations close to the reef in the Coral Sea (up to 1.8 km off the reef). More extensive work in November 1983 and 1984 and January/February 1985 (Leis et al. 1987) indicates that, while there does seem to be temporal variability in the distribution pattern of chaetodontid larvae, larvae are consistently more abundant than expected (if distributions were random) 1.8–5.5 km off the reef, and less abundant 0.5–1.8 km off (Table 3). The very few samples taken between 5.5–11 km off the reef suggest the band of high abundance of chaetodontid larvae may extend further offshore. By 11–19 km off the reef, numbers of chaetodontid larvae again appear to be very low.

On the continental shelf of northwestern Australia where there are fewer coral reefs, a year-long study (Young et al. 1986) captured only 20 chaetodontid larvae (Table 2). Two cross-shelf transects consisting of 10 stations in sum were sampled, and half the chaetodontid larvae were taken at the two inshore stations, with a further four taken at the two stations next to the inshore stations. Ten of these 14 larvae were *Coradion* and *Parachaetodon*. Adults of both genera seem largely confined to coastal reefs and soft bottoms (Steene 1978), which could explain why their larvae were most abundant nearshore.

In the samples from the Lizard Island area, few chaetodontid larvae were identifiable below family. However, distributions for those that were are interesting. Eighteen *Coradion* larvae were captured in the Lagoon, and none in the Coral Sea (this agrees with the NW Australia study cited above, where 7 *Coradion* larvae were captured – all on the continental shelf). Only seven *Forcipiger* larvae were captured, and all were taken in the Coral Sea. It should be noted that most chaetodontid larvae in the Lizard Island area samples were young: only six were postflexion stages, and all of these were taken by bongo nets in the 1984/85 studies. It is quite possible that older larvae have different distributions than do young larvae.

To summarize the available distributional evidence on chaetodontid larvae:

(1) chaetodontid larvae are absent or found in very small numbers in the immediate vicinity of reefs, including small to medium semi-enclosed lagoons;

(2) in more open continental shelf waters including those of the Great Barrier Reef Lagoon, chaetodontid larvae can be found in numbers, although these may be limited to a small number of genera, such as *Coradion*;

(3) there is no indication of a windward/leeward difference in abundance of chaetodontid larvae around either mid-lagoon or outer reefs;

(4) highest numbers of chaetodontid larvae of most (but not all) types were found at some distance seaward of outer reefs (i.e. in oceanic conditions); and very limited evidence suggests that this area of highest abundance may be further offshore off leeward than windward coasts.

Therefore, the limited available evidence seems to largely support the conclusions of earlier, published work. Larvae of most chaetodontid taxa are oceanic and the larvae either avoid or do not survive in areas near shore and reefs. The distance away from shore and reefs at which the larvae are to be found in numbers is not entirely clear and may be less than originally thought (Leis & Miller 1976). Some chaetodontid larvae are apparently capable of completing their pelagic phase in continental shelf waters. For this the best case can be made for *Coradion*. Finally, due to extremely small sample sizes, all these conclusions must be treated with caution.

Vertical distribution

Information on vertical distribution of chaetodontid larvae is extremely limited but seems to indicate that chaetodontid larvae occur in the upper portions of the water-column. Belyanina (1975) concluded that in the Caribbean Sea and Gulf of Mexico chaetodontid larvae were common ('not more than 10 larvae') in the 0–100 m stratum, but were not found at greater depths.

In a study of vertical distribution in water about 25 m deep in the Great Barrier Reef Lagoon (unpublished), samples were taken in 16 day and 6 night 'sets' each of which consisted of a neuston

Table 3. Numbers of chaetodontid larvae (and numbers of samples) in oblique bongo net samples from the Coral Sea and Great Barrier Reef Lagoon near Lizard Island. * not included in analysis: chi square test for combined data, $p<0.005$.

	GBR Lagoon	Coral Sea (distance seaward – km)				
		0–0.5	0.5–1.8	1.8–5.5	5.5–11	11–19
Nov. 1983	0 (3)	4 (2)	3 (5)	6 (3)	6 (4)	1 (4)
Nov. 1984	2 (12)	5 (8)	1 (12)	12 (12)	– (0)	0 (12)
Jan./Feb. 1985	5 (12)	0 (12)	0 (12)	4 (12)	– (0)	5 (12)
Combined	7 (27)	9 (22)	4 (29)	22 (27)	*	6 (26)

tow, and three opening-closing bongo net tows (one each at 0–6, 6–13 and 13–20 m). Sixteen chaetodontid larvae were captured in 10 sets (6 day and 4 night). During the day three larvae (three occurrences) were taken in the 0–6 m stratum, five larvae (five occurrences) were taken at 6–13 m, and none at 13–20 m. At night, five larvae (3 occurrences) were taken at 0–6 m, one larva (one occurrence) was taken at 6–13 m and 2 larvae (one occurrence) were taken at 13–20 m. No chaetodontid larvae were taken in the neuston at any time. Too few larvae were taken to attempt statistical analyses.

In a study of vertical distribution in waters immediately adjacent to Lizard Island reefs, samples were taken in sets, one each at 0–1, 3–4 and 6–7 m (Leis 1986a). Chaetodontid larvae were taken in 4 day sets and the 5 larvae captured were taken from all three strata. At night, chaetodontid larvae occurred in one set, and the 5 larvae captured were again taken from all three strata.

All the larvae taken in the above Australian studies were preflexion-stages as were 29 of the 30 larvae in day oblique bongo net tows taken in the Great Barrier Reef Lagoon and Coral Sea in 1984/85 (unpublished data, Table 2). Only two chaetodontid larvae were found in neuston tows taken at the same time, but both were postflexion-stages. These data suggest the possibility that chaetodontid larvae move into the neuston as they get older.

Size and duration of pelagic phase

Size at settlement of chaetodontid larvae varies widely, from 9–60 mm (Table 4), and even within the genus *Chaetodon*, the entire range for the family is approached (9–50 mm, Table 4). Although some chaetodontids do settle out at a large size, more than half of the species for which there is information settle out at less than 20 mm. It is

Table 4. Duration and size of pelagic stages of chaetodontid fishes. Ages are unverified (**verified) otolith ages except one (indicated with*) which was based on elapsed time between peak in spawning and peak in recruitment. References (1) Ralston (1976, 1981); (2) Burgess (1978); (3) Brothers et al. (1983); (4) Brothers & Thresher (1985); (5) Watanabe (1946); (6) Fowler (1988).

Species	Size at settlement (mm)	Duration of pelagic stages (days)	Reference
Chaetodon			
C. auriga	≤14.5	40–53	4, 5
C. baronessa	–	38	4
C. collare	≤8.5	43	4, 5
C. excelsa	20	–	2
C. kleinii	–	56	4
C. lunula	≤15.5	–	5
C. lineolatus	–	54	4
C. melannotus	≤12	–	5
C. miliaris	25–27	32–38	1
C. miliaris	32–36	about 90*	2
C. octofasciatus	9–18	–	2
C. plebeius	10	21–56**	3, 6
C. rainfordi	11	20–43**	3, 6
C. speculum	–	26	4
C. trifascialis	9–18	30–34	2, 4
C. unimaculatus	50	35	2, 4
C. vagabundus	≤15.6	–	5
Chelmon rostratus	14.7 ± 1.5	19–28	3, 4, 6
Chelmon rostratus	9–18		2
Forcipiger flavissimus	60	41–57	2, 4
Hemitaurichthys spp	60	–	2
Heniochus diphreutes	–	30–34	4

unclear how much variation there is within species because sample sizes are generally small, but it does seem clear that there is less variation within species than within genera.

Otolith age estimates for duration of the pelagic phase vary from 20 to 57 days (Table 4), and once again, the values for *Chaetodon* range by nearly as much as the family range. Available duration ranges within species vary by as much as 83% of the mean value. Not surprisingly, the greatest ranges are for species with the greatest sample size (e.g. those from reference 6, Table 4). For most species listed in Table 4, counts were done on few individuals, so the true variation is probably higher. A mean age at settlement for all chaetodontids would be about 35–40 days. To my knowledge, the daily nature of otolith rings has been confirmed on only two chaetodontids (*C. plebeius* and *C. rainfordi*) and this was done on post-settlement juveniles (Fowler 1988). Verification of otolith ages in other species and in presettlement larvae is urgently required.

Within the family, size at settlement varies more than six-fold, but age at settlement (ignoring one estimate not based on otolith rings, Table 4) varies by only a factor of 2.24 (Table 4). The correlation between mean otolith age and mean size at settlement for the nine species for which there are reported data for both values is not significant. Although this may be due to the small sample size, there is no indication in the available data of a relationship between size and age among species. For example, *Forcipiger flavissimus,* which settles out at 60 mm, has a pelagic duration of 41–57 days, while *Chaetodon plebeius* which settles out at about one-sixth of this size, has a pelagic duration which nearly brackets this (Table 4). Further, the wide range in ages in species for which there are adequate sample sizes is a strong indication that a simple relationship will not be found. Not enough data are available to determine if any correlation exists between age and size within species.

At present, we do not know if chaetodontid larvae are capable of extending the pelagic period if favourable conditions for settlement are not found. We do not know if the variations in size at settlement are genetically determined, dependent on food availability and hence growth rate, or determined by other factors. Nor is there any indication if any of these factors vary regionally or temporally.

Feeding and behaviour

Observations on feeding and behaviour of chaetodontid larvae are extremely limited. In the laboratory, newly-hatched embryos of *Chaetodon nippon* floated up-side-down just below the water surface, and 72 hour old larvae began to swim actively (Suzuki et al. 1980a). The *C. nippon* larvae were offered rotifers and larvae of oysters and sea urchins as food. Only a few oyster larvae were found in the guts of the fish larvae, no growth took place and all *C. nippon* larvae died within eight days of hatching (Suzuki et al. 1980a). At the other end of the larval period, there are a few anecdotal observations of the rapidity of the transformation from pelagic to juvenile colouration (e.g. Burgess 1978). Otherwise, nothing is known of behaviour of larval chaetodontids aside from the observation that settlement-stage larvae are attracted to lights at night.

Seasonality

Although of dubious value, because it is based on few years' data and lumping of all chaetodontid larvae, it is possible to extract some limited information on seasonality from several sources. Leis (1978) sampled in all months close to shore off Leeward Oahu and captured chaetodontid larvae only in March, May, June and July, with maximum numbers per volume in June and July. On the northwest continental shelf of Australia, Young et al. (1986) sampled bi-monthly for 14 months and captured chaetodontid larvae from October to April (Austral spring through autumn), while none were taken during June or August.

At Lizard Island, samples were taken in April, July, October 1979 and January/February 1980 (Leis 1986a), and chaetodontid larvae were taken in spring, and summer, but not autumn or winter.

Within the outer Great Barrier Reef Lagoon

near Lizard Island (samples in 1979/80, 1981/82, 1983, and 1984/85) (Leis & Goldman 1984, 1987, and unpublished, Leis et al. 1987) chaetodontid larvae were most commonly encountered January to April, when they occurred in 25.6% of samples and averaged 1.4 larvae per positive tow. In July, October and November, only 4.2% of tows had chaetodontid larvae, and there was an average of 1.3 larvae per positive tow.

In the Coral Sea (samples in 1979/80, 1983 and 1984/85, Leis & Goldman 1984, and unpublished, and Leis et al. 1987), 9% of tows in January/February had chaetodontid larvae, with 1.3 larvae per positive tow. No chaetodontid larvae were taken in October. In November, 27% of tows contained chaetodontid larvae, and an average of 2.1 larvae were present per positive tow. No sampling was carried out in the Coral Sea in autumn or winter.

Overall, it appears chaetodontid larvae are least abundant in winter. Relatively high abundances may be encountered at any time in spring, summer or autumn. There is not enough information to determine if within this spring to autumn peak, there are differences in temporal patterns of abundance among areas.

Discussion

The rarity of chaetodontid larvae in towed net samples is a major obstacle to further work. In sampling programs to date, the vast majority of samples have not contained chaetodontid larvae, and the average number of larvae per positive sample is low. Perhaps an increase in volume sampled per tow could improve chaetodontid catches. However, the 1984/85 Great Barrier Reef Lagoon/Coral Sea bongo net tows (Table 2), sampled 1000–2000 m^3 each and still caught few chaetodontid larvae. It seems that alternative sampling strategies must be considered if chaetodontid larvae are to be the subject of productive ecological studies. Possible alternatives are light trapping or neuston tows at night.

The distributions of chaetodontid larvae give rise to further questions. The larvae thus far identified as having onshelf distributions (*Coradion, Parachaetodon*) have inshore distributions as adults, so an onshelf larval distribution is not surprising. This should perhaps be expected for larvae of other inshore species. Obviously, this sort of distribution cannot be expected at oceanic islands which lack a continental shelf (Leis 1986b). The oceanic distribution of other chaetodontid larvae, particularly the apparent peaks in abundance some distance seaward of the reef raises questions as to how the larvae arrive at these areas from adult spawning sites on the reef, and how the larvae return (if, indeed, they do) to settle on the reef. It is normally assumed that distributions of larvae are adaptive, and, for example, that larvae found in abundance peaks offshore survive and return to the reef at least as well as, if not better than, larvae found elsewhere. This sort of assumption remains to be tested. Several authors (e.g. Johannes 1978) have speculated that the timing and placement of reef fish spawning are such that larvae would be moved away from the reef, and this is probably involved in producing the observed offshore distribution of larvae. However, the relative contributions of spawning, larval behaviour or other factors to producing larval distributions are unknown. At present, there is only speculation on mechanisms for return to the reef – e.g. current gyres off the lee coast of the Island of Hawaii (Lobel & Robinson 1986). Off a windward reef such as the Great Barrier Reef, larvae could gain an on-reef vector by entering the wind-driven surface layer (neuston), but this type of behaviour would result in offshore movement off a lee reef. Older chaetodontid larvae have been captured in the neuston, but there is no strong evidence that larvae nearing settlement preferentially enter the neuston or utilize it as a return mechanism. Larvae of settlement size (9–60 mm, depending on species, Table 4), particularly those at the upper end of the range, should be very competent swimmers, with the ability to actively move toward a reef. This, of course, begs the question of how the reef is detected and how proper orientation is achieved and maintained (see below). Further work on distribution is clearly necessary, but the very low abundance of chaetodontid larvae make it likely this type of work will be done on larvae of other families. Johannes (1978) has argued that the offshore distribution of many types of larvae is an evolutionary response to high predation pressure by reef-based planktivores. One reviewer suggested that such predation could reduce numbers of chaetodontid larvae nearshore, thereby producing an offshore increase in abundance. Because a number of types of larvae are most abundant nearshore (Leis & Miller 1976, Leis 1986a, Leis & Goldman 1984, 1987), this hypothesis could be correct only if the predation were selectively targeted at larvae with offshore distributions such as labrids and chaetodontids relative to larvae with onshore distributions such as tripterygiids, gobiids and certain pomacentrids and apogonids. This seems unlikely.

If chaetodontid larvae are primarily confined to the upper portion of the mixed layer, sampling for them could be concentrated there with savings in effort both in the field and laboratory. Further, it is possible that, for a given volume sampled, catch rates would be increased. Most larvae of reef fishes occur at greater depths than chaetodontid larvae apparently do (unpublished data), and to the extent that current vectors differ with depth, this could contribute to differences in horizontal distribution. Chaetodontid larvae are heavily pigmented; this is characteristic of shallow-living larvae, and adds some strength to the conclusion that chaetodontid larvae are shallow-living. However, the data base to support this conclusion is very limited and more intensive study of chaetodontid larvae vertical distribution is also necessary.

A positive relationship between size at settlement and duration of the pelagic stage is expected and the fact that a significant relationship was not found suggests differential growth rates among species. An average size at settlement of less than 20 mm and age at settlement of less than 40 days, do not seem exceptional among reef fishes, and it would be of interest to know if larvae with on-shelf distributions differ in these values from larvae with oceanic distributions. The wide range for both size and age within the family means that the potential dispersal ability probably also has a wide range within the family. Most currently available information is based on unverified otolith ages, and verification is required before too much emphasis is placed on these data. In addition, it is highly desirable to obtain more information on variability of age and size for individual species. The wide range in age within the few species with adequate sample sizes combined with lack of evidence that size at settlement varies widely within species, indicates either that there are great variations in growth rate among individuals or that larvae possess the ability to postpone settlement and stop growing. Apparently, labrid larvae have at least the latter ability (Victor 1986a, b), so it would not be surprising if chaetodontid larvae did as well.

The virtual absence of any information on feeding and behaviour of chaetodontid larvae is unfortunate, and moves should be made to fill this gap. Lab rearing is one possible approach (Suzuki et al. 1980a). Settlement-stage larvae can be captured alive with light traps, and this provides an avenue for investigation of behaviour of settlement and the general behaviour of older larvae.

Understanding the distributional ecology of chaetodontid larvae could increase understanding of chaetodontid zoogeography. Chaetodontid larvae with oceanic distributions would be expected to be widely distributed on the Pacific tectonic plate, while those with onshelf distributions would be expected to be largely restricted to localities west of the andesite line (Leis 1986b). This holds true for the onshelf *Parachaetodon* and *Coradion* and oceanic *Forcipiger* (Springer 1982). It will be interesting to see if this holds as taxonomic advances are made and more larvae can be identified and their distributions characterized. Do the larvae of island group endemic chaetodontids (Springer 1982) have distributions closer to their natal reefs than more widely spread confamilials? The answers to such intriguing questions are unlikely to emerge soon.

Acknowledgements

The previously unpublished research cited here was supported primarily by MST Grants 80/2016 to B. Goldman and 83/1357 to me and a Queens Fellowship in Marine Science to me. My thanks to K. Matsuura for assistance with Japanese literature,

P. Young and P. Last for making the CSIRO Northwest Shelf larval fish collections available, A.J. Fowler for criticizing the manuscript and providing information from his otolith studies, and S. Blum for information on chaetodontid phylogeny. S. Bullock and S. Reader provided much needed editorial assistance. T. Goh and L. Spitalieri typed the manuscript.

References cited

Ahlstrom, E.H. 1971. Kinds and abundance of fish larvae in the eastern tropical Pacific, based on collections made on EASTROPAC I. U.S. Fish. Bull. 69: 3–77.

Ahlstrom, E.H. 1972. Kinds and abundance of fish larvae in the eastern tropical Pacific on the second multi-vessel EASTROPAC survey, and observations on the annual cycle of larval abundance. U.S. Fish. Bull. 70: 1153–1242.

Belyanina, T.N. 1975. Preliminary results of the study of ichthyoplankton of the Caribbean Sea and the Gulf of Mexico. Proceedings of the P.P. Shirshov Institute of Oceanology 100: 127–146. [English Translation, 1981 by Al Ahram Center for Scientific Translations].

Bourret, P., D. Binet, C. Hoffschir, J. Rivaton & H. Velayoudon. 1979. Evaluation de 'l'effet d'Ile' d'un Atoll: Plancton et micronecton au large de Mururoa (Tuamotus). Centre Office de la Recherche Scientifique et Technique Outre-Mer de Nouméa, Nouvelle-Calédonie. 124 pp.

Brock, R.E., C. Lewis & R.C. Wass. 1979. Stability and structure of a fish community on a coral patch reef in Hawaii. Mar. Biol. 54: 281–292.

Brothers, E.B., D. McB. Williams & P.F. Sale. 1983. Length of larval life in twelve families of fishes at 'One Tree Lagoon', Great Barrier Reef, Australia. Mar. Biol. 76: 319–324.

Brothers, E.B. & R.E. Thresher. 1985. Pelagic duration, dispersal, and the distribution of Indo-Pacific coral-reef fishes. pp. 53–70. In: M.L. Reaka (ed.) The Ecology of Coral Reefs, Symposia Series for Undersea Research, NOAA's Undersea Research Program, Vol. 3.

Burgess, W.E. 1974. Evidence for the elevation to family status of the angelfishes (Pomacanthidae), previously considered to be a subfamily of the butterflyfish family, Chaetodontidae. Pac. Sci. 28: 57–71.

Burgess, W.E. 1978. Butterflyfishes of the world. T.F.H. Publications, Neptune City. 832 pp.

Chen, Z. & S. Wei. 1978. A preliminary investigation on pelagic fish eggs and larvae from the waters around the Xisha Islands and Zhongsha Islands in the South China Sea. pp. 295–320. In: Investigation Reports of Chinese Studies of the Biology of the Ocean Near the Xisha and Zhongsha Islands, Science Press, Beijing. (In Chinese).

Chen, Z. & S. Wei. 1982. An investigation on pelagic fish eggs and larvae of the central area of South China Sea. pp. 251–268. In: Symposium on Research Reports on the Sea Area of South China Sea, Science Press, Beijing. (In Chinese).

Colin, P.L. & I.E. Clavijo. 1988. Spawning activity of fishes producing pelagic eggs on a shelf edge coral reef, southwestern Puerto Rico. Bull. Mar. Sci. (in press).

Day, F. 1870. Notes on some fishes from the western coast of India. Proc. Zool. Soc. London. 1870: 369–374.

Fourmanoir, P. 1976. Formes post-larvaires et juvéniles de poissons cotiers pris au chalut pélagique dans le sud-ouest Pacifique. Cahiers du Pacifique 19: 47–88.

Fowler, A.J. 1988. Aspects of the population ecology of three species of chaetodonts at One Tree Reef, southern Great Barrier Reef. Ph.D. Thesis, University of Sydney, Sydney. 145 pp.

Fowler, H.W. 1945. A study of the fishes of the southern piedmont and coastal plain. Acad. Nat. Sci. Phila. Mono. 7: 1–408.

Fraser-Brunner, A. 1933. A revision of the chaetodont fishes of the subfamily Pomacanthine. Proc. Zool. Soc. London 1933: 543–598.

Günther, A. 1868. Additions to the ichthyological fauna of Zanzibar. Ann. Mag. Nat. Hist. 4th Ser. 1: 457–459.

Günther, A. 1871. On the young state of fishes belonging to the family of Squamipinnes. Ann. Mag. Nat. Hist. 4th Ser. 8: 318–320.

Günther, A. 1873. Andrew Garrett's Fische der Südsee, part 1. J. Museum Godeffroy 2(3): 1–128.

Houde, E.D., J.C. Leak, C.E. Dowd & S.A. Berkeley. 1979. Ichthyoplankton abundance and diversity in the eastern Gulf of Mexico. Report to the Bureau of Land Management, contract AA550–CT7-28. NTIS-PB-299839. xxxii plus 546 pp.

Houde, E.D., S. Almatar, J.C. Leak & C.E. Dowd. 1986. Ichthyoplankton abundance and diversity in the Western Arabian Gulf. Kuwait Bull. Mar. Sci. 8: 107–393.

Hubbs, C.L. 1963. *Chaetodon aya* and related deep-living butterflyfishes: their variation, distribution and synonymy. Bull. Mar. Sci. Gulf & Carib. 13: 133–192.

Johannes, R.E. 1978. Reproductive strategies of coastal marine fishes in the tropics. Env. Biol. Fish. 3: 65–84.

Johnson, G.D. 1984. Percoidei. pp. 464–498. In: H.G. Moser, W.J. Richards, D.M. Cohen, M.P. Fahay, A.W. Kendall Jr. & S.L. Richardson (ed.) Ontogeny and Systematics of Fishes, Amer. Soc. Ichthyol. Herpetol. Spec. Pub. 1.

Kendall, W.C. & E.L. Goldsborough. 1911. The shorefishes. Mem. Mus. Comp. Zool. 26: 241–344.

Leis, J.M. 1978. Distributional ecology of ichthyoplankton and invertebrate macrozooplankton in the vicinity of a Hawaiian coastal power station. Ph.D. Dissertation, University of Hawaii, Honolulu. 317 pp.

Leis, J.M. 1982. Distribution of fish larvae around Lizard Island, Great Barrier Reef: coral reef lagoon as refuge? Proc. Fourth Intern. Coral Reef Symp. 2: 471–477.

Leis, J.M. 1986a. Vertical and horizontal distribution of fish larvae near coral reefs at Lizard Island, Great Barrier Reef. Mar. Biol. 90: 505–516.

Leis, J.M. 1986b. Ecological requirements of Indo-Pacific larval fishes: a neglected zoogeographic factor. pp. 759–766. *In:* T. Uyeno, R. Arai, T. Taniuchi & K. Matsuura (ed.) Indo-Pacific Fish Biology: Proceedings of the Second International Conference on Indo-Pacific Fishes, Ichthyological Society of Japan, Tokyo.

Leis, J.M. & B. Goldman. 1984. A preliminary distributional study of fish larvae near a ribbon coral reef in the Great Barrier Reef. Coral Reefs 2: 197–203.

Leis, J.M. & B. Goldman. 1987. Composition and distribution of larval fish assemblages in the Great Barrier Reef Lagoon, near Lizard Island, Australia. Aust. J. Mar. Freshwater Res. 38: 211–223.

Leis, J.M., B. Goldman & S. Ueyanagi. 1987. Distribution and abundance of billfish larvae (Pisces: Istiophoridae) in the Great Barrier Reef Lagoon and Coral Sea near Lizard Island, Australia. U.S. Fish. Bull. 85: 757–765.

Leis, J.M., S.E. Reader & B. Goldman. 1989. Composition and abundance of epibenthic fish larvae in inter-reef areas of the Great Barrier Reef Lagoon near Lizard Island, Australia. Jap. J. Ichthyol. 35: 428–433.

Leis, J.M. & J.M. Miller. 1976. Offshore distributional patterns of Hawaiian fish larvae. Mar. Biol. 36: 359–367.

Leis, J.M. & D.S. Rennis. 1983. The larvae of Indo-Pacific coral reef fishes. New South Wales University Press, Sydney and University of Hawaii Press, Honolulu. 269 pp.

Lobel, P.S. & A.R. Robinson. 1986. Transport and entrapment of fish larvae by ocean mesoscale eddies and currents in Hawaiian waters. Deep-sea Res. 33: 483–500.

Lütken, C. 1880. Spolia Atlantica. Vidensk. Selsk. Skr. 5. Raekke, naturvidenska-belig og mathematisk Adf. XII. 6: 413–613. (In Danish, French summary). 1881, English translation of French summary by W.S. Dallas. Ann. Mag. Nat. Hist. 5th Ser. 7: 1–14 and 107–123.

Moe, M.A. jr. 1976. Rearing Atlantic angelfish. Mar. Aquar. 7: 17–26.

Nellen, W. 1973. Fischlarven des Indischen Ozeans. 'Meteor' Forsch.-Ergebnisse, Reihe D, No. 14: 1–66.

Okada, Y. & M. Watanabe. 1947. Juveniles of *Pomacanthus semicirculatus* (Cuvier et Valenciennes). Zool. Mag. (Tokyo) 57: 29–30. (In Japanese).

Ralston, S. 1976. Age determination of a tropical reef butterflyfish utilizing daily growth rings of otoliths. U.S. Fish. Bull. 74: 990–994.

Ralston, S. 1981. Aspects of the reproductive biology and feeding ecology of *Chaetodon miliaris,* an Hawaiian endemic butterflyfish. Env. Biol. Fish. 6: 167–176.

Randall, J.E. 1961. Two new butterflyfishes (family Chaetodontidae) of the Indo-Pacific genus *Forcipiger.* Copeia 1961: 53–62.

Randall, J.E. 1963. An analysis of the fish populations of artificial and natural reefs in the Virgin Islands. Carib. J. Sci. 3: 31–47.

Richards, W.J. 1984. Kinds and abundances of fish larvae in the Caribbean Sea and adjacent waters. NOAA Tech. Rept., NMFS Spec. Sci. Rep. Fisheries 776: 1–54.

Schmidt, P. & G. Lindberg. 1930. A list of fishes, collected in Tsuruga (Japan) by W. Roszkowski. Bulletin l'Académie des Sciences de L'URSS, Classe des Sciences Physico. Math. 1930: 1135–1150.

Springer, V.G. 1982. Pacific plate biogeography, with special reference to shorefishes. Smith. Contrib. Zool. 367: 1–182.

Steene, R.C. 1978. Butterflyfishes and angelfishes of the world, Vol. 1, Australia. Reed, Sydney. 144 pp.

Suzuki, K., Y. Tanaka & S. Hioki. 1980a. Spawning behavior, eggs, and larvae of the butterflyfish, *Chaetodon nippon,* in an aquarium. Jap. J. Ichthyol. 26: 334–341.

Suzuki, K., Y. Tanaka, S. Hioki & Y. Shiobara. 1980b. Studies on reproduction and larval rearing of coastal marine fishes. pp. 53–82. *In:* G. Yamamoto (ed.) Research in Large-Scale Culture of Marine Fisheries Resources, Institute of Oceanic Research and Development, Tokai University, Shimizu.

Thresher, R.E. 1984. Reproduction in reef fishes. T.F.H. Publications, Neptune City. 399 pp.

Victor, B.C. 1986a. Duration of the planktonic larval stage of one hundred species of Pacific and Atlantic wrasses (family Labridae). Mar. Biol. 90: 317–326.

Victor, B.C. 1986b. Delayed metamorphosis with reduced larval growth in a coral reef fish (*Thalassoma bifasciatum*). Can. J. Fish. Aquat. Sci. 43: 1208–1213.

Watanabe, M. 1946. Early life histories of some chaetodontid fishes with special reference to development of color pattern: I. *Chaetodon collaris* No. 11: 5 pp.; II. *Chaetodon melanotus* No. 12: 4 pp.; III. *Chaetodon auriga* No. 13: 4 pp.; IV. *Chaetodon vagabundus* No. 14: 4 pp.; V. *Chaetodon lunua* No. 15: 4 pp.; VI. *Chaetodon pleneius* No. 16: 3 pp.; VII. *Chaetodon trifasciatus* No. 17: 3 pp.; VIII. *Chaetodon strigangulus* No. 18: 3 pp.; Short Report, Resources Exploitation Institute, Tokyo, No. 11–18. (In Japanese).

Williams, D. McB. & A.I. Hatcher. 1983. Structure of fish communities on outer slopes of inshore, mid-shelf and outer shelf reefs of the Great Barrier Reef. Mar. Ecol. Prog. Ser. 10: 239–250.

Young, P.C., J.M. Leis & H.F. Hausfeld. 1986. Seasonal and spatial distribution of fish larvae in waters over the north west continental shelf of western Australia. Mar. Ecol. Prog. Ser. 31: 209–222.

Implications of feeding specialization on the recruitment processes and community structure of butterflyfishes

Mireille L. Harmelin-Vivien
Centre d'Océanologie de Marseille, Station marine d'Endoume, CNRS UA 41, F-13007 Marseille, France, and Centre de l'Environnement d' Opunohu, MNHN-EPHE, BP 1013 Papetoai, Moorea, Polynésie Française

Received 7.4.1988   Accepted 5.9.1988

Key words: Diets, Food specialists, Ontogeny, Settlement, Chaetodontids, Corals, Reef

Synopsis

When settling on coral reefs, fish larvae generally change from zooplanktivory to diverse forms of benthic feeding. Whereas food has not been reported to directly influence settlement, it is hypothesized that food resource might play a major role in the recruitment processes of butterflyfishes. Benthic feeding was found to occur immediatly after settlement, and was related to the degree of specialization of adult diets. Among obligate coral feeders scleractinian polyps were the exclusive diet of new recruits. In non-obligate corallivorous chaetodontids tentacles of sedentarian polychaetes were the preferred prey of juveniles for all the species studied, and represented on average 36.2% of their prey by weight. They formed a transitional food resource for these species which mainly fed on scleratinian polyps when adults (51.6% by weight). Among the chaetodontids studies, some recruited within adult home sites, whereas others exhibited size-specific distributions. The different patterns observed were not closely related to food specialization of the adults. The importance of food resources to the community structure of butterflyfishes on coral reefs is discussed.

Introduction

After a pelagic larval life lasting from one week to several months, with an average of one month (Brothers et al. 1983, Victor 1986a), fish larvae descend to the reef for a more sedentary benthic phase. At this time most species, with the exception of obligate planktivores, change from plankton feeding to benthic feeding. Major factors thought to affect recruitment and early survivorship of newly settled coral reef fishes include the availibity of suitable shelter sites, the density of adults and the encounter rate with predators (Luckhurst & Luckhurst 1977, Williams & Sale 1981, Sale et al. 1984, Shulman 1984, 1985, Victor 1986b). Food resources are not reported to directly influence settlement (Shulman 1984). Even if food is not considered as a limiting factor in recruitment processes questions arise about the way fish cover their energy requirements at settlement, particularly those species exhibiting strong feeding specialization as adults. How and when does the transition take place from the planktonic to the benthic diet? Is the change of diet concomittant with settlement? Is food specialization immediately effective, or is there a transitional diet in juvenile fish? This last hypothesis seems most likely to fit the cases of omnivorous and planktivorous species, but seems irrelevant to highly food selective species. If food specialization occurs immediatly after settlement, the presence of the specific food resource is needed for that particular fish species to recruit, and the distributions of adults and juveniles should coincide. Conversely, the existence of a transitional diet and different food preferences in juvenile fishes may allow a wider spatial dispersion of the recruits and probably ensure a better survival rate. It would also allow for the colonization of a variety of different habitats by the same species during the course of its ontogenetic development.

The chaetodontid family which has evolved within the coral reef ecosystem (Reese 1977) is particularly suited for the study of these problems. Different types of feeding behaviors and various levels of food specializations, from highly specialized feeders to generalists, are observed within the butterflyfishes. Chaetodontids may be broadly divided into coelenterate predators, carnivores preying upon other benthic invertebrates, omnivores, and plankton feeders (Hiatt & Strasburg 1960, Randall 1967, Hobson 1974, Anderson et al. 1981, Harmelin-Vivien & Bouchon-Navaro 1982, 1983). Coelenterate predators predominate, encompassing 31 to 90% of chaetodontid species on Indo-pacific reefs (Bouchon-Navaro 1985). These often display a high degree of feeding specialization for a particular category of prey (scleractinians, alcyonarians, gorgonians, etc.), in some cases at a generic or even a specific level. Obligate scleractinian feeders are numerous in chaetodontids (15 to 44% of the species), more than in any other reef fish family. Diets of adults have been investigated in a number of species in Hawaii (Hobson 1974, Reese 1977, Ralston 1981), in the Marshall Islands (Hiatt & Strasburg 1960, Reese 1975), on the Great Barrier Reef of Australia (Reese 1975, Anderson et al. 1981), in Okinawa, Japan (Sano et al. 1984), in French Polynesia (Harmelin-Vivien & Bouchon-Navaro 1983, Bouchon-Navaro 1986), in Madagascar (Harmelin-Vivien 1979), in Jordan (Harmelin-Vivien & Bouchon-Navaro 1982) and in the Caribbean (Randall 1967, Birkeland & Neudecker 1981, Gore 1984, Lasker 1985). Nothing is presently known about the food of juvenile chaetodontids, especially during the critical phase of their recruitment on the reef. The only study that refers to ontogenetic food changes in butterflyfishes, was performed on a planktivorous Hawaiian species (Ralston 1981).

This paper is a first attempt to consider these problems and to discuss the implications of feeding specialization on the recruitment and community structure of reef fishes.

Methods

To investigate ontogenetic changes in feeding among chaetodontids, specimens from different geographic areas were studied. Fishes were collected at Tulear (S.W. of Madagascar), Moorea (French Polynesia) and Guadeloupe (French West Indies) in various macrohabitats: lagoonal, reef flat and outer coral reef slope formations (Harmelin-Vivien 1979, Harmelin-Vivien & Bouchon-Navaro 1983).

As most of the chaetodontid species are diurnally active (Hobson 1974, Reese 1975, 1977, Harmelin-Vivien 1979) specimens were sampled during daytime between 0900 to 1600 hours, mainly by spearfishing or rotenone poisoning. Some small specimens were collected with quinaldine anaesthetic and hand nets. Standard length (SL) was measured to the nearest millimeter. Fishes were immediately preserved in a 10% buffered formalin solution with abdominal injection for the largest specimens, and later examined for stomach content. Ten species for which both juvenile and adult specimens were sampled, were taken into account in the present study. A total of 292 fishes with food in their stomach were quantitatively analyzed, 34.6% of them being juveniles. According to the maximum standard length of the various species (Burgess 1978), specimens with SL <40 to 60 mm were considered as juveniles (Table 1). Fishes less than 60 mm SL are expected to be less than 6 months old, if one follows the conclusions reached by Ralston (1976) for *Chaetodon miliaris*. They generally were the smallest size-class distinguished when size was considered for visual counts (Bouchon-Navaro 1979, 1981, Lindquist & Gilligan 1986).

Quantitative analysis of food was conducted on stomach contents as prey were more or less fully digested in the intestine. The main prey items were sorted by systematic groups under a binocular microscope, and then weighted to the nearest 0.1 mg. Relative importance of prey in the feeding was determined by their percentages by wet weight (Cw).

Niche breadth in the utilization of food resources was calculated using the Shannon-Wiener index of diversity

$$H' = \sum_{i=1}^{n} p_i \log_2 p_i,$$

where pi is the proportion by weight of a particular prey category, for n prey categories.

Equitability $J' = \dfrac{H'}{H' \max}$

was also calculated since this index is independent of n.

In order to determine the similarity of feeding within – and between – species, a cluster analysis using the mean Chi square distances was performed on the data of juvenile and adult diets of all species.

Results

Specialization level

The total number of prey item taxa eaten by a species was generally equal or a little lower in juvenile than in adult chaetodontid fishes, and the mean number of prey found in one specimen did not significantly differ from juveniles to adults for most species (Table 1). The range of food eaten by juveniles and adults were generally very similar except for *Chaetodon lunula* from Tulear (Table 2). The diversity and equitability of diets were similar for the two size-classes in most species.

All the chaetodontid species analyzed in this study were coelenterate predators (Table 2), but they differed greatly in their degree of feeding specialization. Two groups can be readily distinguished: the obligate coral feeders (*C. trifascialis* and *C. trifasciatus*) preying exclusively upon scleractinian polyps, and the non-obligate or facultative coral feeders (*C. auriga, C. citrinellus, C. kleinii, C. lunula, C. madagascariensis, C. vagabundus* and *C. xanthocephalus*) which fed on coral polyps together with many other benthic organisms, from algae to ascidians (Table 2).

Obligate coral-feeders

Food specialization in *C. trifascialis* and *C. trifasciatus* occurred very early in their benthic life history. All the specimens analyzed from 22 mm in standard length to the largest adults, had exclusively fed on scleractinian polyps. The few smallest specimens of *C. trifasciatus* (19 to 21 mm SL) collected on Tulear reefs presented stomach contents composed half of calanoid copepods, half of coral polyp fragments. These specimens were not fully metamorphosed lacking juvenile pigmentation, and still exhibiting the preopercular spines of the tholichthys larval form. Therefore, in the obligate coral feeders such as *C. trifasciatus*, food specialization appears to occur immediatly after the settlement of early juveniles on the reef, and the transition phase from planktonic to benthic food is very short.

Non-obligate coral feeders

In non-obligate coral feeders, benthic feeding also seemed to occur very early, as stomachs from specimens of 19 mm SL (*C. capistratus*), 21 mm SL (*C. auriga*) or 22 mm SL (*C. lunula*) were already full of benthic organisms.

In spite of a food spectrum very similar to the adult one, juveniles of non-obligate coral feeders clearly exhibited distinct food preferences (Table 2). Whatever the geographic area and the species studied, tentacles of sedentarian polychaetes, mainly from serpulid, sabellid and terebellid worms, were a preferred prey of juvenile fishes. An inverse relationship with size in weight percentages of sedentarian polychaete tentacles and scleractinian polyps was observed within all these chaetodontid species (Table 3). The mean proportion of polychaetes in diets of juveniles (36.2%) averaged over all non-obligate coral feeders was significantly higher than in adult ones (8.8%). Conversely, the mean proportion of scleractinian polyps in adult diets (51.6%) far exceeded that found in juveniles (8.2%).

The cluster-analysis performed on the quantitative data of stomach contents defined three main trophic groups (Fig. 1): group I included only adult size-classes, group II was mainly composed of juvenile size-classes and group III gathered small and large *C. citrinellus*. The last four species were individually very loosely related to the others. Interspecific similarity of feeding was thus higher within each size-class (within adults and within juveniles) than intraspecific similarity between size-classes, *C. citrinellus* excepted.

Distributions of juveniles and adults

Information on the relative distribution of juveniles and adults among chaetodontid populations is sparse in the literature (Table 4). Depending on the species, the patterns of distribution may or may not differ between size-classes. In some species such as *C. paucifasciatus* in the Gulf of Aqaba (Bouchon-Navaro 1979) no difference was observed between juvenile and adult distributions. In others, the different size-classes inhabited different reef zones. Juveniles and sub-adults were generally more abundant in shallow-water areas whether it be coral reefs or seagrass beds, whereas adults predominated in deeper waters (Fricke 1973, Clarke 1977, Bouchon-Navaro 1979, 1981, Lindquist & Gilligan 1986).

Difference or similarity in the distribution patterns of juvenile and adult chaetondontids did not

Table 1. Chaetodontid species studied for ontogenetic change in feeding. Two broad size-classes were distinguished in each species, juvenile individuals (SL <60 mm) and adult individuals (SL >60 mm). Species were collected on Tulear (Madagascar), Moorea (French Polynesia) and Guadeloupe (French West Indies) coral reefs. N = number of specimens analyzed with food in their stomach, XSL = mean standard length of specimens (mm), P = total number of prey types in fish diet. XP = mean number of prey types per individual, s.d. = standard deviation. Significance of t-test was given for the comparison of mean prey per individual (XP) between size-classes in each species (n.s. = non significant, * = p<0.05, ** = p<0.01, *** = p<0.001, - - - = test not calculated, too few data).

Species	Locality	N	XSL (Range)	P	XP (s.d.)
Chaetodon auriga	Tulear	20	39.3 (22–57)	13	3.6 (1.4) n.s.
	Tulear	59	101.4 (65–145)	17	3.5 (1.6)
C. capistratus	Guadeloupe	20	27.5 (19–33)	19	5.3 (1.9) - - -
C. citrinellus	Moorea	5	59.5 (44–60)	9	7.6 (1.2) n.s.
	Moorea	10	84.6 (80–90)	12	7.1 (1.0)
C. kleinii	Tulear	3	35.3 (34–36)	5	3.7 (0.6) - - -
	Tulear	3	57.3 (55–60)	4	2.7 (0.6)
C. lunula	Moorea	4	38.0 (21–60)	9	5.0 (1.4) ***
	Moorea	13	127.8 (106–136)	14	9.0 (1.4)
	Tulear	1	24.0	6	6.0 - - -
	Tulear	4	129.8 (115–135)	2	1.5 (0.6)
C. madagascariensis	Tulear	5	32.4 (25–39)	5	4.7 (0.5) **
	Tulear	10	91.9 (70–131)	4	3.4 (0.5)
C. trifascialis	Tulear	4	40.7 (33–42)	1	1.0 (0.0) n.s.
	Tulear	3	89.3 (84–94)	1	1.0 (0.0)
C. trifasciatus	Moorea	25	38.1 (31–52)	1	1.0 (0.0) n.s.
	Moorea	23	96.3 (78–109)	1	1.0 (0.0)
	Tulear	6	31.2 (19–50)	1	1.0 (0.0) n.s.
	Tulear	35	89.6 (70–106)	1	1.0 (0.0)
C. vagabundus	Moorea	2	43.5 (40–47)	7	6.5 - - -
	Moorea	10	113.8 (104–124)	17	9.4 (1.4)
	Tulear	2	50.1 (43–58)	9	5.5 - - -
	Tulear	13	107.0 (70–130)	15	6.4 (1.3)
C. xanthocephalus	Tulear	4	45.7 (36–60)	7	3.8 (2.5) n.s.
	Tulear	8	129.8 (80–155)	6	3.3 (1.3)

seem to be closely related to their degree of specialization. Distributions were more often similar in highly specialized species, since obligate coral feeders (OCF) and obligate alcyonarian feeders (AL) represented 55.5% of the species with similar patterns, and only 15.8% of the species with different distributions (Table 4). Nevertheless, juveniles of some obligate coral feeders, such as *C. trifascialis* and *C. trifasciatus,* exhibit different patterns of distribution on different reefs.

Discussion

Ontogenetic changes in feeding

The role of larval recruitment in determining the structure and stability of reef fish communities has been often, and sometimes hotly, discussed (Russell et al. 1977, Sale et al. 1980, Williams & Sale 1981, Williams 1983, Victor 1983, 1986b, Eckert 1984, Shulman 1984, 1985, Sale 1985, Doherty & Williams 1988, Mapstone & Fowler 1988, and others). Ecological requirements of larvae and newly settled juveniles generally differ from those of adults, and these differences may explain a number of observed adult distributions (Leis 1986). The importance of shelter, adult density and predation

Table 2. Diets of juveniles and adults in chaetodontids, expressed as weight percentages of prey (Cw) in stomach contents. J = juveniles <60 mm SL, A = adults >60 mm SL, (1) = specimens from 22 to 50 mm SL, (2) = mainly sedentarian polychaete tentacules + juveniles of errant polychaetes, + = Cw <0.1%.

Prey types	C. auriga Tulear		C. citrinellus Moorea		C. kleinii Tulear		C. lunula Moorea		C. lunula Tulear		C. madagascariensis Tulear		C.trifascialis Tulear		C. trifasciatus Moorea		C. trifasciatus Tulear (1)		C. vagabundus Moorea		C. vagabundus Tulear		C. xanthocephalus Tulear	
	J	A	J	A	J	A	J	A	J	A	J	A	J	A	J	A	J	A	J	A	J	A	J	A
Filamentous algae	12.8	0.1	22.6	34.0	–	–	–	–	–	–	32.5	20.0	–	–	–	–	–	–	1.0	22.5	30.0	+	7.5	7.0
Sponges	–	–	0.4	0.8	–	–	–	–	–	–	–	–	–	–	–	–	–	–	–	0.4	–	–	13.0	20.0
Hydroids	0.5	0.4	1.0	1.6	25.0	15.5	13.7	0.1	–	–	7.5	–	–	–	–	–	–	–	–	0.8	–	0.1	–	–
Actinians	5.0	0.6	–	–	–	–	–	–	–	–	–	–	–	–	–	–	–	–	–	–	–	1.6	–	–
Scleractinians	0.5	66.2	7.5	15.4	7.5	70.0	2.5	55.7	–	95.0	26.8	55.0	100.0	100.0	100.0	100.0	100.0	100.0	2.0	18.7	10.0	82.5	1.5	6.0
Alcyonarians	0.8	4.7	29.0	27.5	–	–	–	–	–	–	5.0	–	15.0	–	–	–	–	–	–	3.9	–	9.8	3.0	+
Nemerteans	–	–	–	–	–	–	–	0.2	–	–	–	–	–	–	–	–	–	–	–	0.9	–	+	–	–
Polychaetes (2)	57.8	8.6	30.0	12.6	25.0	–	62.3	20.8	7.0	–	30.5	10.0	–	–	–	–	–	–	50.0	21.2	20.0	1.5	43.0	5.0
Sipunculid introverts	7.5	1.4	1.5	5.6	–	–	9.3	4.9	–	–	–	–	–	–	–	–	–	–	–	1.9	–	–	–	–
Opisthobranchs	–	0.1	–	–	–	–	–	–	–	–	–	–	–	–	–	–	–	–	10.0	5.2	0.2	0.7	–	–
Gastropods	–	–	–	–	–	–	–	5.6	–	–	–	–	–	–	–	–	–	–	–	–	–	–	–	–
Cephalopods	–	–	–	–	–	–	–	0.1	–	–	–	–	–	–	–	–	–	–	–	–	–	–	–	–
Molluscan eggs	6.0	14.8	7.6	–	–	–	2.5	4.7	–	–	–	–	–	–	–	–	–	–	30.0	19.9	26.5	1.1	–	–
Cirriped tentacles	–	–	–	–	–	–	–	–	–	–	2.7	–	–	–	–	–	–	–	–	–	–	–	–	–
Calanoid copepods	–	–	–	–	–	–	–	–	10.0	–	–	–	–	–	–	–	–	–	–	–	1.0	–	–	–
Tanaid isopods	–	–	–	+	–	–	–	1.8	20.0	–	–	–	–	–	–	–	–	–	2.0	+	1.8	+	–	–
Amphipods	+	0.1	–	0.1	15.0	4.5	0.7	+	5.0	–	–	–	–	–	–	–	–	–	5.0	+	0.5	+	0.6	–
Decapod shrimps	–	0.2	–	–	–	–	–	1.0	–	–	–	–	–	–	–	–	–	–	–	0.4	–	–	–	–
Brachyuran larvae	–	0.1	–	–	–	–	–	2.0	–	–	–	–	–	–	–	–	–	–	–	+	–	–	–	–
Echinoid podia	1.8	–	–	0.2	–	–	–	–	–	–	–	–	–	–	–	–	–	–	–	–	–	–	–	–
Holothurian tentacles	–	0.3	–	–	–	–	–	–	–	–	–	–	–	–	–	–	–	–	–	–	–	0.3	–	–
Pterobranchs	3.3	0.9	0.3	–	–	–	–	–	–	–	–	–	–	–	–	–	–	–	–	–	–	1.8	–	–
Ascidians	2.1	0.3	–	1.1	27.5	10.0	1.2	–	10.0	–	–	–	–	–	–	–	–	–	–	1.6	–	+	31.4	60.0
Fish eggs	1.9	0.3	–	–	–	–	–	–	–	–	–	–	–	–	–	–	–	–	–	0.4	–	–	–	–
Mucus + organic matter	–	1.0	–	1.0	–	–	5.0	2.0	48.0	–	–	–	–	–	–	–	–	–	–	2.0	10.0	0.6	–	–
Food niche breadth H'	0.56	1.73	2.30	2.38	1.47	1.30	1.89	2.10	0.92	0.62	1.56	0.83	0.00	0.00	0.00	0.00	0.00	0.00	1.73	2.56	0.64	0.94	1.65	1.58
J'	0.16	0.42	0.72	0.69	0.64	0.65	0.59	0.56	0.37	0.38	0.67	0.36	0.00	0.00	0.00	0.00	0.00	0.00	0.62	0.69	0.19	0.31	0.55	0.68

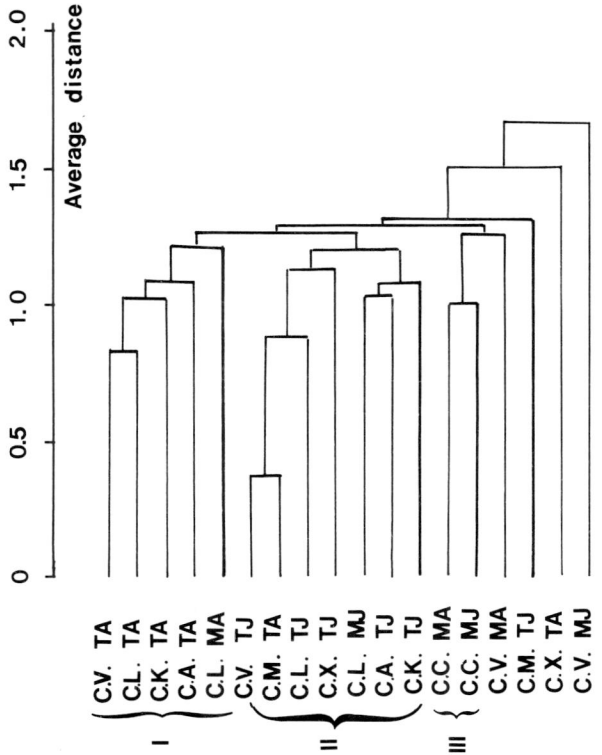

Fig. 1. Cluster phenogram of diets of 18 non-obligate corallivorous butterflyfishes using average distance coefficients. Group I gathered adult size-classes, group II gathered juvenile size-classes, and group III juveniles and adults of *Chaetodon citrinellus* (C.A. = *Chaetodon auriga*, C.C. = *Chaetodon citrinellus*, C.K. = *Chaetodon kleinii*, C.L. = *Chaetodon lunula*, C.M. = Chaetodon madagascariensis, C.V. = *Chaetodon vagabundus*, C.X. = *Chaetodon xanthocephalus*, T. = Tulear, M = Moorea, A = Adults, J = Juveniles).

has been more thoroughly investigated than the role of the food requirements of early juveniles (Luckhurst & Luckhurst 1977, Williams & Sale 1981, Sale et al. 1984, Shulman 1984, 1985). Based on experiments, Shulman (1984) concluded that food had little effect on recruitment of reef fishes in the Virgin Islands, but did not eliminate 'the possibility that prey availability was one of the selective forces involved in settlement site selection by recruits'.

Prey composition diets of larval fishes are very homogeneous and size dominated prey selection patterns. Naupliar through adult stages of copepods are the staple food of most marine fish larvae (Hunter 1981). This is probably the case for the diets of chaetodontid larvae, as calanoid copepods were found in the smallest recruits analyzed.

Table 3. Relative importance of polychaetes, mainly tentacles of sedentarian polychaetes andscleractinian polyps in the feeding of juveniles and adults of non-obligate corallivorous chaetodontid species, expressed as weight percentages of the total food ingested. T = Tulear, M = Moorea, G = Guadeloupe, J = juveniles, A = adults, – = no data.

Species	Locality	% Polychaetes J	% Polychaetes A	% Scleractinians J	% Scleractinians A
Chaetodon auriga	T	57.8	8.6	0.5	66.2
C. capistratus	G	36.8	–	23.8	–
C. citrinellus	M	30.0	12.6	7.5	15.4
C. kleinii	T	25.0	0.0	7.5	70.0
C. lunula	M	62.3	20.8	2.5	55.7
	T	7.0	0.0	0.0	95.0
C. madagascariensis	T	30.5	10.0	26.8	55.0
C. vagabundus	M	50.0	21.2	2.0	18.7
	T	20.0	1.5	10.0	82.5
C. xanthocephalus	T	43.0	5.0	1.5	6.0
Mean		36.2	8.8	8.2	51.6
s.d.		17.3	8.1	9.6	31.4
Significance of t-test on means			*** p<0.001		*** p<0.001

Change from planktonic to benthic feeding in butterflyfish seems to take place at the very moment of settlement on the reef. The transitional phase between the two life-styles appears to be very short, but needs to be studied more closely. No break in feeding was noticed from the planktonic to the benthic life in the butterflyfish species analyzed, nor in the planktivorous *C. miliaris* studied by Ralston (1981) in Hawaii. In contrast, an interruption in feeding activity during the transformation from the larval to the juvenile state has been observed in other reef fishes such as acanthurids (Randall 1961, Harmelin-Vivien personal observations). For example, during metamorphosis from the acronurus larva, *Acanthurus triostegus* ceases to feed, while largely subsisting on fat reserves stored in its abdominal cavity (Harmelin-Vivien personal observation).

Among chaetodontids, feeding specialization on scleractinian polyps begins at the time of settlement in obligate coral feeders (*C. trifascialis* and *C. trifasciatus*). In these species dietary requirements are probably as important (if not more important) than shelter availability for successful recruitment.

In non-obligate coral feeders, a narrowing of food niche breadth in juveniles, as generally reported for fish (Ross 1986), is hardly noticeable. The food spectrum of the juveniles we studied was similar to that of adults but the relative importance of the main prey eaten differed with size-class. Juveniles of most species took a high proportion of tentacles of sedentarian polychaetes (sabellids, ser-

Table 4. Difference or similarity in the distribution patterns of juvenile and adult chaetodontid fishes. Indication of diets are abbreviated after each species name. OCF = obligate coral feeders, CF = non-obligate coral feeders, AL = alcyonarian feeders, BI = other benthic invertebrate feeders, PK = plankton feeders.

Locality	Similar distributions of juveniles & adults	Different distributions of juveniles & adults	Reference
Guadeloupe		*C. capistratus* (CF)	Bouchon Navaro pers. comm.
Bahamas		*C. capistratus* (CF, BI) *C. ocellatus* (BI) *C. sedentarius* (BI) *C. striatus* (BI)	Lindquist & Gilligan 1986
Jordan (Aqaba)	*C. austriacus* (OCF) *C. paucifasciatus* (CF)	*C. trifascialis* (OCF)	Bouchon-Navaro 1979
Comoro Islands (Anjouan)	*C. melannotus* (AL) *C. trifasciatus* (OCF)	*C. auriga* (CF) *C. kleinii* (BI) *C. trifascialis* (OCF) *C. unimaculatus*	Fricke 1973
Madagascar (Tulear)	*C. blackburni* (AL) *C. guttatissimus* (CF) *C. kleinii* (BI) *C. madagascariensis* (CF) *C. melannotus* (AL) *C. trifascialis* (OCF)	*C. auriga* (CF) *C. lunula* (CF) *C. trifasciatus* (OCF) *C. vagabundus* (CF) *C. xanthocephalus* (BI)	Harmelin-Vivien 1979
Hawaii	*C. miliaris* (PK)		Ralston 1981
French Polynesia (Moorea)	*C. citrinellus* (CF) *C. ornatissimus* (OCF) *C. pelewensis* (CF) *C. quadrimaculatus* (CF) *C. reticulatus* (OCF) *C. trifascialis* (OCF) *C. trifasciatus* (OCF)	*C. auriga* (CF) *C. ephippium* (CF) *C. lunula* (CF) *C. vagabundus* (CF)	Bouchon-Navaro 1981

pulids, terebellids), whereas adults mainly fed on scleractinians or other sessile invertebrates (alcyonarians, ascidians, sponges). Polychaetes probably are more nutritious than coral tissue, and better supply the higher energy demand of juveniles. The high dietary overlap between juveniles of different species which might result from similar prey preferences does not necessarily imply any competitive interaction, as juvenile density is generally low and their main prey is very abundant on coral reefs. Greater similarity of diets among juveniles than among adults has been observed in other fish species (Duka 1976, Ross 1986), and has been interpreted as a consequence of the exploitation of some abundant types of prey. How prey utilization is related to ontogenetic changes in behavior and/or morphology of the feeding apparatus has not been determined in this study. In a study of the functional morphology of the feeding apparatus of butterflyfishes, Motta (1988) observed that in many cases the polychaete worms retracted in their tube, and escaped predation faster than fish was able to catch them. A faster jaw protrusion time in juvenile butterflyfish might explain why polychaete worms are mostly preyed upon by juvenile individuals. An alternative hypothesis is that the iron content of the teeth of butterflyfish, which is correlated with their diet (Motta 1987) is much lower in juvenile fishes and prevents them from feeding intensively on hard-bodied prey like corals. In the latter case, the iron content of the teeth caps should be much higher in juvenile specimens of obligate coral feeders than in non-obligate ones. We were unable to investigate this point.

Implication on community structure

The described patterns of butterflyfish communities are based on the distribution and interactions of adult individuals; and generally show a positive relationship between cover of living scleractinians, fish species richness and fish abundance (Bouchon-Navaro 1979, 1981, Anderson et al. 1981, Bell & Galzin 1984, Bell et al. 1985, Bouchon-Navaro et al. 1985, Findley & Findley 1985). Some authors have suggested there is little evidence that butterflyfish assemblages are resource-limited (Bell et al. 1985, Findley & Findley 1985). Others, however, have emphasize the possible importance of competitive interactions among chaetodontid species (Reese 1975, Anderson et al. 1981, Bouchon-Navaro 1986).

It is often claimed that reef fishes are relatively site-attached after recruitment (Sale 1980). In fact, the site of settlement and the site of adult-life differ in many reef fish species, and one can notice a differential distribution of size-class cohorts across a reef in the same species (Galzin 1985, Shulman & Ogden 1987). Among chaetodontids, where adults are reported to be highly sedentary (Reese 1975, 1977, 1981), it seems that settlement occurs preferentially in shallow-water habitats for at least half of the species studied (Table 4). The patterns of size-class distribution may vary with geographical areas within the same fish species, and do not seem to be closely related to feeding specialization, contrary to what hypothesized. The importance of shallow waters for reef fish recruits is well known in species such as *Acanthurus triostegus* (Randall 1961, Sale 1969). Particular hydrological conditions, food resources or avoidance of predators are put forward to account for the high rate of fish recruitment observed in shallow waters.

When a fish settles in a reef zone different to that of adults of the same species, its recruitment is independent of local adult density, and no intraspecific overlap in food and microhabitat exists during the early life-history of the species. When recruitment occurs within adult home ranges, differences in food requirements between size-classes reduces the potential intraspecific competition for that resource.

The present study demonstrates that among chaetodontids specific food and spatial requirements of adults and juveniles may range from identical to completely different. Recruitment studies to date have emphasized species with similar juvenile and adult habits. It is suggested that the dynamics of species in which food and habitat requirements differ significantly between juveniles and adults may differ fundamentally from these species. In particular, post-settlement events, especially at times of transition between juvenile and adult behavior, may be of greater importance in determining recruitment to adult populations than initial densities of post-larvae.

Feeding requirements of juvenile butterflyfishes – and of coral reef fishes in general – need to be thoroughly investigated for a better understanding of post-settlement events. The timing of the transition between planktonic feeding to benthic feeding should be determined by analysis of otoliths and stomach contents for the same individuals. It is necessary to know the food preferences of recruits, and to determine if there is any consistency within fish families or within feeding guilds. The caloric value and chemical composition of prey should be determined, and related to the energetic requirements of juvenile fishes. These analyses should be conducted on syntopic species exhibiting different level of feeding specialization. For species with very specific feeding requirements, distribution and abundance of suitable food resources may greatly restrict the availability of potential settling sizes. The importance of food resources on settlement may be investigated by appropriate experiments where the real specific food resources of recruits are modified.

Acknowledgement

Many thanks are expressed to Y. Bouchon-Navaro for providing juvenile specimens of *Chaetodon capistratus* from Guadeloupe seagrass beds; and to F. Bourlière for critical comments on a draft of the manuscript. I am particularly grateful to E.S. Reese, D. McB. Williams and P.J. Motta for reviewing the manuscript.

References cited

Anderson, G.R.V., A.H. Ehrlich, P.R. Ehrlich, J.D. Roughgarden, B.C. Russell & F.H. Talbot. 1981. The community structure of coral reef fishes. Amer. Nat. 117: 476–495.
Bell, J.D. & R. Galzin. 1984. The influence of live coral cover on coral reef fish communities. Mar. Ecol. Prog. Ser. 15: 265–274.
Bell, J.D., M.L. Harmelin-Vivien & R. Galzin. 1985. Large scale spatial variation in abundance of butterflyfishes (Chaetodontidae) on Polynesian reefs. Proc. Fifth Internat. Coral Reef Symp. 5: 421–426.
Birkeland, C. & S. Neudecker. 1981. Foraging behavior of two Caribbean chaetodontids: *Chaetodon capistratus* and *C. aculeatus*. Copeia 1981: 169–178.
Bouchon-Navaro, Y. 1979. Quantitative distribution of the Chaetodontidae on a fringing reef of the Jordanian coast (Gulf of Aqaba, Red Sea). Téthys 9: 247–251.
Bouchon-Navaro, Y. 1981. Quantitative distribution of the Chaetodontidae on a reef of Moorea Island (French Polynesia). J. Exp. Mar. Biol. Ecol. 55: 145–157.
Bouchon-Navaro, Y. 1985. Ecologie des Chaetodontidae des récifs coralliens d'Aqaba (Mer Rouge) et de Moorea (Polynésie française). Diplome EPHE, Paris. 193 pp.
Bouchon-Navaro, Y. 1986. Partitioning of food and space resources by chaetodontid on coral reefs. J. Exp. Mar. Biol. Ecol. 103: 21–40.
Bouchon-Navaro, Y., C. Bouchon & M.L. Harmelin-Vivien. 1985. Impact of coral degradation on a chaetodontid fish assemblage (Moorea, French Polynesia). Proc. Fifth Internat. Coral Reef Symp. 5: 427–432.
Brothers, E.B., D.Mc.B. Williams & P.F. Sale. 1983. Length of larval life in twelve families of fishes at 'One Tree Lagoon', Great Barrier Reef, Australia. Mar. Biol. 76: 319–324.
Burgess, W.E. 1978. Butterflyfishes of the world. T.F.H. publications, Neptune City. 832 pp.
Clarke, R.D. 1977. Habitat distribution and species diversity of chaetondontid and pomacentrid fishes near Bimini, Bahamas. Mar. Biol. 40: 277–289.
Doherty, P.J. & D.Mc.B. Williams. 1988. The replenishment of coral reef fish populations. Oceanogr. Mar. Biol. Annu. Rev. 26: 487–551.
Duka, L.A. 1976. Feeding and food relationships of the larvae and young of the family Labridae. J. Ichtyol. 16: 398–407.
Eckert, G.J. 1984. Annual and spatial variation in recruitment of Labroid fishes among seven reefs in the Capricorn (Bunker Group, Great Barrier Reef). Mar. Biol. 78: 123–127.
Findley, J.S. & M.T. Findley. 1985. A search for pattern in butterflyfish communities. Amer. Nat. 126: 800–816.
Fricke, H.W. 1973. Behaviour as part of ecological adaptation. In situ studies in the coral reef. Helgoländer wiss. Meeresunters. 24: 120–144.
Galzin, R. 1985. Ecologie des poissons récifaux de Polynésie Française: variations spatio-temporelles des peuplements, Dynamique de populations de trois espèces dominantes des lagons nord de Moorea, Evaluation de la production ichtyologique d'un secteur récifo-lagonaire. Ph. D. Dissertation, Univ. Sciences et Technique du Languedoc, Montpellier. 195 pp.
Gore, M.A. 1984. Factors affecting the feeding behavior of a coral reef fish, *Chaetodon capistratus*. Bull. Mar. Sci. 35: 211–220.
Harmelin-Vivien, M.L. 1979. Ichtyofaune des récifs coralliens de Tuléar (Madagascar): Ecologie et relations trophiques. Ph. D. Dissertation, Univ. Aix-Marseille II, Marseille. 165 pp.
Harmelin-Vivien, M.L. & Y. Bouchon-Navaro. 1982. Trophic

relationships among chaetodontid fishes in the Gulf of Aqaba (Red Sea). Proc. Fourth Internat. Coral Reef Symp. 2: 537–544.

Harmelin-Vivien, M.L. & Y. Bouchon-Navaro. 1983. Feeding diets and significance of coral feeding among chaetodontid fishes in Moorea (French Polynesia). Coral Reefs 2: 119–127.

Hiatt, R.W. & D.W. Strasburg. 1960. Ecological relationships of the fish fauna on coral reefs of the Marshall Islands. Ecol. Monogr. 30: 65–127.

Hobson, E.S. 1974. Feeding relationships of teleostean fishes on coral reefs in Kona, Hawaii. U.S. Fish. Bull. 72: 915–1031.

Hunter, J.R. 1981. Feeding ecology and predation of marine fish larvae. pp. 33–77. In: R. Lasker (ed.) Marine Fish Larvae: Morphology, Ecology and Relation to Fisheries, University of Washington Press, Seattle.

Lasker, H.R. 1985. Prey preferences and browsing pressure of the butterflyfish Chaetodon capistratus on Caribbean gorgonians. Mar. Ecol. Prog. Ser. 21: 213–220.

Leis, J.M. 1986. Ecological requirements of Indo-Pacific larval fishes: a neglected zoogeographic factor. pp. 759–766. In: T. Uyeno, R. Arai, T. Taniuchi & K. Matsuura (ed.) Indo-Pacific Fish Biology: Proc. Second Internat. Conf. Indo-Pacific Fishes, Ichtyol. Soc. of Japan, Tokyo.

Lindquist, D.G. & M.R. Gilligan. 1986. Distribution and relative abundance of butterflyfishes and angelfishes across a lagoon and barrier reef, Andros Island, Bahamas. Northeast Gulf Science 8: 23–30.

Luckhurst, B.E. & K. Luckhurst. 1977. Recruitment patterns of coral reef fishes on the fringing reef of Curaçao, Netherlands Antilles. Can. J. Zool. 55: 681–689.

Mapstone, B.D. & A.J. Fowler. Recruitment and the structure of assemblages of fish on coral reefs. TREE 3: 72–77.

Motta, P.J. 1987. A quantitative analysis of ferric iron in butterflyfish teeth (Chaetodontidae, Perciformes) and the relationship to feeding ecology. Can. J. Zool. 65: 106–112.

Motta, P.J. 1988. Functional morphology of the feeding apparatus of ten species of Pacific butterflyfishes (Perciformes, Chaetodontidae): an ecomorphological approach. Env. Biol. Fish. 22: 39–67.

Ralston, S. 1976. Age determination of a tropical reef butterflyfish utilizing daily growth rings of otoliths. U.S. Fish. Bull. 74: 990–994.

Ralston, S. 1981. Aspects of the reproductive biology and feeding ecology of Chaetodon miliaris, a Hawaiian endemic butterflyfish. Env. Biol. Fish. 6: 167–176.

Randall, J.E. 1961. A contribution to the biology of the convict surgeon fish of the Hawaiian Islands, Acanthurus triostegus sandvicensis. Pac. Sci. 15: 215–272.

Randall, J.E. 1967. Food habits of reef fishes of the West Indies. Stud. Trop. Oceanogr. 5: 665–847.

Reese, E.S. 1975. A comparative field study of the social behavior and related ecology of reef fishes of the family Chaetodontidae. Z. Tierpsychol. 37: 37–61.

Reese, E.S. 1977. Coevolution of corals and coral feeding fishes of the family Chaetodontidae. Proc. Third Internat. Coral Reef Symp. 1: 267–274.

Reese, E.S. 1981. Predation on corals by fishes of the family Chaetodontidae: implications for conservation and management of coral reef ecosystems. Bull. Mar. Sci. 31: 594–604.

Ross, S.T. 1986. Resource partitioning in fish assemblages: a review of field studies. Copeia 1986: 352–388.

Russel, B.C., G.R.V. Anderson & F.H. Talbot. 1977. Seasonality and recruitment of coral-reef fishes. Aust. J. Mar. Freshw. Res. 28: 521–528.

Sale, P.F. 1969. Pertinent stimuli for habitat selection by the juvenile manini, Acanthurus triostegus sandvicensis. Ecology 50: 616–623.

Sale, P.F. 1980. The ecology of fishes on coral reefs. Oceanogr. Mar. Biol. Ann. Rev. 18: 367–421.

Sale, P.F. 1985. Patterns of recruitment in coral reef fishes. Proc. Fifth Internat. Coral Reef Symp. 5: 391–396.

Sale, P.F.. P.J. Doherty & W.A. Douglas. 1980. Juvenile recruitment strategies and the coexistence of territorial pomacentrid fishes. Bull. Mar. Sci. 30: 147–158.

Sale, P.F., W.A. Douglas & P.J. Doherty. 1984. Choice of microhabitats by coral reef fishes at settlement. Coral Reefs 3: 91–100.

Sano, M., M. Shimizu & Y. Nose. 1984. Changes in structure of coral reef fish communities by destruction of hermatypic corals: observational and experimental views. Pac. Sci. 38: 51–79.

Shulman, M.J. 1984. Resource limitation and recruitment patterns in a coral reef fish assemblage. J. Exp. Mar. Biol. Ecol. 74: 85–109.

Shulman, M.J. 1985. Recruitment of coral reef fishes; effects of distribution of predators and shelter. Ecology 66: 1056–1066.

Shulman, M.J. & J.C. Ogden. 1987. What controls tropical reef fish populations: recruitment or benthic mortality? An example in the Caribbean reef fish Haemulon flavolineatum. Mar. Ecol. Prog. Ser. 39: 233–242.

Victor, B.C. 1983. Recruitment and population dynamics of a coral reef fish. Science 219: 419–420.

Victor, B.C. 1986a. Duration of the planktonic larval stage of one hundred species of Pacific and Atlantic wrasses (family Labridae). Mar. Biol. 90: 317–326.

Victor, B. 1986b. Larval settlement and juvenile mortality in a recruitment-limited coral reef fish population. Ecol. Monogr. 56: 145–160.

Williams, D.Mc.B. 1983. Daily, monthly and early variability in recruitment of a guild of coral reef fishes. Mar. Ecol. Prog. Ser. 10: 231–237.

Williams, D.Mc.B. & P.F. Sale. 1981. Spatial and temporal patterns of recruitment of juvenile coral reef fishes to coral habitats within One Tree Lagoon, Great Barrier Reef. Mar. Biol. 65: 245–253.

Sexual differentiation, gonad development, and spawning seasonality of the Hawaiian butterflyfish, *Chaetodon multicinctus*

Timothy C. Tricas[1] & Joy T. Hiramoto
Department of Zoology, University of Hawaii at Manoa, Honolulu, HI 96822, U.S.A.
[1] *Present address: Washington University School of Medicine, Department of Otolaryngology, 517 S. Euclid Ave., Box 8115, St. Louis, MO 63110 and Marine Biological Laboratory, Woods Hole, MA 02543, U.S.A.*

Received 24.2.1988   Accepted 7.9.1988

Key words: Chaetodontidae, Coral reef fishes, Fish reproduction, Gonad histology

Synopsis

The reproductive biology of the coral reef butterflyfish, *Chaetodon multicinctus,* was investigated by histological examination of gonads sampled over an 18 month period from a shallow inshore population on Oahu, Hawaii. Most gonads developed directly from previously undifferentiated tissue. Ovarian development (the structural formation of lamellae and primary oocytes) was observed in fish ≥ 44 mm and testicular development (the formation of spermatogenic crypts) in fish ≥ 62 mm standard length (SL). In addition, testis formation was identified within the ovarian lamellae of several differentiated but immature fish. It is hypothesized that prematurational sex change may facilitate monogamy within the highly competitive social structure of this site attached species. Oocyte development in mature females was marked by distinct phases of primary growth, the formation of yolk vesicles, and vitellogenesis. Spawning activity was histologically identified by the maturation and hydration of fully yolked oocytes, and presence of postovulatory follicles. Recently spawned females from field collections and experimental gonadotropin-treatments exhibited postovulatory follicles that were estimated to persist at least 24 h after ovulation. Atresia of yolked oocytes was classified into four stages of cell degeneration and resorption. Monthly analyses of oocyte development and atresia within the sample population show that *C. multicinctus* has a protracted annual spawning season with a major peak during the early spring and evidence of spawning activity among some individuals in the fall. Histological analyses of spawning activity provide more accurate and unambiguous information than do traditional gonadosomatic assays in this and probably other coral reef fishes.

Introduction

Sexual maturation and spawning are important aspects of the life history of coral reef fishes, yet our understanding of the reproductive biology of this group is remarkably limited (reviewed by Thresher 1984, Walsh 1987). The majority of studies on reproductive activity have employed gonadosomatic indices (GSI) (e.g. Ralston 1981), macroscopic classification of gonad ripeness (e.g. Munro et al.1973, Nzioka 1979), or direct observations of spawning (e.g. Neudecker & Lobel 1982, Robertson 1983, Moyer 1984). With the exception of studies exclusively devoted to sex change (e.g. Moyer & Nakazono 1978, Ross 1984) only a few provide histological analyses of gamete development (e.g. Bouain & Siau 1983, Hourigan & Kelley 1985). This latter method may be preferred over more traditional GSI analysis for the study of spawning seasonality since the relationship between ovarian weight and body size often changes with the stage of oocyte development (de Vlaming et al. 1982). Histological classification should be especially advantageous to identify reproductive activity in tropical reef species during the non-peak periods of a protracted spawning season.

Butterflyfishes of the family Chaetodontidae are among the most conspicuous members of coral reefs. However, detailed reproductive data exist for only one species, *Chaetodon miliaris*, studied by Ralston (1976, 1981), and no published analyses of sexual differentiation, gonad maturation, or oocyte development are available for any chaetodontid. The banded butterflyfish, *Chaetodon multicinctus* Garrett, is endemic to the Hawaiian Islands and Johnston Atoll (Burgess 1978) where it occurs on most coral reefs. It is a small obligate coral feeder and forms monogamous pairs that defend permanent feeding territories (Reese 1975, Tricas 1985, 1988, 1989). It is closely allied with the congeners *Chaetodon punctatofasciatus*, *Chaetodon guttatissimus*, and *Chaetodon pelewensis* that occur regionally in the Indo-Pacific (Burgess 1978) and thus is an excellent model species for comparative study.

This paper presents the first histological analysis of the reproductive biology of a member of the family Chaetodontidae. Patterns of gonad differentiation, oocyte development, and atresia are described. Histological analyses are shown to be preferred over GSI methods, and reveal that *C. multicinctus* spawns over a protracted annual period with a major peak in the spring and activity among some fishes in the fall.

Study site and methods

Adult and juvenile pairs, and unpaired juveniles of *C. multicinctus* were collected over an 18-month period in 1981–1982 on shallow coral reefs (3–10 m deep) near Kahe Point on the west shore of Oahu, Hawaii (Fig. 1). All but one of the 1981 collections were taken semi-monthly from July through December, while those in 1982 were made every week. Fifty-eight separate samples were taken. Fish were speared between 0900 and 1200 h by

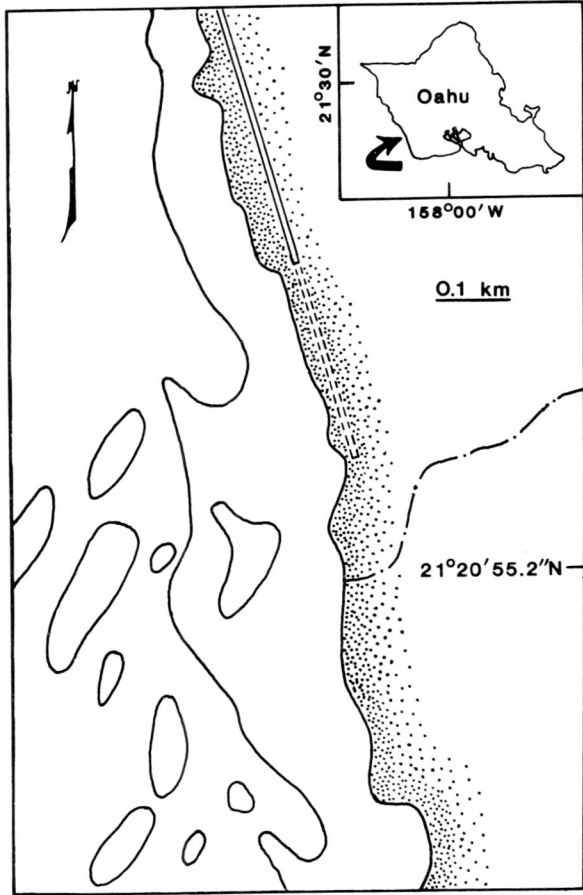

Fig. 1. Patch reef collection site at Kahe Point on the leeward coast of Oahu, Hawaii. Some collections made on patch reefs 0<0.5 km to the north of area shown. Submarine relief shows approximate 4 m contour of reef. After U.S. Geological Survey Map: Ewa, Hawaii 1983.

divers using scuba, placed in plastic bags, and transported on ice to the laboratory. Whole individuals were weighed (nearest mg), measured (nearest mm) for standard length (SL), and their gonads removed then weighed (nearest 0.1 mg). In fish ≤60 mm SL, tissues along the medial mesentery between the gut and dorsal wall of the coelomic cavity were dissected when gonads were not visible. The GSI, computed as the ratio of ovary weight to gonad-free body weight × 100, was determined for sexually mature females.

Gonads were fixed in 10% buffered formalin for at least 5 days, washed, and preserved in 70%

isopropanol. Tissues were dehydrated and cleared in ethanol and xylene, respectively, or by a single series of ethanol-butanol washes. Gonad tissues were embedded in Paraplast, and 7 μm-thick sections cut on a microtome. Hematoxylin/eosin, hematoxylin/triosin, Mallory's, and hematoxylin/triosin/Mallory's were used to study tissue morphology. Feulgen's stain confirmed the presence of DNA-rich material in cell aggregates within ovaries suspected of spermatogenic activity. Carbohydrate inclusions characteristic of cortical alveoli in oocytes (Aketa 1954) were identified by aqueous PAS treatment. Ooplasmic lipids were identified by either a 1% osmium tetroxide stain (and hematoxylin/eosin counterstain) of preserved tissue or a Sudan Black B stain of 40–45 μm fresh-frozen sections.

Ovarian sections were examined from fish in all size classes collected during all months of the study. Oocytes were staged and measured on a compound microscope with an ocular micrometer. Postovulatory follicles were measured along their major axis. Transverse histological sections were taken from the midregion of ovaries to maximize cross sectional area. Comparison of primary growth and vitellogenesis among oocytes from anterior, mid, and posterior regions of mature females showed that oocyte development was homogeneous throughout the ovary ($P>0.975$, 3-sample Smirnov test, n = 5 ovary pairs, Conover 1971).

Postovulatory follicles in recently spawned ovaries were examined by induced ovulation in captive fish. During February, March, and May 1982, and March, 1983, 16 adult females (71–94 mm SL) were collected with handnets by divers and transported to the Waikiki Aquarium. Each fish was given a single intramuscular injection of human chorionic gonadotropin (HCG Sigma) (approximately 10 IU/gbw) and held in large tanks with fresh flowing seawater (23.5–25.0° C). After 32–36 h, fish were stripped of ovulated oocytes by applying gentle finger pressure along the abdomen. Gonads were processed and stained with hematoxylin/eosin.

Results

A total of 628 fish were collected in the study. Histological sections of gonads were examined for 416 individuals between 29 and 95 mm SL. Ovary development (the formation of primary oocytes and ovarian lamellae) was recognized in fish as small as 44 mm SL (Table 1). Sections from the dorsal mesentery of smaller fish revealed only beds

Table 1. Sex ratios by size class for *Chaetodon multicinctus*. Fully differentiated but immature ovaries that contained spermatogenic crypts were classified as male and are indicated in parentheses.

Size class (mm SL)	n	Percent		
		Females	Males	Undifferentiated
<30	2	0	0	100
31–35	5	0	0	100
36–40	2	0	0	100
41–45	6	17	0	83
46–50	5	100	0	0
51–55	12	67	0	33
56–60	9	56	0	44
61–65	25	72	20 (4)	8
66–70	22	64	36 (14)	0
71–75	70	52	47 (1)	1
76–80	125	66	34	0
81–85	225	56	44	0
86–90	100	35	65	0
91–95	20	25	75	0

of connective tissue and primordial germ cells. The smallest fish with testis organization (formation of spermatogenic crypts in a structurally indifferent gonad) was 62 mm SL. All smaller fish either lacked differentiated gonads or showed structural ovarian development.

Ovaries

The onset of ovarian morphogenesis in *C. multicinctus* was marked by the proliferation of the stroma and gonia in the dorsal mesentery fold. This medial layer thickened and formed cell columns that developed into lamellae which projected laterally into the ovary lumen. Concurrent with gross structural development of the ovary was the appearance of primary oocytes (Fig. 2A). These cells formed in the lamellar folds and basal tissue layers, and were characterized by enlarged basophilic cells with large nuclei. Proliferation of the stromal layer abated as lamellae formation completed. Mature ovaries were paired, medially fused, and surrounded by a layer of connective tissue. Ovaries distended with yolked oocytes were anteriorly bilobed. During spawning eggs pass through an oviduct continuous with the lumen on the posterior aspect of the ovary.

Testes

In the majority of juveniles where testicular morphogenesis was observed, a thick bilateral stromal layer was formed within the undifferentiated dorsal-medial mesentery. Spermatogonia gave rise to aggregations of spermatocytes which formed spermatogenic crypts(Fig. 2B). Mature testes were elongate, paired, fused along the medial axis, and of the unrestricted spermatogonial testis-type (Grier 1981). Spermatozoa were only observed in crypts within fully-formed radially branched tubules. Sperm are released from crypts into the tubule lumen and collect into a large posteriorly projecting *vasa deferens*. Fully developed lobule luminae, sperm ducts, and liberated spermatozoa were observed in males as small as 65 mm SL.

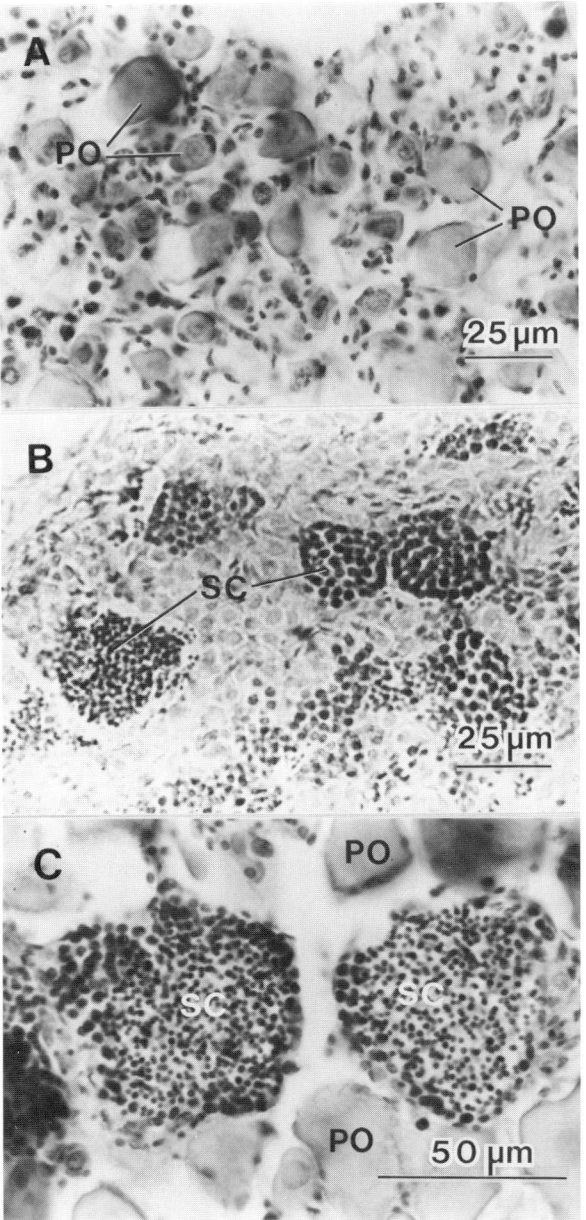

Fig. 2. Differentiation of gonad tissue in *Chaetodon multicinctus:* A – early ovary development from previously undifferentiated tissue in a 44 mm SL fish; B – early testis development from previously undifferentiated tissue in a 73 mm SL fish; C – spermatogenic crypt formation within lamellae of immature ovary from a 67 mm SL fish. All tissues stained with hematoxylin/eosin. PO = primary oocytes; SC = spermatogenic crypts.

A second distinct but less frequent pattern of testes formation was also observed in juveniles. Spermatogenic tissue secondarily developed within structurally differentiated ovaries that contained primary oocytes of the central nucleoli or more advanced perinucleolar stage (described below) (Fig. 2C, 3). Epithelial, stroma, and gonia (presumably spermatogonia) proliferated along the surface of the ovarian lamellae to form groups of mitotic cells that organized into basophilic crypts within the lamellae. Newly formed crypts contained aggregations of basophilic spermatocytes (Fig. 2C) that gave rise to central concentrations of spermatids. There was no difference in mean cell diameter of either primary spermatocytes or spermatids between adult testes and those of transitional juveniles (Two-way analysis of variance, $F = 0.054$, $P < 0.90$). Like that observed in testes that developed directly from primordial germ cells in the dorsal mesentery, there was a predominance of spermatids but a lack of spermatozoa prior to tubule formation indicating that spermatozoa are not formed during testis differentiation. Although many immature individuals 61–70 mm SL showed stroma proliferation in the lamellae which might indicate spermatogenic activity, only five non-vitellogenic fish collected between February and May showed clear evidence of organized spermatogenic tissue. No remnant of an ovarian lumen or secondary development of *vasa deferentia* on the gonad surface, as occurs in many protogynous hermaphroditic fishes (e.g. Reinboth 1962, Dipper & Pullin 1979, Ross 1984), was observed in any mature testes. No evidence of degenerate testicular tissue, or oocytes in association with fully differentiated testes was found to indicate prematurational protandry. The existence of spermatogenic tissue within a structurally developed ovary appears to be strictly prematurational since no crypt formation was observed in females with vitellogenic ova.

Oocyte development

Oogenesis involved distinct nuclear and cytoplasmic events, and rapid cell growth. Dimensions of

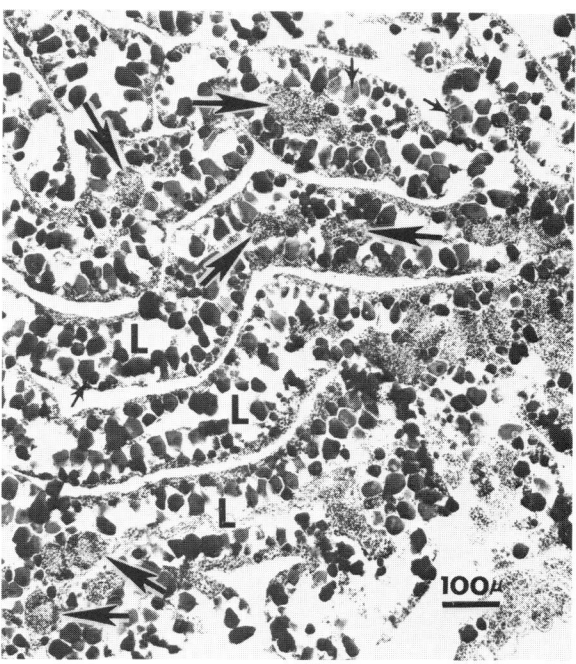

Fig. 3. Transitional gonad in a 67 mm SL *Chaetodon multicinctus*. Primary oocytes (small arrows) and spermatogenic crypts (large arrows) in ovary lamellae (L).

various stages and associated structures are summarized in Table 2.

Primary growth (Fig. 2A, 4A). – Primary growth involved the enlargement of primary oocytes and appearance of nucleoli in a centrally located nucleus. The basophilic stain of cytoplasm adjacent to the nucleus indicated the presence of a 'yolk nucleus' (reviewed by Wallace & Selman 1981). During subsequent growth the cytoplasm appeared more heterogeneous and less basophilic. Lampbrush chromosomes were visible within the germinal vesicle (egg nucleus). Oil inclusions appeared as small transparent droplets under hematoxylin/eosin stain and black when fixed with osmium tetroxide. A perinucleolar stage (c.f. Kuo et al. 1974) was identified following migration of nucleoli to the nuclear membrane, and persisted in oocytes throughout vitellogenesis.

Cortical alveoli (yolk vesicle) stages (Fig. 4B). – Prior to vitellogenesis, PAS positive yolk vesicles (the precursors of cortical alveoli) formed between the egg nucleus and oolemma. Under hematoxylin/eosin stain, previtellogenic oocytes with yolk vesicles were distinguished from primary oocytes by the former's transparent inclusions within the basophilic cytoplasm. Vitellogenic oocytes contained numerous cortical alveoli concentrated adjacent to the oolemma.

Vitellogenesis (Fig. 4A, 4B). – Yolk uptake rapidly accelerated oocyte enlargement. In early vitellogenesis, the vitelline membrane appeared as a thin eosinophilic band around the oocyte. Yolk spheres first appeared near the ooplasm periphery and migrated towards the cell interior. As cells enlarged, the vitelline membrane thickened and developed radial striations to form the zona radiata. In later stages of vitellogenesis, enlarged yolk spheres filled the oocyte interior and lipid droplets aggregated near the nucleus.

Maturation (Fig. 4C) *and hydration*. – Early oocyte maturation involved the coalescence of lipid droplets followed by migration of the nucleus to the cell periphery. The nuclear membrane then degenerated and its contents were released into the cytoplasm. Yolk fusion variably occurred during nuclear migration or after the breakdown of the nuclear membrane with the zona radiata still intact. Maturation and only early stages of hydration were observed among females taken during regular field collections. Later phases of hydration were found among females collected between 1500 h and dusk. Hydration involved the uptake of large amounts of water into the cell, yolk fusion, and cytoplasmic clearing. In late hydration, cells exhibited central concentrations of eosinophilic yolk, and peripheral concentrations of basophilic cytoplasm and liberated nucleoplasm. Striations of the zona radiata became less distinct prior to ovulation.

Ovulation and postovulatory follicles (Fig. 4D, 5A, 5B). – At ovulation, hydrated eggs are discharged from their follicular envelope into the ovary lumen. Newly evacuated follicles, termed postovulatory follicles, consisted of a hollow multilayered band of follicle cells attached to the lamellar surface (Fig. 4D). Older postovulatory follicles (probably ≥40 h

Table 2. Cell diameters and structural dimensions for stages of oocyte development in *Chaetodon multicinctus*. Measurements in microns. n = number of cells or structures measured.

Stages/Structure	n	Range	Mean ± SD
Primary growth			
Central nucleoli stage	201	31– 92	58 ± 12
Perinucleolar stage	220	51–154	92 ± 20
Cortical alveoli	23	123–179	142 ± 14
Early vitellogenesis	180	113–299	207 ± 34
Zona radiata (width)	32		3 ± 6
Yolk spherule (diam)	42		5 ± 1
Late vitellogenesis	331	227–453	338 ± 39
Zona radiata (width)	46		8 ± 2
Yolk globule (diam)	56		13 ± 2
Maturation	36	350–453	399 ± 25
Hydration	61	391–618	498 ± 56
Postovulatory follicles	84	52–206	101 ± 28
Ovulated eggs	4	464–639	525 ± 78
Atretic oocytes			
Alpha	21	258–381	316 ± 41
Beta	53	206–412	298 ± 50
Gamma	83	103–381	213 ± 75
Delta	162	21–484	104 ± 63

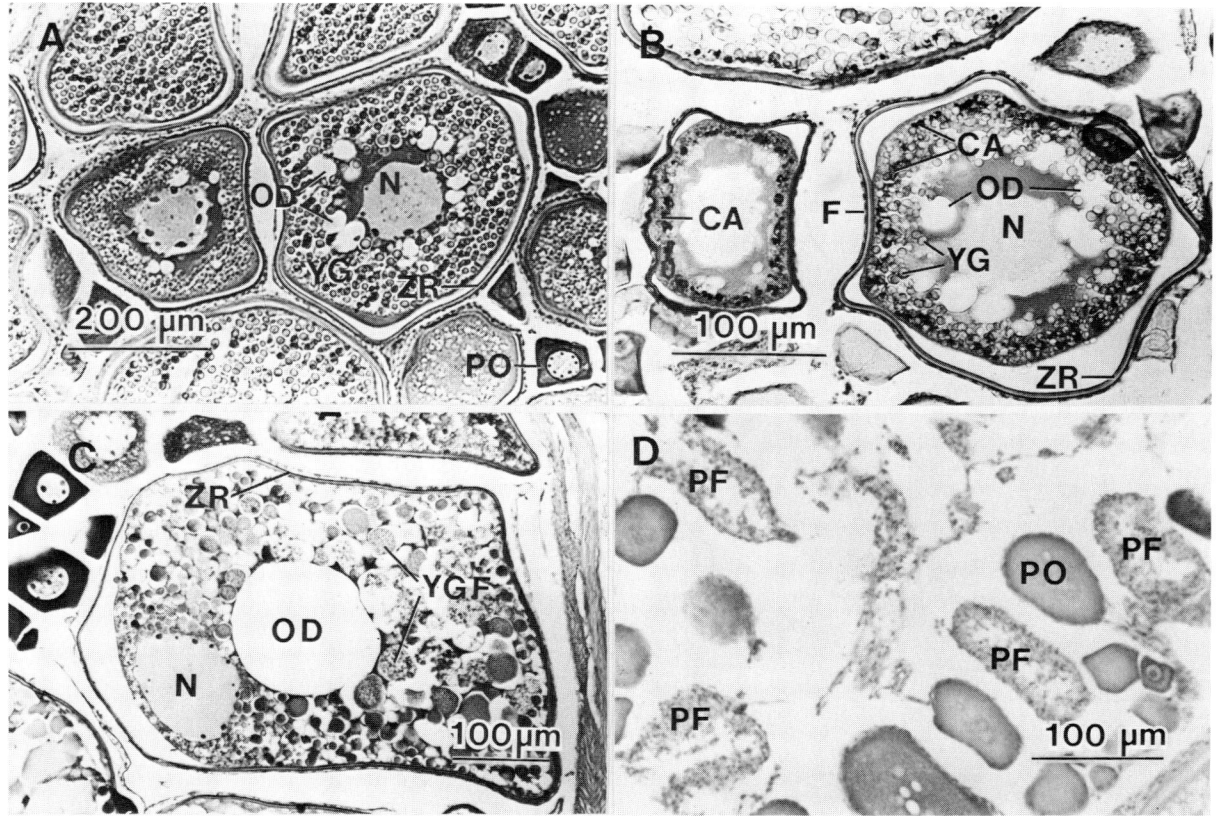

Fig. 4. Stages of oocyte development in *Chaetodon multicinctus:* A – primary growth and vitellogenesis (note lampbrush chromosomes and nucleoli in nuclei of oocytes); B – yolk vesicle stage (cell at left) and early vitellogenesis (cell at right) showing cortical alveoli (dark cytoplasmic inclusions); lighter eosinophilic bodies are yolk granules, PAS stain; C – maturation; D – postovulatory follicles from ovary of female (79 mm SL) collected in April 1982. All tissues except Fig. 4B stained with hematoxylin/eosin. CA = cortical alveoli; F = follicle; OD = oil droplet; N = nucleus; PF = postovulatory follicle; PO = primary oocyte; YG = yolk granule; YGF = yolk granule fusion; ZR = zona radiata.

of age) from field collections appeared as solid masses of degenerate cells of smaller size (Table 2) than atretic stage oocytes (described below). However, only fresh postovulatory follicles were used in analyses of spawning activity. Vitellogenic females injected with HCG exhibited postovulatory follicles (Fig. 5A, 5B) and produced clear buoyant eggs that could be fertilized. Recently formed postovulatory follicles were identified in gonad sections 36 h following injection of HCG.

Oocyte atresia

Not all fully developed oocytes were spawned. Some vitellogenic ova arrested development and showed distinct degenerative stages. Atresia of yolked oocytes in *C. multicinctus* was qualitatively similar to that described for other fishes (Bretschneider & Duyvene de Wit 1947, Lambert 1970, Khoo 1975, Saidapur 1978, Hunter & Macewicz 1985) although some differences exist.

Alpha atresia (Fig. 6A). – Alpha atresia was marked by convolutions of the zona radiata, nuclear disruption, early breakdown of yolk globules,

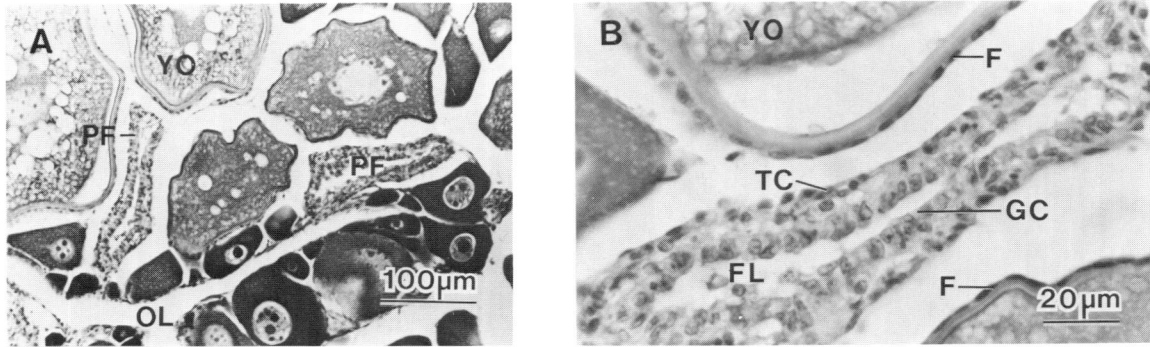

Fig. 5. Postovulatory follicles in ovary after induced ovulation in 94 mm SL *Chaetodon multicinctus:* A – photomicrograph showing two postovulatory follicles attached to lamellae wall; B – close up of follicle in Figure 5A. F = follicle; FL = follicle lumen; GC = granulosa cell layer; OL = ovarian lumen; PF = postovulatory follicle; TC = theca cell layer; YO = yolked oocyte.

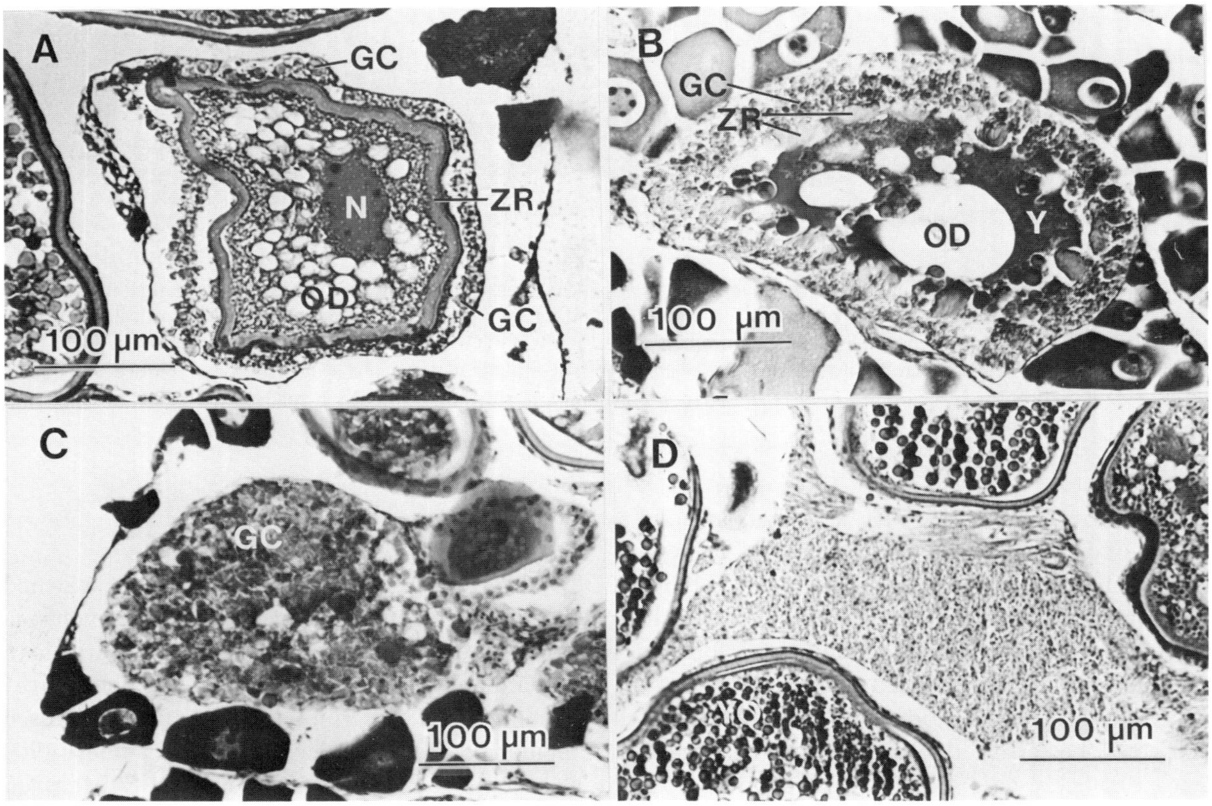

Fig. 6. Stages of atresia in yolked oocytes of *Chaetodon multicinctus:* A – alpha stage showing hypertrophy of follicle granulosa cells and coalescence of oil droplets; B – beta atresia indicated by disruption of the zona radiata; C – gamma stage in advanced resorption of yolk and cytoplasm by granulosa cells; D – delta atresia characterized by large brown body after full resorption of cytoplasm and yolk. GC = granulosa cells; N = nucleus; OD = oil droplets; Y = yolk; YO = yolked oocyte; ZR = zona radiata.

and hypertrophy of granulosa cells peripheral to the zona radiata. The end of alpha atresia was defined by disruption of the zona radiata following that described for the goldfish, *Carassius auratus* (Khoo 1975). This uncommonly observed stage apparently is of relatively short duration and signals the onset of subcellular reorganizations and breakdown prior to resorption.

Beta atresia (Fig. 6B). – During beta atresia, granulosa cells migrate into the cell to phagocytose yolk. Theca cells were not observed to invade the oocyte interior. The end of beta atresia was characterized by the full disintegration of the zona radiata (sensu Khoo 1975). Pronounced neovascularization associated with oocyte resorption often reported for other species (e.g. Bretschneider & Duyvene de Wit 1947, Lambert 1970, Hunter & Macewicz 1985, Saidapur 1978) was not observed.

Gamma atresia (Fig. 6C). – Resorption of oocyte contents continued during gamma atresia until completed. Small irregular shaped follicles (possibly lutein cells) were formed that stained orange-yellow in hematoxylin/eosin (Bretschneider & Duyvene de Wit 1947, Hunter & Macewicz 1985, Saidapur 1978). Theca cells of the follicle still surrounded the remnant oocyte.

Delta atresia (Fig. 6D). – The delta stage of atresia was identified by the residual cell components after resorption of yolk and cytoplasm. These oocytes contained light brown material under hematoxylin/eosin stain. No evidence for epsilon atresia, where cells of the corpus luteum differentiate into new oogonia cells (Khoo 1975), was found.

Seasonality of oocyte development

Gonadosomatic data show an increase in relative gonad weight during the spring months compared to that of the summer and fall (Fig. 7). The lowest monthly mean was 0.6% in September 1981 with an almost steady increase through May at 4.0%. This was followed by a dramatic decline in June to 1.5% which remained at about 1% for the rest of 1982. A

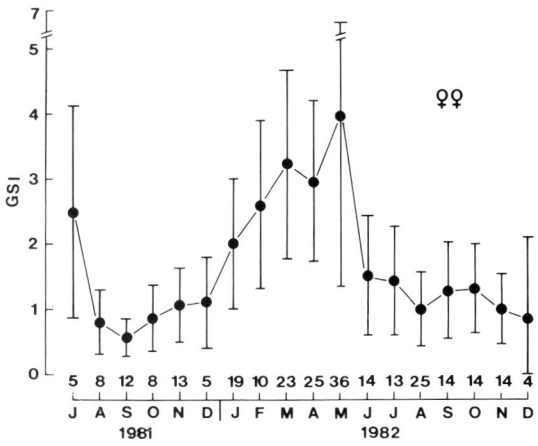

Fig. 7. Monthly gonadosomatic index (GSI) for female *Chaetodon multicinctus*. Vertical bars = 95% confidence limits. Numbers indicate monthly sample sizes.

similar drop is seen after July 1981. Visual inspection of these data suggest spawning occurred from January through May 1982, although the high variability of gonad weight among females within monthly samples precludes a clear statistical boundary for this period.

The uncertainties of spawning activity from the gonadosomatic data can be more directly examined by histological analyses of ovaries. The majority of females sampled in every month of the study contained yolked oocytes (minimum = 58%, August 1981). Females spawned through most of 1982 with the most active period in late winter and early spring (Fig. 8). The broad spawning activity observed in the first half of 1982 peaked during March and was followed by a steady decline through August. The proportion of females spawning during the months of January–July (42%) was greater than females sampled in August–December (15%) ($P<0.01$, Mann-Whitney U-test) which indicates more reproductive activity for the population in the former period. The 1982 spring peak was followed by increases in the proportion of females that showed no vitellogenic activity during July (15%) and August (24%), indicating subsidence of vitellogenic activity. Further evidence for a decline in spawning activity was seen in the cycle of oocyte atresia that peaked in May and June 1982 at over

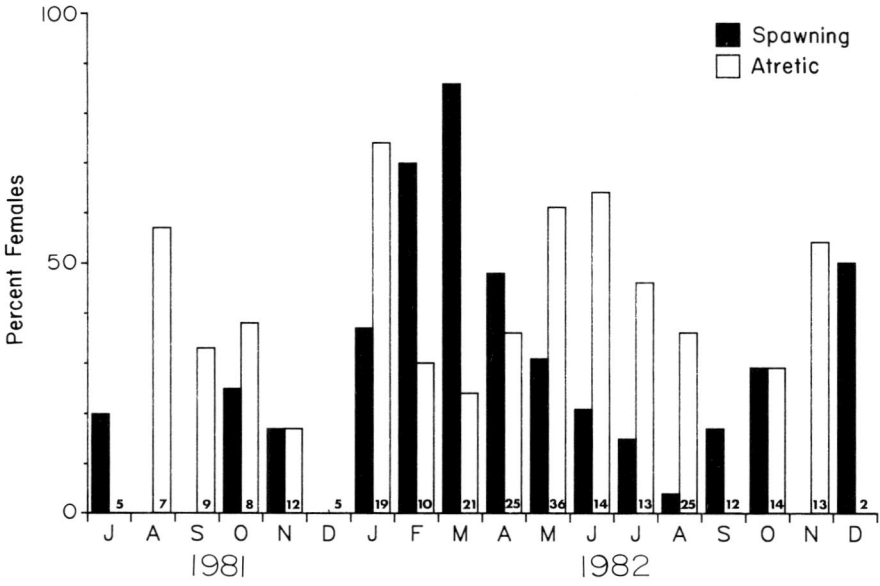

Fig. 8. Monthly summary of oocyte development and atresia in *Chaetodon multicinctus*. Ovaries with histological evidence of impending (oocyte maturation or hydration) or recent (postovulatory follicles) spawning activity indicated by dark bars. Atresia (alpha, gamma, or beta stages) indicated by open bars. Numbers indicate monthly sample sizes.

60% and lagged spawning peaks by 2–3 months (Fig. 8). Additional spawning is indicated during October–November 1981, and September–October 1982. Fall spawning differed from the spring cycle by its relatively short duration and low proportion of spawning females. Both bimonthly periods were followed by one month of spawning inactivity and indications of subsequent oocyte atresia.

Although atresia of yolked oocytes occurred seasonally, the percentage of atretic ova per female was highly variable over the study period and never reached 100% (range = 0 to 83%, n = 250). Advanced atresia (greater than 50% of oocytes in alpha, beta, or gamma atresia) is often used as an indicator of spawning cessation, but in *C. multicinctus* consistently occurred among only a low proportion of females (July–December 1981: $\bar{X}$ = 9.5%, SD = 8.6, January–December 1982; $\bar{X}$ = 2.3%, SD = 3.2). Further, atretic oocytes often occurred with postovulatory follicles within the same ovary. Thus, spawning may be reduced in frequency or intensity but not fully terminated during peak periods of oocyte atresia.

Discussion

Sexual differentiation

Juvenile *C. multicinctus* recruit to the reef at a small size and with undifferentiated gonads that may develop along one of three pathways recorded in this study. First, all ovaries differentiated from a structurally indifferent gonad. Similarly, the majority of differentiating testes also developed directly from previously undifferentiated tissue. Third, some testes developed within previously differentiated ovarian tissue. Unfortunately, we can not link the structural development of ovaries with sex determination especially since in many fishes all individuals pass through a female-like phase (Atz 1964, Yamamoto 1969, Takahaski 1977, Shimizu & Takahashi 1980, Matsuyama et al. 1988). However, since most males and females developed directly from previously undifferentiated tissue, the development of testes in structurally differentiated ovaries may present a case of prematurational sex change. Prematurational sex change was reported in the parrotfish *Sparisoma* (Robertson & Warner 1978) but is apparently uncommon among coral reef fishes. Juvenile intersexes are generally thought to be a passive phenomenon and their potential function within the mating system of a species is unknown. One possible advantage unrelated to adjustments in relative measures of reproductive effort between the sexes (Ghiselin 1969, Warner 1975) is the assurance of obtaining a mate under severe ecological conditions (sensu Liem 1968) or existing social constraints. New *C. multicinctus* recruits spend their first few months on the reef at a restricted site within reef interstices to avoid fish predators and vigorous aggression from conspecific adult pairs. At sizes as small as 50 mm SL, fish with structurally differentiated ovaries often form pairs and begin to establish small feeding areas in suboptimal habitats or between existing adult territories. Because a mate is essential for defense of a feeding area and suitable coral habitat is often limited (Tricas 1985, 1988), homosexual 'protopairing' of juvenile females may represent an adaptation to enhance the early establishment of successful feeding territories in a highly competitive environment. Under appropriate conditions the subsequent differentiation of one individual into a functional male may occur after pair formation. This phenomenon may function to ensure a heterosexual mate in a monogamous mating system where social contact between sexes is constrained by strong selection for early site attachment.

Oocyte development and atresia

Oocyte development in *C. multicinctus* follows general patterns described for other marine and freshwater teleosts (reviewed by Wallace & Selman 1981, Nagahama 1983) but also exhibits notable differences that are important when using histological criteria to assess spawning activity. For example, in primary growth centrally arranged nucleoli migrate to a peripheral distribution within the nucleus similar to that in the grey mullet, *Mugil cephalus* (Kuo et al. 1974). Unlike grey mullet, however, nucleoli in *C. multicinctus* persist in that arrangement throughout vitellogenesis and on into final maturation until nuclear degeneration (Fig. 3C). Early or late vitellogenesis can be identified by the thickness of the zona radiata and diameter of yolk elements (Table 2). The striations of the zona radiata present throughout vitellogenesis form from associations between follicle cells and oocyte microvilli for transport of protein yolk precursors into the oocyte (reviewed by Laale 1980, Ng & Idler 1983) and increase in width during vitellogenesis in *C. multicinctus*. Measurement of the zona radiata and yolk spheres can eliminate the need to measure maximum oocyte diameters to identify vitellogenic stages.

The dynamics of oocyte development that precede spawning in *C. multicinctus* occur rapidly. The onset of oocyte maturation and rare occurrence of hydrated oocytes in the morning collections indicate that final maturation begins approximately 6–8 h before dusk, and is consistent with the diel spawning periodicity of *C. multicinctus* (Lobel 1978). Final maturation also began in the morning for the related Caribbean pomacanthid angelfish, *Holacanthus tricolor*, and was completed in the afternoon for dusk spawning (Hourigan & Kelley 1985). The hydration and fusion of yolk globules increase egg transparency and buoyancy, characteristics that respectively reduce egg predation by planktivores and enhance offshore dispersal of zygotes.

The relatively uncommon observation of alpha atresia was due in part to early and rapid breakdown of the zona radiata. Beta and gamma atresia involve a longer process of cell resorption and therefore were more commonly observed. 'Brown bodies' of delta atresia occurred in >60% of ovaries and in all months which indicates a long-term persistence of that stage. For comparative purposes, alpha, beta, and gamma atresia in *C. multicinctus* can be pooled and related to the alpha stage described for some temperate species (e.g. Bretscheider & Duyvene de Wit 1947, Lambert 1970, Hunter & Macewicz 1985). Oocyte atresia in *Chaetodon* further differs from the above studies in that pronounced neovascularization was not observed during resorption. This may reflect the comparatively short duration and low intensity of oocyte resorption in a tropical species with a protracted spawning period in which existent vascularization (or only minimal neovascularization undetected by our histological preparations) is sufficient.

Spawning seasonality

Although gonadosomatic data indicate spawning activity during the first 5 months of 1982 and a peak in May, the high variability among females within monthly samples make statistical confirmation difficult. The insensitivity of the GSI is due to the asynchronous and multiple spawning of females within monthly samples of *C. multicinctus* (Tricas 1986) and perhaps in part to the change in relationship between ovary weight and body size with stage of development (de Vlaming et al. 1982). Furthermore, the GSI method alone can not provide information on possible low levels of spawning activity that may occur in the summer and fall months. In contrast, histological analyses of oocyte development indicate that *C. multicinctus* has a peak in March and a protracted annual spawning period. The occurrence of oocytes undergoing primary growth and vitellogenesis during all months of the year indicates a constant recruitment of oocytes among females and lack of a distinct end to spawning in the population.

Although the spring spawning peak was followed by an increase in atretic females, the low proportion of females in advanced atresia does not indicate a clear postspawning period for the population as in temperate fishes with more gross postspawning atresia (e.g. Crossland 1977). Furthermore, alpha or beta atresia often co-occurred in ovaries with maturing oocytes or fresh postovulatory follicles. Thus, atresia in tropical species with protracted spawning periods may better indicate a decline in reproductive output by individuals or termination of spawning among a portion of females rather than the end of the spawning for the entire population.

Unfortunately, the static sampling of females makes it impossible to assess the spawning activity of individuals over the year. For example, a second but less pronounced spawning peak may also exist during the fall. From the data it is difficult to conclude that the fall spawning activity is a separate and distinct cycle rather than variability in reproductive output in which some females begin spawning earlier in the fall than others. Ralston (1981) reported peak spawning from January–May (based upon a GSI analysis) for the planktivorous Hawaiian butterflyfish, *C. miliaris*, but found no gravid or spawning ovaries in the summer or fall. Similarly Lobel (1989) reported peak GSI for *C. multicinctus* at Kona Hawaii from May–June. Spring spawning appears to be the general pattern for Hawaiian reef species although multimodal annual spawning peaks of unequal amplitude were reported for other Hawaiian species (reviewed by Walsh 1987).

Although the histological staging of oocytes provided important indications of spawning activities, the best evidence of recent spawning activity were postovulatory follicles. Postovulatory follicles, however, degenerate rapidly and can be difficult to distinguish from the latter stages of normal oocyte atresia (Khoo 1975, Hunter & Macewicz 1985). Furthermore, degeneration rates are probably an increasing function of water temperature and their value as indicators of recent spawning is limited to collections made immediately following spawning. In the temperate northern anchovy, *Engraulis mordax*, identifiable postovulatory follicles persist up to 48 h after spawning (Hunter & Goldberg 1980). Takita et al. (1983) reported the presence of postovulatory follicles in ovaries of aquarium held *Callionymus enneactis* (a tropical species) up to about 15 h after spawning. Assuming that *C. multicinctus* ovulated the day following HCG-treatment, 24 h would be a conservative estimate of the persistence of postovulatory follicles in this species since ovaries were fixed approximately 36 h after injection. The presence of identifiable postovulatory follicles in field collections (all collected before 1200 h) supports this minimum estimate since *C. multicinctus* spawns at dusk (Lobel 1978).

The coexistence of oocytes in different developmental stages indicates that females spawn multiply over a reproductive season. This dynamic distribution of oocytes termed 'group synchronous' development (Wallace & Selman 1981) indicates the existence of multiple batches that are shed serially over time. A more detailed analysis of the size-frequency distribution of ova shows that *C. multicinctus* spawns, perhaps multiply, during weeks prior to new and full moon phases (Tricas 1986). Asynchronous spawning activity undoubtedly introduces variability in the monthly proportions of spawning females by multiple sampling in the lunar month. In the present study this was partially controlled by the regular 7-day collection interval which sampled each lunar quarter within each calendar month. Further application of histological techniques to the reproduction of tropical species could provide valuable information on the growth rates of oocytes to help determine time intervals between spawnings.

Acknowledgements

Thanks to J. Peck (Hawaiian Electric Company) and J. Bond for their assistance at depth with field collections, and S. Kraul (Waikiki Aquarium) for help with the induced-spawning experiments. S.R. Haley provided facilities, equipment, and guidance during the histological analyses. Discussions with R. Haley, T. Hourigan, J. Hunter, C. Kelley, B. Macewicz, E. Reese, R. Ross, and R. Warner improved this manuscript. This work was possible only through the laboratory assistance, support, and endorsement of Helen and Nicole Tricas, and Glenn Young. The paper is drawn from a dissertation submitted to the University of Hawaii by the senior author in partial fulfillment of the requirements for the Ph.D. degree in Zoology, and from a directed undergraduate research project by JTH.

References cited

Aketa, K. 1954. The chemical nature and the origin of the cortical alveoli in the egg of the medaka, *Oryzias latipes*. Embryologia 2: 63–66.

Atz, J.W. 1964. Intersexuality in fishes. pp. 145–232. *In*: C.N. Armstrong & A.J. Marshall (ed.) Intersexuality in Vertebrates Including Man, Academic Press, New York.

Bouain, A. & Y. Siau. 1983. Observations on the female reproductive cycle and fecundity of three species of groupers (*Epinephelus*) from the southeast Tunisian seashores. Mar. Biol. 7: 211–220.

Bretschneider, L.H. & J.J. Duyvene de Wit. 1947. Sexual endocrinology of non-mammalian vertebrates. Monogr. Prog. Res. Holland during the war, Vol. 2, Elsevier, Amsterdam.

Burgess, W.E. 1978. Butterflyfishes of the world. T.F.H. Publications, Neptune City. 832 pp.

Conover, W.J. 1971. Practical nonparametric statistics. Wiley, New York. 462 pp.

Crossland, J. 1977. Seasonal reproductive cycle of snapper *Chrysophrys auratus* (Forster) in the Hauraki Gulf. N. Z. J. Mar. Freshwater Res. 11: 37–60.

de Vlaming, V., G. Grossman & F. Chapman. 1982. On the use of the gonadosomatic index. Comp. Biochem. Physiol. 73A: 31–39.

Dipper, F.A. & R.S.V. Pullin. 1979. Gonochorism and sex inversion in British Labridae (Pisces). J. Zool. Lond. 187: 97–112.

Ghiselin, M.T. 1969. The evolution of hermaphroditism among animals. Quart. Rev. Biol. 44: 189–208.

Grier, H.J. 1981. Cellular organization of the testis and spermatogenesis in fishes. Amer. Zool. 21: 345–357.

Hourigan, T.F. & C.D. Kelley. 1985. Histology of the gonads and observations on the social behavior of the Caribbean angelfish *Holacanthus tricolor*. Mar. Biol. 88: 311–322.

Hunter, J.R. & S.R. Goldberg. 1980. Spawning incidence and batch fecundity in northern anchovy, *Engraulis mordax*. U. S. Fish. Bull. 77: 641–652.

Hunter, J.R. & B.J. Macewicz. 1985. Rates of atresia in the ovary of captive and wild northern anchovy, *Engraulis mordax*. U. S. Fish. Bull. 83: 119–136.

Khoo, K.H. 1975. The corpus luteum of goldfish (*Crassius auratus* L.) and its functions. Can. J. Zool. 53: 1306–1323.

Kuo, C., C.E. Nash & Z.H. Shehaded. 1974. A procedural guide to induce spawning in grey mullet. Aquaculture 3: 1–14.

Laale, H.W. 1980. The perivitelline space and egg envelopes of bony fishes: a review. Copeia 1980: 210–226.

Lambert, J.G.D. 1970. The ovary of the guppy, *Poecilia reticulata*. The atretic follicle, a corpus atreticum or a corpus luteum praeovulationis. Z. Zellforsch. 107: 54–67.

Liem, K.F. 1968. Geographical and taxonomic variation in the pattern of natural sex reversal in the teleost fish order Synbranchiformes. J. Zool. Lond. 156: 225–238.

Lobel, P.S. 1978. Diel, lunar, and seasonal periodicity in the reproductive behavior of the pomacanthid fish, *Centropyge potteri*, and some other reef fishes in Hawaii. Pac. Sci. 32: 193–207.

Lobel, P.S. 1989. Ocean current variability and the spawning season of Hawaiian reef fishes. Env. Biol. Fish. 24: 161–171.

Matsuyama, M., R.T. Lara & S. Matsuura. 1988. Juvenile bisexuality in the red sea bream, *Pagrus major*. Env. Biol. Fish. 21: 27–36.

Moyer, J.T. 1984. Reproductive behavior and social organization of the pomacanthid fish, *Genicanthus lamarck* at Mactan Island, Phillipines. Copeia 1984: 194–200.

Moyer, J.T. & A. Nakazono. 1978. Population structure, reproductive behavior, and protogynous hermaphroditism in the angelfish *Centropyge interruptus* at Miyake-jima, Japan. Jap. J. Ichthyol. 25: 25–39.

Munro, J.L., V.C. Gaut, R. Thompson & P.H. Reeson. 1973. The spawning seasons of Caribbean reef fishes. J. Fish Biol. 5: 69–84.

Nagahama, Y. 1983. The functional morphology of teleost gonads. pp. 223–275. *In:* W.S. Hoar, D.J. Randall & E.M. Donaldson (ed.) Fish Physiology, Vol. IXA, Academic Press, New York.

Neudecker, S. & P.S. Lobel. 1982. Mating systems of chaetodontid and pomacanthid fishes at St. Croix. Z. Tierpsychol. 59: 299–318.

Ng, T.B. & D.R. Idler. 1983. Yolk formation and differentiation in teleost fishes. pp. 373–404. *In:* W.S. Hoar, D.J. Randall & E.M. Donaldson (ed.) Fish Physiology, Vol. IXA, Academic Press, New York.

Nzioka, R.M. 1979. Observations on the spawning seasons of East African reef fishes. J. Fish Biol. 14: 329–342.

Ralston, S. 1976. Anomalous growth and reproductive patterns in populations of *Chaetodon miliaris* (Pisces, Chaetodontidae) from Kaneohe Bay, Oahu, Hawaiian Islands. Pac. Sci. 30: 395–403.

Ralston, S. 1981. Aspects of the reproductive biology and feeding ecology of *Chaetodon miliaris,* a Hawaiian endemic butterflyfish. Env. Biol. Fish. 6: 167–176.

Reese, E.S. 1975. A comparative field study of the social behavior and related ecology of reef fishes of the family Chaetodontidae. Z. Tierpsychol. 37: 37–61.

Reinboth, R. 1962. Morphologische und funktionelle Zweigeschlechtlichkeit bei marinen Teleostiern (Serranidae, Sparidae, Centracanthidae, Labridae). Zool. Jb. Abt. Allg. Physiol. 69: 405–480.

Robertson, D.R. 1983. On the spawning behavior and spawning cycles of eight surgeonfishes (Acanthuridae) from the Indo-Pacific. Env. Biol. Fish. 9: 193–223.

Robertson, D.R. & R.R. Warner. 1978. Sexual patterns in the labroid fishes of the western Caribbean, II: The parrotfishes (Scaridae). Smiths. Contrib. Zool. 255: 1–26.

Ross, R. 1984. Anatomical changes associated with sex reversal in the fish *Thalassoma dupperey* (Teleostei: Labridae). Copeia 1984: 245–248.

Saidapur, S.K. 1978. Follicular atresia in the ovaries of non-mammalian vertebrates. Inter. Rev. Cytol. 54: 225–244.

Shimizu, M. & H. Takahashi. 1980. Process of sex differentiation of the gonad and gonoduct of the three-spined stickleback, *Gasterosteus aculeatus* L. Bull. Fac. Fish. Hokkaido Univ. 31: 137–148. (in Japanese).

Takahashi, H. 1977. Juvenile hermaphroditism in the zebra fish, *Brachydanio rerio.* Bull. Fac. Fish. Hokkaido Univ. 28: 57–65.

Takita, T., T. Iwamoto, S. Kai & I. Sogabe. 1983. Maturation and spawning of the Dragonet, *Callionymus enneactis,* in an aquarium. Jap. J. Ichthyol. 30: 221–226.

Thresher, R.E. 1984. Reproduction in reef fishes. T. F. H. Publications, Neptune City. 399 pp.

Tricas, T.C. 1985. The economics of foraging in coral-feeding butterflyfishes of Hawaii. Proc. 5th Int. Symp. Coral Reefs. 5: 409–414.

Tricas, T.C. 1986. Life history, foraging ecology, and territorial behavior of the Hawaiian butterflyfish, *Chaetodon multicinctus.* Ph.D. Dissertation, University of Hawaii, Honolulu. 247 pp.

Tricas, T.C. 1988a. Determinants of feeding territory size in the corallivorous butterflyfish, *Chaetodon multicinctus.* Anim. Beh. (in press).

Tricas, T.C. 1988b. Prey selection by coral-feeding butterflyfishes: strategies to maximize the profit. Env. Biol. Fish. (in press).

Wallace, R.A. & K. Selman. 1981. Cellular and dynamic aspects of oocyte growth in teleosts. Amer. Zool. 21: 325–343.

Walsh, W.J. 1987. Patterns of recruitment and spawning in Hawaiian reef fishes. Env. Biol. Fish. 18: 257–276.

Warner, R.R. 1975. The adaptive significance of sequential hermaphroditism in animals. Amer. Nat. 109: 61–82.

Yamamoto, T. 1969. Sex differentiation. pp. 117–175. *In:* W.S. Hoar & D.J. Randall (ed.) Fish Physiology, Vol III, Academic Press, New York.

Spawning behavior of *Chaetodon multicinctus* (Chaetodontidae); pairs and intruders

Phillip S. Lobel
Woods Hole Oceanographic Institution, Woods Hole, MA 02543, U.S.A.

Received 25.3.1988   Accepted 12.8.1988

Key words: Butterflyfish, Fish social behavior, Mating strategy, Hawaii

Synopsis

Spawning by the banded butterflyfish, *Chaetodon multicinctus* (Chaetodontidae) was observed on coral reefs off Kona, Hawaii. These fish occurred in male-female pairs during normal daytime activities, a behavior which is typical for the family. Courtship is also a paired male-female activity. During spawning, however, other individuals (males?) may intrude on the spawning pair. Spawning typically takes place at least a meter or two above the bottom. The spawning position consists of the male below and behind the female with his snout against the female's ventral flank or anal fin area. Intruding individuals may join in when the pair is in position and about to spawn. Intruders line-up against the male in the same position as he is against his female. Underwater photographs are included to illustrate these behaviors.

Introduction

A long-term pair-bond is believed to exist in most species of chaetodontid fishes. The general pattern is that most individuals of many species are found in pairs (Reese 1975). This behavior has been a focus for discussions about the behavioral ecology of chaetodontids (Reese 1975, Neudecker & Lobel 1982, Driscoll & Driscoll 1988).

The spawning behaviors of several *Chaetodon* species, including *C. multicinctus*, have been described with the fishes in a particular spawning posture when positioned to release gametes (Lobel 1978, Susuki et al. 1980, Neudecker & Lobel 1982, Thresher 1984). Although potential spawning intruders have been seen nearby mating pairs in the field (Lobel 1978, Neudecker & Lobel 1982), intruders actually co-spawning with a pair have been seen only in a large aquarium (Susuki et al. 1980). Other aspects of the aquarium fishes behavior appeared normal for wild chaetodontids (Susuki et al. 1980). The question remained as to whether intruder co-spawning was an aquarium artifact or a natural behavior. This is a field report of spawning *Chaetodon multicinctus*, Garret 1863, and the occurrence of 'spawning intruders' in the wild.

Chaetodon multicinctus, is endemic to the Hawaiian Islands and Johnston Atoll (Randall et al. 1985). Reese (1975) classified it as a strongly paired species (see also Driscoll & Driscoll 1988). It is one of the most numerous of the chaetodontids on Kona reefs, feeding mostly on scleractinian corals during the day and inactive at night (Hobson 1974, Tricas 1985). The primary spawning period is during March to July with occasional spawning at other times (Lobel 1989, Tricas & Hiramoto 1989). Recruitment of *C. multicinctus* larvae from the open sea to the reef habitat is greatest during summer months (Walsh 1987).

Materials and methods

Study site and schedule

The dive site was Kamoa reef, situated about midway between Kailua-Kona Bay and Keahou Bay, Kona Coast of the Island of Hawaii. The reef ranged in depth from 8 to 20 m. Between September 1980 and October 1982, about 400 scuba dives were done by the author during all times of day and night; 52 of these dives during dusk (1730 to 1915 h).

Fish sex ratio

Collections were taken by spear from an adjacent reef. Specimens were caught prior to dusk throughout the year for gonadosomatic index measurement (see Lobel 1989). This provided some consistency in ovary status, wherein the eggs become hydrated a few hours before mating. Squash mounts of gonads were examined using a light microscope.

Results

Sex ratio

A total of 467 specimens were collected for study of the fish's spawning season (Lobel 1989). The sex ratio of mature fish collected in pairs was 1:1 (N = 400); 14% (N = 67) of individuals collected were immature and not sexually identifiable using a light microscope.

Chaetodon multicinctus was found in pairs during the daytime throughout the year. Selected pairs of fish (N = 36 pairs) were carefully watched for about 10 min to determine if the individuals maintained a pair-bond or were just temporarily associated. This was a reasonable observation duration to determine pairing given the behavior of this species (see Reese 1975, Driscoll & Driscoll 1988). The pair was collected after observation. Pairs were observed during November (N = 8), December (N = 5), March (N = 7) and May (N = 18). Pairs were composed of a male and a female in all cases where sex could be determined (N = 24 pairs). Twelve pairs of *C. multicinctus* possessed sexually immature gonads. The mean size ($\pm$ SD) of sexually mature paired fish was 67.5 $\pm$ 4.4 mm SL (N = 48 individuals). The mean size ($\pm$ SD) of the paired immature fish was 50.4 $\pm$ 6.7 mm SL (N = 24 individuals).

Courtship and spawning behavior

Mating behavior of *C. multicinctus* was observed in the evening period (1730 to 1915 h) during 7 of a total of 52 dives in the same reef area. Thirteen spawning events were seen on 7 dates (Table 1). Courtship activity was seen once during the months of October, November, December and the first spawning for a year was seen 5 January. The greatest degree of courtship activity and all other spawning occurred from March until mid July.

Courtship activity became apparent about 45 to 30 min before spawning occurred. In the early stages, a pair of fish swam along a track of reef at moderate speed with the female followed closely by the male. The male would periodically rush upon the female, approaching her from behind and placing his snout to her anal fin area. This behavior progressed and later appeared as false starts before actual spawning. If intruding fish were present at this time, they followed several body lengths behind the pair (see photo in Lobel 1978). Courtship usually involved a single male and a female (N = 14), but on five occasions several other *C. multicinctus* were seen closely following a mating pair. On one occasion a group of 8 intruders were seen a few minutes earlier as members of a group of about 20 *C. multicinctus* which was aggregated in one section of the reef. These intrusions took place once in January, twice in March and twice in May. Further description of the courtship and spawning behavior of *C. multicinctus* is in Lobel (1978).

Release of gametes occurs with the fish assuming a stereotyped spawning position. In this position, the fish are a meter or more above the bottom, the male is below and behind the female with his snout against her anal fin area (Fig. 1). Once a pair is in spawning position, intruders may rush in line. Intruders follow behind the first male and mimic the spawning position of the pair (Fig. 2). A pair was frequently disturbed by the intruders and sometimes broke away. In such cases, the pair swam to a nearby location and were followed by the intruders. Although I cannot be absolutely certain, it appeared in most cases that the intruders released gametes at the same instant the leading pair spawned. At the time of gamete release the fish made a quick sharp body movement and then darted back to the bottom.

Time of spawning

The time of spawning relative to sunset was calculated for spawning events consisting of pairs alone (N = 7) and pairs plus intruders (N = 5). In all cases, spawning occured after the time of sunset and before nautical twilight (Table 1). The limited data obtained to date suggests that intruders are more likely to spawn with a pair earlier in the evening. The mean (± SD) time of spawning by pairs plus intruders was 24 ± 6 (range 15 to 29) min after sunset. The mean time of spawning of pairs alone was 34 ± 4 (range 29 to 39) min after sunset.

Discussion

The behavior of intruder fish rushing upon a breeding pair and releasing gametes simultaneously is not unusual among fishes. The behavior has been described as 'streaking' among labroid fishes and as 'sneakers' among salmonid and centrarchid fishes. Labroid fishes (Labridae and Scaridae) are broadcast spawning species. Pair spawning occurs between a terminal phase male and a single female and as the spawning pair peaks in its ascent to spawn, initial phase males rush in (i.e., 'streaking') and release sperm at the same instant (Warner et al. 1975, Robertson & Warner 1978, Warner & Robertson 1978, Warner & Hoffman 1980). Pair and group spawning has also been seen among acanthurid species (Robertson 1983). Salmonid and centrarchid freshwater fishes lay eggs in nests on cleared space on the bottom. Intruder or 'sneaker' males rush upon the spawners and release sperm simultaneously with the spawning pair (Keenleyside 1979, Gross 1984). Fertilization stealing has also been described for several other groups of freshwater fishes. The possible evolutionary implications of this spawning strategy have been recently reviewed by Gross (1984).

The pair spawning position described for *C. multicinctus* (Lobel 1978) has since been referenced for its similarity to the pair spawning position in other chaetodontids (Susuki et al. 1980, Neudecker &

Table 1. Dates and times of spawning *Chaetodon multicinctus*.

Date	Pair only	Pair & intruder(s)	Time of spawning	Sunset	Nautical twilight
1. 3 Mar 81	x		1835	1806	1853
2. 10 Apr 81	x		1852	1817	1906
3. 10 Apr 81	x		1855		
4. 10 Apr 81	x		1856		
5. 17 Jul 81	x		not recorded		
6. 5 Jan 82		x + 1	1750	1735	1826
7. 6 Mar 82		x + 2	1828	1807	1854
8. 21 Mar 82		x + 3	1840	1811	1859
9. 5 May 82		x + 8	1850	1825	1915
10. 5 May 82	x		1853		
11. 5 May 82		x + 2	1853		
12. 5 May 82	x		1859		
13. 5 May 82	x		1900		

Fig. 1. A spawning pair of *Chaetodon multicinctus,* a moment before gamete release (1835 h, 3 March 1981). The male approaches the female from below and places his snout against her anal fin area immediately prior to spawning. Notice the swollen belly of the female bearing eggs.

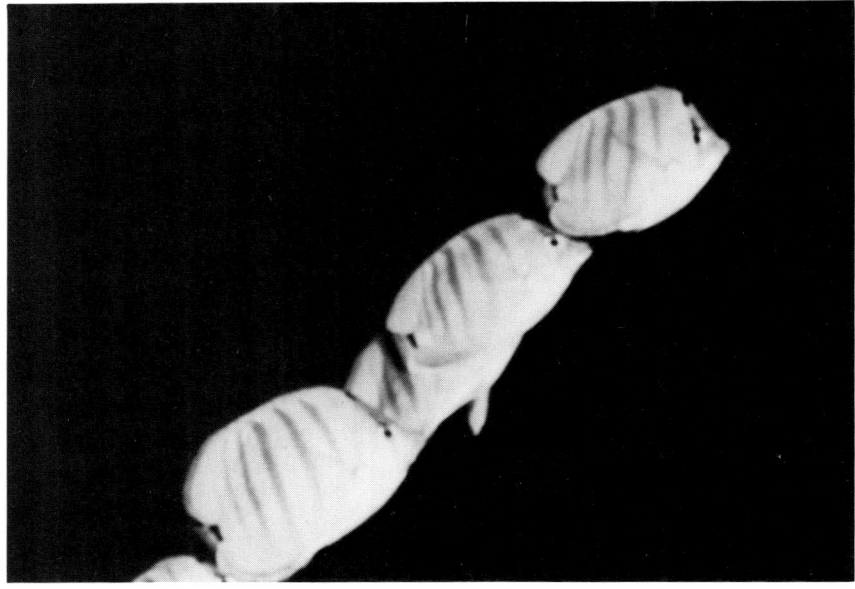

Fig. 2. A spawning pair of *Chaetodon multicinctus* accompanied by 3 intruders (1840 h, 21 March 1982). The intruders rushed into line behind the primary male at the time when the original pair were about to spawn.

Lobel 1982, Thresher 1984). The exception to the pair spawning behavior was seen among *C. nippon* in an aquarium by Susuki et al. (1980). They described co-spawning by intruding males. These males joined a pair about to spawn. The intruder males released sperm simultaneously when the leading pair spawned (Susuki et al. 1980). The same intruder behavior has now been observed among *C. multicinctus* in the field (this study). The intruder fish I saw were also possibly males; females would have been noticeably swollen with eggs and females have not been observed assuming a male's position during courtship. The limited observations reported herein indicate that intruder co-spawning was more common earlier than later during the dusk period.

These observations suggest an added complexity to the social relationships of chaetodontids on coral reefs. The question is, are these spawning intruders unattached marauding males or are they males who have temporarily abandoned their mates? At this point, both possibilities are plausible. First, every individual *C. multicinctus* on a reef is not necessarily associated in a pair; $83 \pm 10\%$ of individuals on a reef were confirmed in pair association leaving about 17% unattended (Reese 1975). Second, sexual selection theory can argue for mate desertion after spawning planktonic offspring (Trivers 1972). So far, female chaetodontids have been observed to spawn only once per evening (Lobel 1978, Susuki et al. 1980, Neudecker & Lobel 1982, Thresher 1984). It may be that male chaetodontids could benefit by temporarily deserting their female mates after spawning to seek additional matings. One constraint on this behavior is the temporal synchronization of spawning throughout a local population. There is a narrow time window which closes shortly after sunset during which a male is required to court and mate his partner, desert her and then locate and co-spawn with another pair. This time constraint may be one factor favoring pair-bonding among fish with spawning restricted to the dusk crepuscular period (Neudecker & Lobel 1982).

Acknowledgements

This research was supported by the NSF (OCE-8009554 and OCE-8117891) while the author was a postdoctoral fellow in Oceanography, Center for Earth and Planetary Physics, Harvard University. Further support provided by grants from the U.S. Department of Energy's program for O.T.E.C. (DE-A503-83CE89302), administered by the Marine Science Group, University of California at Berkely, the U.S. Army (DACA83-83-C0049) and Sea Grant (NA86AA-D-SG090 WHOI No. R/A-26-PD). Contribution No. 6774 from the Woods Hole Oceanographic Institution.

References

Driscoll, J.W. & J.L. Driscoll. 1988. Pair behavior and spacing in butterflyfishes (Chaetodontidae). Env. Biol. Fish. 22: 29–37.

Gross, M.R. 1984. Sunfish, salmon, and the evolution of alternative reproductive strategies and tactics in fishes. pp. 55–76. *In:* G.W. Potts & R.J. Wootton (ed.) Fish Reproduction Strategies and Tactics, Academic Press, New York.

Hobson, E.S. 1974. Feeding relationships of teleostean fishes on coral reefs in Kona, Hawaii. U.S. Fish. Bull. 72: 915–1031.

Keenleyside, M.H.A. 1979. Diversity and adaptation in fish behavior. Springer–Verlag, New York. 208 pp.

Lobel, P.S. 1978. Diel, lunar and seasonal periodicity in the reproductive behavior of the pomacanthid fish, *Centropyge potteri,* and some other reef fishes in Hawaii. Pac. Sci. 32: 193–207.

Lobel, P.S. 1989. Ocean current variability and the spawning season of Hawaiian reef fishes. Env. Biol. Fish. 24: 161–171.

Neudecker, S. & P.S. Lobel. 1982. Mating systems of chaetodontid and pomacanthid fishes at St. Croix. Z. Tierpsychol. 59: 299–318.

Randall, J.E., P.S. Lobel & E.H. Chave. 1985. Annotated checklist of the fishes of Johnston Island. Pac. Sci. 39: 24–80.

Reese, E.S. 1975. A comparative field study of the social behavior and related ecology of reef fishes of the family Chaetodontidae. Z. Tierpsychol. 37: 37–61.

Robertson, D.R. 1983. On the spawning behavior and spawning cycles of eight surgeonfishes (Acanthuridae) from the Indo-Pacific. Env. Biol. Fish. 9: 193–223.

Robertson, D.R. & R.R. Warner. 1978. Sexual patterns of the labroid fishes of the western Caribbean, II. The parrotfishes (Scaridae). Smithsonian Contrib. Zool. 255: 1–26.

Susuki, K., Y. Tanaka & S. Hioki. 1980. Spawning behavior, eggs and larvae of the butterflyfish, *Chaetodon nippon* in an aquarium. Jap. J. Ichthyol. 26: 334–341.

Thresher, R.E. 1984. Reproduction in reef fishes. T.F.H. Publ., Neptune City. 399 pp.

Tricas, T.C. 1985. The economics of foraging in coral-feeding butterflyfishes of Hawaii. Proc. Fifth Internat. Coral Reef. Symp. 5: 409–414.

Tricas, T.C. & J.T. Hiramoto. 1989. Sexual differentiation, gonad development, and spawning seasonality of the Hawaiian butterflyfish, *Chaetodon multicinctus*. Env. Biol. Fish. 25: 111–124.

Trivers, R.L. 1972. Parental investment and sexual selection, pp. 136–179. *In:* B. Capbell (ed.) Sexual Selection and the Descent of Man 1871–1971, Aldine, Chicago.

Walsh, W.J. 1987. Patterns of recruitment and spawning in Hawaiian reef fishes. Env. Biol. Fish. 18: 257–276.

Warner, R.R. & D.R. Robertson. 1978. Sexual patterns in the labroid fishes of the western Caribbean, I. The wrasses (Labridae). Smithsonian Contrib. Zool. 254: 1–27.

Warner, R.R. & S. Hoffman. 1980. Local population size as a determinant of mating system and sexual composition in two tropical marine fishes (*Thalassoma* spp.). Evolution 34: 508–518.

Warner, R.R., D.R. Robertson & E.G. Leigh, Jr. 1975. Sex change and sexual selection. Science 190: 633–638.

Aspects of the spawning of western Atlantic butterflyfishes (Pisces: Chaetodontidae)

Patrick L. Colin
Caribbean Marine Research Center, 100 East 17th Street, Riviera Beach, FL 33404, U.S.A.
Mailing address: CMRC, c/o Florida State University Marine Laboratory, Rt. 1, Box 456, Sopchoppy, FL 32358, U.S.A.

Received 5.5.1988   Accepted 9.11.1988

Key words: Eggs, Lunar periodicity, Seasonality, Predation, Coral reefs, Aggression, Western Atlantic

Synopsis

The status of knowledge of spawning among the five shallow water *Chaetodon* species in the western Atlantic is reviewed. Spawning has been observed for three species in Puerto Rico, St. Croix and the Bahamas, with possible courtship in a fourth. *Chaetodon aculeatus* spawned near the time of sunset over objects on the reef as single female/male pairs or as two females and one male, with pair spawning in rapid succession. Spawning occurred during much of the lunar month from February to April and it is uncertain whether any lunar periodicity to spawning exists. Male-male aggression was noted. Spawning sites (coral heads) were alternated daily and it is likely that females spawn only once every two days. A single female produced as many as 2090 eggs in a single spawning. *Chaetodon capistratus* spawned during much of the lunar month from February to April. It spawned about 5 min after *C. aculeatus*, occasionally using the same sites, and alternated sites daily. A female produced as many as 3710 eggs in one spawning. *Chaetodon striatus* spawned from February to April but it is unknown if it has any lunar spawning cycle. No predation attempts by piscivores on spawning adults were seen. Predation by *Melichthys niger* on eggs of *C. striatus* occurred. No egg predation was observed for *C. aculeatus* and *C. capistratus*. With an assumed four month reproductive season, alternate day spawning and observed egg production values, *C. aculeatus* and *C. capistratus* produce respectively about 100 000 and 200 000 eggs per large female per year. The reproductive strategy of smaller species may be to produce moderate numbers of eggs per day over a spawning season of at least a few months while larger species may produce more eggs per day for a shorter period.

Introduction

Despite their diversity and conspicuous presence on most coral reefs, the reproductive biology of the butterflyfishes (Chaetodontidae) is poorly known. Thresher (1984) has summarized recent information on the family. There are five shallow water species of *Chaetodon* in the western Atlantic and aspects of their reproduction discussed in this paper are summarized in Table 1. Aiken (1983) provided data on gonadal condition of various *Chaetodon* spp. near Jamaica, based on trap collections, but did not observe actual spawning. Neudecker & Lobel (1982) described spawning of *Chaetodon aculeatus* and *C. capistratus* while Colin & Clavijo (1988) described spawning in the two preceeding species as well as *C. striatus*. This paper reports additional information concerning spawning of western Atlantic species and attempts to discuss spawning strategies of chaetodontids in comparison with other reef fishes.

Materials and methods

Observations and collections were made in Puerto Rico, St. Croix, U.S. Virgin Islands and the Bahamas. Puerto Rican observations were made on a shelf edge reef at 18 m depth off southwestern Puerto Rico in an area described in detail by Colin & Clavijo (1988). Observations in St. Croix were made from the Hydro-Lab habitat on the east wall of the Salt River submarine canyon at 18 m depth. In the Bahamas work was carried out from the Caribbean Marine Research Center on Lee Stocking Island, Exumas, on a reef ledge at 12–18 m depth facing Exuma Sound.

Table 1. Spawning and courtship occurrences of shallow water western Atlantic species of Chaetodon. FM = full moon, AFM = after full moon, BFM = before full moon, NM = new moon.

Species	Location	Months	Lunar phase	Temp. (°C)	Source
C. aculeatus	Puerto Rico	April 78', 79'	FM-8AFM	25.5	Colin & Clavijo (1988)
	St. Croix	Sept. 1978	4 AFM (probable spawn)	28–29	pers. obs.
	St. Croix	Feb. 1980	1–3 BFM		Neudecker & Lobel (1982)
	St. Croix	Oct. 1980	11 BFM (courtship only)		I.E. Clavijo, pers. comm.
	Bahamas	Feb. 1988	3–4 AFM	24.3	pers. obs.
	Bahamas	April 1988	1 AFM–14 AFM (NM)	23.5–24.2	pers. obs.
	Bahamas	May 1988	11 AFM, 15 BFM	25.0	pers. obs.
C. capistratus	Puerto Rico	April 1979	5 BFM	25	Colin & Clavijo (1988)
	St. Croix	Feb. 1980	9, 8, 4, 1 BFM		Neudecker & Lobel (1982)
	Jamaica	much of year	(gonads)		Aiken (1983)
	Bahamas	April 1988	2 AFM–14 AFM (NM)	23.5–24.2	pers. obs.
	Bahamas	May 1988	11 AFM, 15 BFM	25.0	pers. obs.
C. ocellatus	Jamaica	Jan, May (ND March, Apr., June), some other months.			Aiken (1983)
	Bahamas	May 1988 (ripe)	12 AFM	25.5	pers. obs.
	N. Carolina	May 1978 (ripe)	1 week BFM, FM	18–20	S.W. Ross (pers. comm.)
C. striatus	Puerto Rico	Feb–March 1979	9–11 BFM; 3 AFM	24–25	Colin & Clavijo (1988)
		Feb. 1988 (courtship only)	10–11 AFM	26.7	pers. obs.
	Jamaica	Jan. Feb. peak, some active all months			Aiken (1983)
	Bahamas	May 1988 (ripe)	13 AFM	25.5	pers. obs.
C. sedentarius	Puerto Rico	Mar. Apr, June	1–5 BFM (courtship)		
	Jamaica	Jan., May, Sept. (a few ripe)			Aiken (1983)
	N. Carolina	May 1978 (ripe)	1 week BFM, FM	18–20	S.W. Ross (pres. comm.)

Observations were made while SCUBA diving. Spawning behavior was recorded using an 8 mm video camera. Temperatures were measured using calibrated recording thermographs in Puerto Rico and the Bahamas. Time of events was determined to 2 min accuracy using calibrated watches.

Eggs were collected using a short handled dip net 15 cm in diameter with 210 micron mesh. This was used to strain the water in the area of egg release within 5–10 seconds of release. The net with captured eggs was then everted inside a plastic bag to retain the eggs. Eggs were returned to the laboratory in the plastic bags, filtered from the water using 250 micron mesh and preserved in alcohol. Once in alcohol, the eggs became opaque and were easily counted in a petri dish.

Results

Accounts of individual species

Chaetodon aculeatus. The longsnout butterflyfish is the chaetodontid most often seen to spawn. Colin & Clavijo (1988) observed spawning in Puerto Rico in April 1978 and 1979 while Neudecker & Lobel (1982) saw it from 13 February to 1 March 1980 in St. Croix. More recently, I observed this species spawning in the Bahamas during February, April and May 1988 (n = 34 spawns). Spawnings were observed 1 and 3 days before the new moon (Neudecker & Lobel 1982) and daily from the day of the full moon to at least the day after the new moon (Colin & Clavijo 1988, personal observation). Neudecker & Lobel (1982) reported spawning by *C. aculeatus* not to occur between 9 and 11 days before the full moon in February and 2 days after the full moon in early March. No attempt has yet been made to determine if individuals are spawning during the entire period from after the new moon until near the time of the full moon, but this should be done before any firm statements can be made regarding lunar periodicity of its spawning.

Spawning was not observed on three evenings in July 1988 at Lee Stocking Island on phases of the moon and at sites where it had been observed in April and May. There may be some limited spawning occurring during the summer or fall periods. I observed one probable spawning by *C. aculeatus* in St. Croix in September 1978 and I.E. Clavijo (personal communication) also saw two instances of courtship without apparent spawning there in October 1980.

Spawning behavior also appears to be variable. Neudecker & Lobel (1982) reported an ascent of less than one meter with the male underneath the female and both fish quivering for about one sec while releasing gametes. My observations indicate that pairs of *C. aculeatus* often circle several times with the fish oriented laterally, the snout of the male touching the caudal fin of the female before he moves slightly underneath her for gamete release. The pair did not always move upward with the male pushing the female at gamete release, but rather the male orients behind the female with his snout touching her caudal fin. They circle laterally with the males' snout continuing to touch the female until gamete release. During one spawning the pair 'made a short slow "dash" into the water column to release gametes' (A. Gronell in Thresher 1984, personal observation). In the spawnings I observed there was considerable variation in vertical ascent, from less than 10 cm to as much as 50–60 cm, with much of the actual rise coming during the circling phase prior to gamete release. At gamete release the female is stationary, angled upwards with the male below her with his snout near her vent area. He moves forward while the eggs are released, leaving the female in the water column. There were no false starts to spawning as occurs in some other reef fishes (Colin & Clavijo 1988). In all cases where spawning was positively observed, the female was visibly swollen with eggs before spawning.

Chaetodon aculeatus typically spawned near or above objects on the bottom. One pair in Puerto Rico spawned above a large sponge *Xestospongia muta*. In the Bahamas it spawned along a steep rocky reef face which extended from 12 m depth to a sandy plain at 18 m. This sand extended offshore to the shelf break another 300–400 m offshore, so the actual spawning site was well inside the edge of the insular shelf. Small coral heads, ranging from only about 50 cm to 1.5 m in height, projected up from both the edge and upper surface of the ledge. Fish spawned both along the edge of the face and over heads located 10–15 m inside the reef edge at depths of 10–12 m.

Spawning occurred just before or at the time of sunset, generally about 5 minutes before *Chaetodon capistratus* spawned. Fish spawned once per day, as both male-female pairs and in several instance as two females spawning consecutively with one male.

Without exception, the spawning sites at the Lee Stocking Island study area were utilized only on alternate days. Whether individual fish were utilizing different sites on alterate days or if spawning by individuals took place only every other day could not be positively determined, but based on anecdotal evidence, I feel it is likely alternate day spawning by an individual females is probably the case. There were four coral heads, all within 15 m of each other, that were utilized for spawning and each was monitored continuously during the period when spawning would occur from the day of the full moon to the new moon in April 1988. In addition, spawning sites nearby were monitored intermitently. Fish appeared at the spawning site about 10–15 minutes prior to spawning and remained within a few meters of that site until after spawning had occurred. The spawning sites from the previous night were not utilized by any fish or visited during the half hour prior to spawning.

One coral mound was utilized only by a trio of fish, a single male and two females, with one female considerably smaller than the other (estimated standard length 45 mm vs. 55 mm). This trio was observed spawning on seven occasions, on a two day cycle, and unless disturbed by egg collection, spawnings of the two females occurred within one minute. On 4 days the large female spawned first and on 3 days the small one was first. There was no aggressive interaction between the two females. The small female, the only individual which confidently could be identified, was never seen to spawn elsewhere on nights she did not spawn amongst the trio.

In a nearby area, about 40 m from the primary observation sites, three *C. aculeatus* (presumed to be 2 males and 1 female) were found in which one male fought for a prolonged period with the second, who was with an associated female. Within the hour before sunset the presumed intruding male attempted to court the female while the second male kept trying to drive him from the vicinity. On the first day observed, the female was visibly swollen with eggs and fighting between the two males was fierce. The male with the female repeatedly attempted to drive the intruding male away from the vicinity of the female, and occasionally the female also directed an attack at the intruding male. More aggressive actions were used by the male when the intruder did not retreat. This involved raising the dorsal spines, angling the body forward to orient the spines toward the facing male and charging forward. The two males often ended up spinning, their dorsal fins meeting while both in this posture, until one fish was spined and broke the exchange off. This aggressive behavior was observed for over 30 min with intervals where the males faced one another about 1 m apart without attacking. Both males were battered with wounds to the body and dorsal fin membranes, with the intruder appearing worse. The next day, a presumed non-spawning day for the female, both males were present in the same area, but beyond weak attempts to drive one another away, there was little aggression between them. The following day, 2 days after the initial fierce encounter, the female was again swollen with eggs (supporting alternate day spawning by individuals) and the males fighting as aggressively as two days before. On this evening the defending male was seen to successfully mate with the female.

The eggs were 0.74–0.76 mm in diameter, clear, with a single oil globule, 0.16 mm diameter. The embryos hatched between 26 and 36 h after spawning at 25°C.

Chaetodon capistratus. Spawning by the foureye butterflyfish has been reported previously by Neudecker & Lobel (1982) in St. Croix in February 1980 from 9–4 days before the full moon and by Colin & Clavijo (1988) in Puerto Rico in April 1979, 5 days before the full moon. Additional spawnings have been seen in the Bahamas in April and May 1988 from the day after the full moon until the day after the new moon (n = 14 spawns). Spawning was not observed at the Lee Stocking Island site in July 1988 at locations and on lunar phases where spawning had been seen in April and May.

There are differences reported in spawning behavior of *C. capistratus*. Neudecker & Lobel(1982) reported it to forage in monogamous pairs and spawn in those pairs only. Colin & Clavijo (1988) found that in Puerto Rico it occurs in social groups of as many as 15 individuals with no consistent pairing, indicating some flexibility in social activity and spawning. A social system similar to that observed by Colin & Clavijo (1988) was reported by Gore (1982) from Jamaica and Grand Cayman island. These differences in social system in different areas could be due to density of adults on the reef which cause the pairing of fish to break down. At Lee Stocking Island the social system of *C. capistratus* was similar to that described by Neudecker & Lobel (1982) in St. Croix with pairs foraging together before spawning, and spawning occurring only as isolated pairs.

There was little courtship prior to spawning. Like *C. aculeatus*, at gamete release the female remained nearly stationary, while the male moved forward and downward, possibly serving to mix the eggs and sperm. Close observation of the eggs on release indicate they emerge as a dense stream several centimeters in length, rather than as a single burst or cloud. They quickly begin to diffuse away from the center of the stream so that within about 10 sec the dense stream of eggs is no longer identifiable. Within a few seconds more the clear eggs are nearly invisible to a human observer.

Spawning by *C. capistratus* occasionally occurred at the same sites as *C. aculeatus*, but unless disturbed by an observer, took place about 5 minutus after the latter. At the same sites, *C. capistratus* spawned slightly higher (50–100 cm above objects) than *C. aculeatus*. Like *C. aculeatus*, this species alternated spawning sites each evening and it believed this is due to females spawning only on alternate evenings. Just before the time of spawning some pairs of *C. capistratus* were found in the study area in which neither fish was visibly swollen with eggs. These occurred almost alongside pairs in which one fish was noticeably swollen with eggs and adds some support to the argument that during the spawning season females spawn only every other night. In all cases in which pairs spawned, the female was easily identified by her egg-swollen condition. Despite possible alternate-night spawning by individual females, there were different pairs of *C. capistratus* spawning each night. The eggs were 0.76–0.77 mm in diameter, clear, with a single oil globule 0.18 mm diameter.

Chaetodon ocellatus. The spawning of the spotfin butterflyfish has not been observed. Robins et al. (1986) reported coloration differences between sexes with males having a dark spot at the posterior edge of the dorsal fin, but other works (Allen 1979, Burgess 1978) do not report any sexual dichromatism. In most mature pairs I have seen, assumed to be male and female, plus actual male/female pairs speared and sexed, both individuals possess the black spot on the dorsal fin edge.

During April 1988, when I was regularly observing spawning by *C. aculeatus* and *C. capistratus*, there were several *C. ocellatus* present in the Lee Stocking Island study area and no courtship behavior or visibly swollen females were seen. I am fairly certain *C. ocellatus* was not spawning during this period from the full moon to new moon. At south Cat Island, Bahamas a slightly swollen female *C. ocellatus* was observed on 13 May 1988, two days before the new moon. This fish, 113 mm SL, was collected (along with a 116 mm SL male) at sunset and approximately 19 000 partially hydrated eggs were hand stripped with light pressure from the fish. Certainly this pair was near the time of spawning, but the partially hydrated state of the eggs indicates that spawning was still some hours away. It is possible that spawning may occur at dawn due to the incomplete hydration of these eggs at sunset.

S.W. Ross (personal communication) found gonad development indicating imminent spawning among *C. ocellatus* collected at 38–62 m depth in Onslow Bay, North Carolina during the week before the full moon in May 1978. He found females 111–137 mm SL had ovaries comprising 1.1–8.1% of body weight. Moe (1976) reported the eggs of *C. ocellatus* to be 0.6–0.7 mm in diameter based on eggs taken in the laboratory.

Chaetodon sedentarius. A few instances of possible courtship by a large pair of the reef butterflyfish were seen in Puerto Rico. One fish of probable male/female pairs chased the other, placing its snout on the posterior margin of the anal fin. These instances occurred 1–5 days before the new moon in March, April and June (Colin & Clavijo 1988). S.W. Ross (personal communication) collected female *C. sedentarius* at 38–62 m depth off Onslow Bay, North Carolina with ovaries comprising 3.1–8.3% of body weight during the week before the full moon in May 1978.

Chaetodon striatus. The courtship and spawning of the banded butterflyfish was observed on three occasions on a shelf edge coral reef off southwestern Puerto Rico (Colin & Clavijo 1988). There was no elaborate courtship with pairs of nearly equal-sized fish progressively swimming closer to one another over several minutes. Prior to this, they ranged widely over the reef, covering as much as 50 m in one direction within a few minutes, swimming 50–100 cm above the bottom. The male would occasionally approach the visibly swollen female and touch his snout to her caudal fin. A pair would undergo a few false starts in which the spawning ascent was initiated but broken off prior to gamete release. During the spawning ascent the male was behind the female, his snout touching her caudal fin, while the fish rose at a slight angle. After gradually rising off the bottom, they spawned as much as 7–8 m above the bottom. They were not oriented with nor did they start their ascent from any projection above the surface of the reef. At the release of gametes, the male gave a flip of his caudal fin which may have helped to mix the gametes. After release the fish swam down to the substratum.

Spawnings in Puerto Rico were seen 3 days after the full moon in February 1979 and 9 and 11 days before the full moon in March 1979. In addition, courtship similar to that seen previously was observed 10 and 11 days after the full moon in February 1988. Females were also heavy with eggs, but observers had to leave the site before spawning was observed.

During April 1988 a single pair of *C. striatus* were checked daily for any signs of courtship or spawning activity at the Lee Stocking Island study site from the time of the full moon to the new moon. Neither fish was ever found to be swollen with eggs and near the time of dusk no courtship activity was seen. Almost certainly this pair was not spawning during the period of observation. On 14 May 1988, one day before the new moon, a moderately swollen female *C. striatus* was seen at south Cat Island, Bahamas and collected along with a male. The female produced approximately 5 500 partially hydrated eggs by hand stripping.

In Puerto Rico the eggs were collected, but not measured, and the embryos hatched in about 30 h at 26°C. At 72 h after hatching the eyes were pigmented and the yolk absorbed.

Hybridization

Among western Atlantic species, only a single probable hybrid individual has been reported (Clavijo 1985). It probably represents a cross between *C. ocellatus* and *C. striatus* with the color pattern containing elements of both probable parent species. Since spawnings of chaetodontid species that have been observed have been well separated in time and space, it is unlikely that hybrid *Chaetodon* are the result of chance mixing of eggs and sperm from separate spawnings. More likely they are the result of a suitable mate of the same species not being available to an isolated individual.

Egg numbers per spawning

Eggs were collected using a hand net from individual females immediately after release and the numbers captured from each spawning varied considerably. When the gamete cloud was released by the pair, collection would commence, if possible, within 5 sec so the net could be brought through the area of concentrated eggs while they were still visible. In such cases, most (if not all) released eggs were collected.

Among *C. aculeatus* spawns, the eggs collected per spawn ranged up to 2090 (n = 13) with 9 collections over 500 eggs (531, 545, 654, 692, 1109, 1540, 1630, 1933 and 2090). Four collections numbered less than 200 eggs and since definite egg clouds were seen on release, it is assumed that most of the eggs were not collected from these spawns. The smallest female (approximately 45 mm standard length, SL) observed spawning produced at least 692 and 225 eggs on two nights when her eggs were collected. A full-size female (approximately 55 mm SL) produced 1933 eggs and two nights later 2090 eggs (presumed to be the same fish based on size, general appearance and spawning location). The question of whether a single female spawns nightly (as opposed to alternate nights) cannot be answered based on my data, but I am confident the same females were spawning at least on alternate evenings.

Counts of *C. capistratus* eggs collected from large females (approximately 80 mm SL) were 110, 450, 610 and 3710 (n = 4) with the lower numbers, again, almost certainly due to a large part of the eggs released not being collected.

Aiken (1983) reported estimated total egg numbers from ovaries of four species of *Chaetodon* from Jamaica. For *C. capistratus* total egg numbers ranged between 2 900 and 12 900. The differences with actual numbers of eggs collected above imply that a significant number of eggs are retained in the ovaries and are probably not yet ready for release. Larger species, such as *C. ocellatus* had roughly one order of magnitude more eggs (up to 64 000) in its ovaries than *C. capistratus*.

Predation on eggs and spawning adults

No attempts by piscivores to prey on spawning adults have been seen. Predation on the eggs of *C. striatus* by the black durgeon, *Melichthys niger*, was seen on two occasions immediately after release of the eggs (Colin & Clavijo 1988). The *M. niger* stationed themselves close behind the pair of *C. striatus* who were gradually rising into the water column to spawn and made no effort to escape or deter the egg predators. At the instant of gamete release, the *M. niger* immediately swam to the cloud of gametes and picked at what were assumed to be the eggs for a few minutes. The *C. striatus* made no effort to protect their eggs and the observers remained far away from the spawning so they had negligible effect on the fishes.

No instances of egg predation were noted for 'undisturbed' spawns (where no attempt was made to collect eggs, otherwise egg collection would have interrupted egg predators) by either *C. aculeatus* (n = 17) or *C. capistratus* (n = 8) at the Lee Stocking Island site.

Discussion

Seasonality and lunar periodicity

The seasonality and lunar periodicity of spawning by western Atlantic butterflyfish are not well defined. From available information it does seem likely there are seasonal spawning peaks. Aiken (1983) implied seasonal peaks from his data while Colin & Clavijo (1988) over nearly three years found spawning by chaetodontids only during the winter and spring in Puerto Rico (February to April). At other times, though, a probable spawning by *C. aculeatus* in September (personal observation) and courtship in October (I.E. Clavijo personal communication) in St. Croix were seen, implying a small amount of spawning may occur during the 'off' season.

The alternation of spawning sites and/or bidaily spawning by *C. aculeatus* and *C. capistratus* signal caution in determining accurately when spawning is occurring. For example, Neudecker & Lobel (1982) reported *C. aculeatus* in St. Croix to spawn on two evenings (two days apart) at the same site. Subsequently they reported the fish not to spawn on three evenings, 5–7 days later. If each female spawns every other night, as suspected, then this would explain the lack of spawning by the pair on two of the three nights when spawning was reportedly not seen (5 and 7 days after). Whether or not spawning occurred on the remaining evening would then become the sole basis for making the statement that spawning does not occur during the first quarter of the moon. Similarly another instance of non-spawning by *C. aculeatus* was reported after the full moon in March (Neudecker & Lobel 1982) and could have been due again to either lack of spawning or alternate night spawning. Consequently the data here (and most reef fish spawning data based on observations) can only be considered positive data and negative spawning data must be viewed with caution. Ideally, comparative data, such as gonad indices, should support any conclusions concerning negative spawning. Similarly, spawning observations on a particular phase of the moon do not imply lunar periodicity unless, using consistent and sensitive techniques, significant variation in spawning activity was detected over the course of the lunar month.

A few comparisons can be made regarding spawning patterns between western Atlantic and Indo-Pacific species. Hawaiian and western Atlantic *Chaetodon* do seem to share a similar reproducive season with spawning during the winter and spring. Lobel (1978) reported three species of Hawaiian *Chaetodon* (*C. fremblii*, *C. multicinctus* and *C. unimaculatus*) to spawn during the week before the full moon in February and March. Ralston (1981) found *Chaetodon miliaris*, an Hawaiian endemic, to spawn, based on gonad samples, between January and May with a peak during February and March and no apparent lunar periodicity. Based on gonadal evidence, Tricas (1986) found *C. multicinctus* to spawn between October and May with a semilunar periodicity.

There is little information for chaetodontids in the southern hemisphere. Thresher (1984) briefly described spawning by *Chaetodon rainfordii* and *Heniochus acuminatus* on the Great Barrier Reef, but did not give details of timing. Near Port Moresby, Papua New Guinea, I found *Chaetodon unimaculatus* and *Chaetodon kleinii* swollen with eggs and undergoing typical chaetodontid courtship at the new moon in June when water temperatures were at their yearly minimum (24–25° C). Spawning was never observed, but based on the swollen condition of the females, almost certainly occurred.

Most observations of butterflyfish spawning have been during periods of winter low water temperatures (Ralson 1981, Tricas 1986, Neudecker & Lobel 1982, Colin & Clavijo 1988, this paper) of around 23–26° C which tends to correlate with the period of most active spawning by many reef fishes (Walsh 1987, Colin & Clavijo 1988, Munro et al. 1973). *C. aculeatus* and *C. capistratus* at the Lee Stocking island site were spawning in May when water temperatures were 24–25.5° C but were no longer spawning during July at which time water temperatures were 27.5–29.5° C. It would be interesting to see how well the spring rise in water temperatures correllates with the cessation (or great reduction) in spawning by these two species.

S.W. Ross (personal communication) found ripe females of *C. ocellatus* and *C. sedentarius*, which were almost certainly near the time of spawning, at 38–62 m depth off North Carolina in May 1978 when water temperatures measured at the collection sites were 18–20° C. Winter minimums in this area are near 8–12° C so the observed temperatures represent a significant rise over the yearly minimums. During summer, water temperatures at these depths may not rise significantly or may actually become lower than those observed during May due to upwelling of colder water.

Fricke (1986) reports spawning by *Chaetodon chrysurus* in the Red Sea to first be noticed in June and July during the period of warmest water temperatures. These are only 24–26° C (Fishelson et al. 1987), similar to temperatures recorded in areas where chaetodontids are spawning during winter and spring.

It seems that chaetodontids do not actively spawn at water temperatures above about 26° C and the seasonality of their spawning may be different in response to particular temperature regimes. In areas where water temperatures do not reliably reach the minimums found in true tropical areas, such as along the North Carolina coast, spawning may still proceed, possibly during the warmest period of the year.

Various reasons have been suggested to account for prevalence of winter spawning seasonality in many reef fishes in tropical areas, including increased primary production during winter, occurrence of particular current regimes to either disperse or retain the eggs and larvae near their point of origin and the influence of glacial conditions adapting spawning to present day low water temperatures (see Thresher 1984 for review plus Colin & Clavijo 1988). An additional hypothesis is that spawning by many reef fishes with pelagic eggs at the yearly minimum water temperatures of most reef areas insures embryos have developed to the point at which they are able to initiate feeding (eyes pigmented, mouth and gut functional) at a time where they have a full day ahead (morning) rather than when first feeding would have to be deferred or halted prematurely (night or late afternoon). Aquiring sufficient food and continuing to feed for an entire day may be crucial in allowing larvae to survive the first night after yolk is absorption.

Spawning styles

There may be a change in the reproductive patterns of butterflyfishes with increasing size. Small species may spawn often, perhaps as often as every other day, producing modest numbers of eggs throughout a lengthy period of the year. Large species produce much greater numbers of eggs per spawning but spawn less often, perhaps only a few times a year. Between large and small there is a variation with frequency of spawning smaller numbers of eggs being traded for more eggs produced at greater intervals.

The presence of visibly swollen females is the surest sign of imminent spawning aside from actual observation of the spawning (Fricke 1986). If data on swollen females are acceptable to indicate probable spawning occurrence (in addition to actual observations of spawning) there is even stronger evidence for differences in spawning styles with size among chaetodontids. In the western Atlantic, the two smallest species, *C. aculeatus* and *C. capistratus,* spawned regularly during the period of coolest water temperatures. The three larger species, though, were not observed as often to spawn or have visibly swollen females.

At the Lee Stocking Island only *C. aculeatus* and *C. capistratus* were seen to spawn and to have visibly swollen females. A few pairs of *C. ocellatus* and *C. striatus* were present at this study site, but none were ever seen be swollen. Almost certainly they were not spawning on the same days as *C. aculeatus* and *C. capistratus*. *C. striatus* has been seen to spawn and to have swollen females only in Puerto Rico in February and early March. Colin & Clavijo (1988) saw it spawning on only three occasions in a portion of their study area that was under regular observation for surgeonfish spawning and it is unlikely the species was spawning there regularly (daily or alternate days). Qualitatively the Puerto Rico study indicates that the frequency of spawning by *C. striatus* there is less than what was observed in the Bahamas for the two small species.

The numbers of eggs produced per year by *C. aculeatus* and *C. capistratus* can be estimated based on the limited data available. These species have a four to six month active spawning season with observed spawning occurring from February and to at least May. Each female probably spawns every other day and a full-size adult female of *C. aculeatus* and *C. capistratus* produces approximately 1500 and 3000 eggs per spawning respectively. This results, for each species respectively, in about 100 000 and 200 000 eggs per year per individual. The spawning style of these two small species appears to be to produce gametes on a regular basis (spawning occurred every day for both species whether individuals spawned each day or not) for a period of one third to one half of the year.

The remaining shallow-water western Atlantic *Chaetodon* may have a different style. In the Bahamas in April 1988 *C. striatus* and *C. ocellatus* (no *C. sedentarius* pairs present) were not actively spawning in the same area as *C. aculeatus* and *C. capistratus*. The data available indicate the three larger species produce more gametes per spawning than the two smaller species. The eggs released by a large *C. ocellatus* in a single spawning number in the tens of thousands and it seem likely these larger species produce a larger number of eggs per day on fewer days per year.

Tricas (1986) examined egg numbers produced by Hawaiian *C. multicinctus*. After actively spawning in October and November there was a possible reduction in spawning activity for about a month in December or January. This is followed by a yearly peak in March. He found apparent semilunar periodicity to spawning based on occurrence of postovulatory follicles in female specimens with spawning occurring about nine days prior to one day after the new moon and again nine days before to the day of the full moon. The data indicate that females do not spawn on successive nights and in fact may spawn only a single to a few times during each of the semi-lunar periods. Tricas (1986) found evidence of increased egg production per female during the peak of the spawning period. During the spawning season seven females injected with gonadotropins, held for 24 h in aquaria and then hand stripped of ova, produced between about 2 000 and 20 000 eggs with about 10 000 eggs coming from average size adult females (85–90 mm SL). Assuming that (1) the spawning season of *C. multicinctus* runs from mid-October to mid-May with a brief hiatus or reduction in activity in December or January, (2) spawning occurs on a semilunar cycle with two spawnings per female during each semilunar period, (3) there are reductions in spawning activity and egg numbers produced per spawn early and late in the spawning season and (4) the average females produces about 10 000 eggs per spawning during the peak of the reproductive season, the annual egg production is very roughly estimated at 200 000 to 300 000 eggs per female.

Bauer & Bauer (1981) found female angelfishes (Pomacanthidae) of the genus *Centropyge* in the aquarium to daily produce 50 to over 100 eggs. For six species this resulted in an estimated production of 20 000–50 000 eggs per female per year. Members of *Centropyge* are smaller than most chaetodontids and produce perhaps only about 20% the number of eggs per year as the smaller Atlantic chaetodontids. They may, however, spawn more regularly than small chaetodontids and have a more extended spawning season resulting in a style where even fewer eggs are produced over a more extended period.

Almost a continuum can be seen among fishes producing pelagic eggs from small species which produce their modest number of eggs over a much longer period to large species, such as groupers, which spawn once or a few times per year yet produce eggs numbering two orders of magnitude greater than species such as those of *Centropyge* at that time. Variation of spawning style with size appears to be the case even among the limited size range of chaetodontids, although data are still very limited.

Suggestions for future research

There is an obvious need for a detailed long term study of spawning activity by chaetodontids to both determine patterns within and among species. With five readily accessible species in the western Atlantic, it would be very feasible to include all of them in such a study and a wide range of spawning styles would probably be present within those.

Many aspects to chaetodontid spawning biology are suitable for investigation if the proper techniques are used. It is relatively easy to obtain eggs of some species, opening possibilities to study early development and conduct experimental work on the embryos and larvae which would be impossible using ichthyoplankton collections due to the relative rarity of chaetodontids in such collections.

Acknowledgements

Field work in the Bahamas was made possible by a grant from the Office of Undersea Research, National Oceanic and Atmospheric Administration, U.S. Dept. of Commerce and by the Perry Foundation, Inc. Work in St. Croix at the Hydrolab was made possible by a grant for habitat use from the office of Undersea Research, NOAA. Work in Puerto Rico was carried out as part of a program funded by the National Geographic Society and utilized equipment purchased under a grant from the Division of Biological Oceanography of the National Science Foundation. Steve W. Ross provided valuable information from his work off North Carolina. I would like to thank A. Charles Arneson, Lori J. Bell, Ralf Boulon, Jr., Ileana E. Clavijo, Ann Gronell and Ed Wishinski for their help in the field. I also thank Lori J. Bell, Bori Olla, Robert Wicklund and two anonymous reviewers for comments on the manuscript.

References cited

Aiken, K. 1983. The biology, ecology and bionomics of the butterfly and angelfishes, Chaetodontidae. pp. 155–165. *In:* J.L. Munro (ed.) Caribbean Coral Reef Fishery Resources, Inter. Cent. Living Aquatic Resour. Manag. Stud. and Rev. No. 7, ICLARM, Manila.

Allen, G.R. 1979. Butterflyfishes and angelfishes of the world. Vol. 2. Wiley-Interscience Publ., New York. 352 pp.

Bauer, J.A., Jr. & S.E. Bauer. 1981. Reproductive biology of pigmy angelfishes of the genus *Centropyge* (Pomacanthidae). Bull. Mar. Sci. 31: 495–513.

Burgess, W.E. 1978. Butterflyfishes of the world. T.F.H. Publications, Neptune City. 832 pp.

Clavijo, I.E. 1985. A probable hybrid butterflyfish from the western Atlantic. Copeia 1985: 235–238.

Colin, P.L. & I.E. Clavijo. 1988. Spawning activity of fishes producing pelagic eggs on a shelf edge coral reef, southwestern Puerto Rico. Bull. Mar. Sci. 43: 249–279.

Fishelson, L., L.W. Montgomery & A.A. Myrberg, Jr. 1987. Biology of surgeonfish *Acanthurus nigrofuscus* with emphasis on changeover in diet and annual gonadal cycles. Mar. Ecol. Prog. Ser. 39: 37–47.

Fricke, H.A. 1986. Pair swimming and mutual partner guarding in monogamous butterflyfish (Pisces, Chaetodontidae): a joint advertisement for territory. Ethology 73: 307–333.

Gore, M.A. 1982. The effect of a flexible spacing system on the social organization of a coral reef fish, *Chaetodon capistratus*. Behaviour 85: 118–145.

Lobel, P.S. 1978. Diel, lunar and seasonal periodicity in the reproductive behavior of the pomacanthid fish, *Centropyge potteri,* and some other reef fishes in Hawaii. Pac. Sci. 32: 193–207.

Moe, M.A., Jr. 1976. Rearing Atlantic angelfish. Mar. Aquar. 7: 17–26.

Munro, J.L., V.C. Gaut, R. Thompson & P.H. Reeson. 1973. The spawning seasons of Caribbean reef fishes. J. Fish Biol. 5: 69–84.

Neudecker, S. & P.S. Lobel. 1982. Mating systems of chaetodontid and pomacanthid fishes at St. Croix. Z. Tierpsychol. 59: 299–318.

Ralston, S. 1981. Aspects of the reproductive biology and feeding ecology of *Chaetodon miliaris*, a Hawaiian endemic butterflyfish. Env. Biol. Fish. 6: 167–176.

Robins, C.R. & G.C. Ray. 1986. A field guide to Atlantic coast fishes of North America. Houghton Miflin Co., Boston. 354 pp.

Thresher, R.E. 1984. Reproduction in reef fishes. T.F.H. Publications, Neptune City. 399 pp.

Tricas, T.C. 1986. Life history, foraging ecology, and territorial behavior of the Hawaiian butterflyfish, *Chaetodon multicinctus*. Ph.D Dissertation, University of Hawaii, Honolulu. 248 pp.

Walsh, W.J. 1987. Patterns of recruitment and spawning in Hawaiian reef fishes. Env. Biol. Fish. 18: 257–276.

Chaetodon capistratus from the Carribean. Photo by P.S. Lobel.

Eye camouflage and false eyespots: chaetodontid responses to predators

Stephen Neudecker
Bayfront Conservancy Trust, 1000 Gunpowder Point Drive, Chula Vista, 92010, U.S.A.

Received 6.4.1988   Accepted 14.11.1988

Key words: Butterflyfishes, *Chaetodon*, Communication, Coloration hypotheses, Aposematic coloration, Mullerian mimicry

Synopsis

The roles of eye camouflage and eyespots are examined within the genus *Chaetodon* as are the various theories explaining the evolutionary significance of the brilliant colors. While eye camouflage is not common among reef fishes, 91% of the 90 species of *Chaetodon*, have eyemasks (82) or black heads (4). Eye camouflage occurs concomitantly with diurnal false eyespots in 45.5% (41 of 90) of the species. Diurnal false eyespots serve to misdirect attacks by predators and/or to advertise unpalatability. False eyespots are located on areas of the body which allow escape and survival following an attack. Data suggesting that predators learn about the undesirability of butterflyfishes are presented. Butterflyfishes are inactive at night, forage during the day and spawn at dusk. It is unlikely that nocturnal color changes are useful in conspecific interactions and are therefore believed to provide visual cues to potential predators. Nocturnal eyespots probably function to intimidate potential predators but could also remind them of unpalatability. The aggression release hypothesis (Lorenz 1962, 1966) to explain the brilliant coloration of chaetodontids is not supported because butterflyfish coloration changes and few species are territorial. The species recognition hypothesis (Zumpe 1965) is not supported by results of field experiments. The disruptive coloration hypothesis (Longley 1917) is rejected as a general explanation for poster coloration but does explain the prevalence of eyebars of *Chaetodon* spp. The aposematic hypothesis (Gosline 1965) is supported by morphology, behavior, a lack of predation and field observations. The possibility of Mullerian mimicry is suggested. It is concluded that the primary selective force behind chaetodontid coloration, particularly eyespots, has been predation and color patterns have evolved to minimize this threat.

Introduction

Butterflyfishes are laterally compressed, deep-bodied fishes with bright colors patterns. Lorenz (1962, 1966) explained the functional significance of butterflyfish coloration as an intraspecific stimulus to reduce aggression accompanying territorial behavior. This explanation began a controversy over the purpose of butterflyfish coloration which is still not resolved (Ehrlich et al. 1977, Allen 1980).

Field observations of chaetodontids reveal that a large amount of information is communicated through presentations of the fishes' coloration to conspecifics and to potential predators (Fricke 1973, Hamilton & Peterman 1971, Neudecker & Lobel 1982). Chaetodontids do not flee large potential predators, they turn sideways and present their lateral aspect (Motta 1984). They face threats with lowered heads and erect dorsal spines (Burgess 1978, Ehrlich et al. 1977, personal observation). These fishes are extremely easy to collect because they pause and present their sides to the spear fisherman before fleeing. This behavior suggests that visual cues are important in predator interactions and is further supported by the fact that most piscivorous predators depend mainly upon vision to capture prey (Hobson 1974, 1979).

This paper attempts to explain the significance of eyebars, false eyespots and nocturnal changes in color and pattern within the genus *Chaetodon* and reviews various hypotheses explaining the brilliant coloration of butterflyfishes.

Eyebars

Many authors agree that dark eyebars serve to camouflage the eye from predators. When the eyes are obliterated a primary target for predators is hidden (Cott 1957, Barlow 1972, Burgess 1978). Wickler (1968) demonstrated that the blenny *Runula* targeted the eyes of prey. Rasa (1969) has shown that eyes of conspecifics are the foci of agonistic attacks by the herbivorous pomacentrid *Eupomacentrus jenkinsi*. It could be that many predators are more attuned to movement of the eye than to the eye itself.

Some alternative explanations for eyebars as antiglare devices and for sighting lines have been offered by Hailman (1977). Since many of the fishes here occur at depths where light penetration is reduced, the importance of the antiglare hypothesis is questionable. The lines do not appear to be properly configured to be useful for sighting prey. Hailman (1977) has pointed out that there are probably multiple purposes for eyebars.

The attitude of eyebands is correlated with relative body depth. Deep-bodied fishes have predominantly vertical eyelines whereas fusiform species typically have longitudinal ones (Barlow 1972). Eyebar configuration is tailored to morphology such that the camouflage effect is maximized. After presenting their sides to threats, butterflyfishes are known to zigzag back and forth on their way to cover (Hamilton & Peterman 1971, Ehrlich et al. 1977). The protective function of eyebars is through visual confusion of potential predators especially when flashed against a complex background (Barlow 1972).

Thresher (1977) analyzed 85 species of Caribbean reef fishes and found that only 21% (18 species) had obliterated eyes, of which only six species concealed their eyes with masks whereas the majority possessed extremely prominent or ornamented eyes. The scarcity of eye camouflage among most Caribbean reef fishes may result from the fact that interspecific agonism is widespread there (Myrberg & Thresher 1974) and conspicuousness of eyes has been evolved as a signal during agonistic encounters (Rasa 1969, Thresher 1977).

Diurnal false eyespots

Eyespots occur at or around the eyes of the fish whereas false eyespots look like eyes but occur in other regions of the body. Dark spots with a ring around them are target-like structures which possess maximum visibility (Cott 1957). Cott (1957) suggested that *Chaetodon plebeius* and *C. capistratus* used their false eyespots to produce the impression of a head at the wrong end to misdirect attacks by predators. This paper also uses the term ocellus, for these eye-like spots. Hailman (1981) suggested that a false eyespot on *Forcipiger flavissimus* and *F. longirostris* was a type of mimicry, which he called symmetry-deception, used to deceive and escape from predators. The species have a black, eye-sized spot in the ventral posterior position below a clear caudal fin and black heads that conceal the real eyes. When foraging in a typical 45° head-down position, its head and anal fins tend to disappear and the form of a fish facing in the opposite direction is visible. Presumably the deception works best when the fish is most susceptible to attack with its head pointed down while foraging. My observations and photographs of *Forcipiger* spp. in the field tend to support Hailman's contentions.

False eyespots are believed to protect juveniles of some fishes from predation by their adults. Ocelli on the caudal fins of some juvenile cichlids and piranhas are thought to advertise their identities to conspecific adults and thereby protect them from cannibalism (Leccia 1970, Zaret 1977). These ocelli are mimicked by other species of potential prey of the adults and are thought to deter predation.

Nocturnal eyespots

At night, several species of *Chaetodon* change not only their coloration but also their patterns. Often the nocturnal coloration phase includes eyespots in species which lack them during the day. A nocturnal spot is a dark or light spot which appears at night but is absent during the day. Whereas some butterflyfishes only display spots at night, several species which have diurnal spots highlight them at night.

Fricke (1973) suggested that eyespots which appear on *C. melannotus* at night are adaptations against predators. These spots could be a way of mimicking large predators in an attempt to scare potential predators away. Thresher (1977) reported that pelagic piscine predators had black pupils surrounded by silver irises which is known to be a highly visible pattern (Cott 1957, Levine et al. 1980). Wickler (1968) stated that only cephalopod eyes correspond to the ocellated, vertebrate-like pattern of false eyespots. Since cephalopods tend to hunt nocturnally, they might be an appropriate model.

The aggression release hypothesis

Lorenz (1962, 1966) called the flamboyant coloration of butterflyfishes poster coloration because they looked like highly visible advertisements designed by man to be seen from a distance. While it is correct that chaetodontids are advertising with their dazzling colors and patterns, Lorenz' (1962, 1966) explanation of this coloration as permanent, intraspecific signals used by highly territorial fishes to release aggression and thereby space individuals over the reef is incorrect.

The coloration of chaetodontids is not static but changes in social situations and at night. Hamilton & Peterman (1971) pointed out that chaetodontids can rapidly change the amounts of contrast and color and use these color changes to communicate social information. Fricke (1973) reported that different shading and color patterns in *Chaetodon melannotus* communicated a wide variety of social information from a readiness to attack to a readiness to flee. I have also observed many species of *Chaetodon* communicating inter- and intraspecific information through dynamic shading of areas of their bodies, particularly when spawning (Neudecker & Lobel 1982).

Territorial behavior is not typical for butterflyfishes and has been described in the field for only four of 90 species: *Chaetodon baronessa* and *C. trifascialis* (Reese 1973, 1975), *C. fasciatus* (Fricke 1966) and *C. larvatus* (Allen 1980). Intra- and interspecific agonistic behavior in general is noticeably absent among chaetodontids during the day (Reese 1973, 1975, Ehrlich et al. 1977), although some aggression does accompany spawning at dusk (Neudecker & Lobel 1982) and during crepuscular changeover when chaetodontids move to and from resting sites (Ehrlich et al. 1977, personal observation). Because the color and patterns of these fishes are not static and few species are territorial, this theory does not explain the coloration of chaetodontids, although the term poster coloration is descriptive and should be retained. The question then becomes: What are chaetodontids advertising?

The species recognition hypothesis

Zumpe (1965) suggested that an important function of poster coloration was species recognition over distances. Individuals must be able to recognize conspecifics on coral reefs that are typically inhabited by multitudes of multicolored fishes. Foraging patterns of several *Chaetodon* species in the Pacific show that separated mates of monogamous pairs relocate each other visually (Reese 1975). However, the benefit of such high visibility to mates is countered by the cost of being seen by potential predators.

When conspecific and heterospecific models were introduced into the home ranges of several *Chaetodon* spp. the intruder elicited no agonistic behavior from the resident. Most residents were curious and attracted to both types of models. The introduced models did not illicit the predicted responses if species recognition was important (Ehrlich et al. 1977). Consequently, the species recognition hypothesis is also rejected.

The disruptive coloration hypothesis

Disruptive coloration is a form of deception that simply disrupts the telltale outline of the animal and thereby reduces its chance of being seen (Hailman 1977). Longley (1917) stated that no matter how apparent the bright colors of reef fishes are to man, they coincide and blend in with the background colors of the reef. This repetition of background by irregular patches of contrasted colors tends to catch the observer's eye and draw attention away from the animals and is called disruptive coloration (Cott 1957). Since chaetodontids are benthic browsers, grazers and corallum feeders (Neudecker 1979), they do not need to conceal themselves from their nonelusive prey.

For a pattern to be conspicuous, its components must differ from themselves and the background in brightness, wave length distribution or both (Levine et al. 1980). Black and white patterns are highly conspicuous regardless of illumination (Levine et al. 1980) and are included in the patterns of many *Chaetodon* spp.

False eyespots are an integral part of the coloration of many of these fishes. They are visual cues, a form of blatant advertising, and their purpose is to be seen not to hide the fish. Consequently, the disruptive coloration hypothesis as a general explanation poster coloration is not supported. However, eyebars are a type of disruptive coloration used to camouflage the eyes of butterflyfishes as discussed below.

The aposematic hypothesis

Since most piscivorous fishes depend mainly on vision to capture prey (Hobson 1974, 1979), it is reasonable to believe that colors and patterns communicate information to potential predators. Gosline's (1965) aposematic hypothesis suggests that chaetodontid coloration functions to advertise that these fishes are a dangerous, low-quality meal because they are deep-bodied, bony and difficult for predators to swallow. Hobson & Chave (1972) also felt that strong fin spines and exceptionally deep bodies restricted predation on adult butterflyfishes.

Model insects which are poisonous and advertise these qualities to their potential predators are said to be aposematic (Brower 1988). Warning coloration is a synonym used to describe this phenomenon. Throughout the animal kingdom, there is a widespread coincidence between conspicuous coloration and the presence of noxious characters (Edmunds 1974). The erection of anterior dorsal fin spines is the typical aggressive and defensive display of butterflyfishes (Burgess 1978, Ehrlich et al. 1977, Motta 1984).

Vertebrate predators such as birds are known to be very conservative in attacking potential prey which are aposematically colored (Brower 1984). Experimental work has shown that fishes can learn to avoid conspicuous unpalatable prey (Gawlik 1984, Kruse & Stone 1984, Guilford 1988). At least some insectivorous birds retain the learned ability to avoid aposematic insects for long periods of time (Rothschild 1964, Waldbauer 1988). Predators of warningly colored species are encouraged to generalize because they benefit by recognizing similar patterns as indicating unpalatability (Huheey 1988).

Materials and methods

The color patterns of all species of the genus *Chaetodon* were examined. Patterns were compared and quantified from over 600 hours of field observations and photographs taken in the Atlantic, Caribbean, Pacific and Indian Oceans. In addition, photographs contained in Burgess (1978), Steene (1978) and Allen (1980) were also analyzed.

Once all of the color data were collected, the fishes were grouped by subgenera following the classification of Allen (1980), which followed Burgess (1978), to compare and contrast coloration among related species. Together with Meyers (1980), these publications account for a total of 90 valid species of the genus *Chaetodon*.

The location of diurnal and nocturnal spots was recorded as occurring in one of ten sectors of the body. The body was divided into nine approximately equal sized sectors from the posterior to anterior and from dorsal to ventral. Because so many eyespots occur on the caudal peduncle it was treated as a separate location. The presence or absence of a ring around the spot was also noted.

The diurnal coloration of adults and juveniles was analyzed and compared within and between subgenera. Adults of all 90 species were examined as were the diurnal color patterns of all but 23 species of juveniles. The presence or absence of nocturnal eyespots was determined from field observations and photographs of 34 of the 90 species because the nighttime color of the remaining 56 species is not known.

The presence or absence of eyebars was determined in the same manner. Since concealment of the eye could depend upon concomitant camouflage of the iris, the amount of iris covered by the eyebar was quantified. Species were classified as having either no eyebar, eyebar completely hiding the iris, iris blackened except for the front, iris blackened except for the back, iris blackened except for the front and back, iris not blackened at all or iris a different color than the surrounding body color. When part of the iris was not blackened, it was noted whether or not the uncovered portion of the iris matched the surrounding body color of the fish.

Results

Eyebars

Within the genus *Chaetodon*, 82 of 90 species (91%) have eyebars. Four of the eight species which lack an eyebar are black or have black heads and their eyes are thereby camouflaged (*C. dichrous, C. litus, C. nigropunctatus* and *C. smithi*). Two species have no eyebars or eye camouflage at all as adults (*C. fremblii* and *C. xanthocephalus*). Two other species do not have eyebars but they have a colored patch around the eyes (*C. larvatus* – orange patch and *C. semilarvatus* – blue patch) which may serve to hide the eye.

While nearly all species which exhibit an eyebar as an adult also have them as juveniles, a few do not. Both *Chaetodon ephippium* and *C. xanthocephalus* only have eyebars as juveniles.

The eyebars of most butterflyfishes are black (74/82, 90%), six are yellow, one is yellow with a black outline and one is orange. To effectively hide the eye, the iris should also be black (Hailman personal communication). I recorded the number of species in which the iris was totally blackened or mostly blackened (only the front or back edge uncovered) as well as the number which had an iris color different than the surrounding body color. Figure 1 shows that 30 species have totally blackened eyes, 16 species leave the back portion of the eye uncovered and four species have the front portion uncovered. Therefore 60% (54/90) of the *Chaetodon* spp. have completely or mostly darkened eyes whereas neither the front nor rear portion is not covered in 31% (28 species). Only four species with eyebars have completely exposed irises (*C. bennetti, C. dichrous, C. flavocoronatus* and *C. tinkeri*). When the iris is partially covered, it is the same color as the surrounding body except in three species.

Diurnal false eyespots

A total of 42 of 90 *Chaetodon* spp. (46.7%) have diurnal false eyespots as adults (Fig. 2). Of those species exhibiting diurnal spots, 37 species have one spot and five have two spots. Diurnal eyespots only occur in six of ten body sectors. The most common location for diurnal spots is on the caudal peduncle (13 species), in the dorsal posterior sector (8) and in the dorsal anterior sector (7). Species with a spot on the caudal peduncle generally have a translucent caudal fin.

Of the 42 species with false eyespots, 19 (45.2%) are surrounded by a ring. The ocellus is thought to represent an iris around a pupil and thereby mimic a vertebrate eye. While not measured, false eyespots appear to be two to three times larger than the fishes' eye.

The majority of the species display diurnal false eyespots all day long. But two related species (*C. mertensii* and *C. paucifasciatus*, subgenus *Citharoedus*) display their diurnal spots only some of the time during the day.

More juvenile *Chaetodon* spp. have false eye-

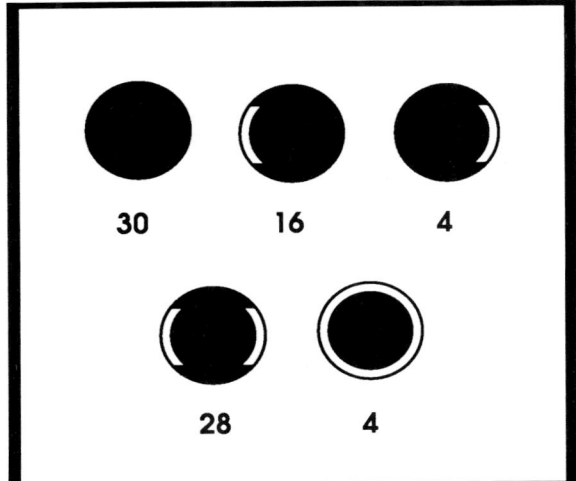

Fig. 1. The amount of iris covered by eyebars in the genus *Chaetodon*. A total of 82 of 90 species (91%) have eyemasks. Of the eight species which do not have eyebars, four of them have black faces and the eye is hidden. In 54 species (60%), the eye is completely or mostly blackened (top row + four black) whereas the iris is not blackened in four species. When the iris is not completely blackened, it is the same color as the body color adjacent to the eye except in three species which have an iris color different than the surrounding body color. Most eye bars are black (74/82, 90%), six are yellow, one is yellow with a black outline and one is orange.

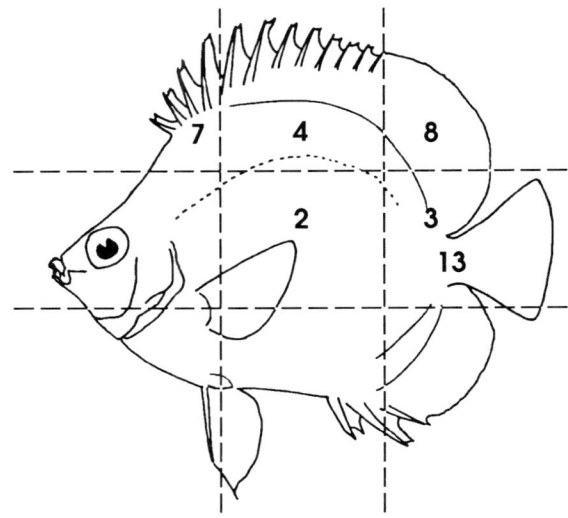

Fig. 2. Location of diurnal eyespots in adult *Chaetodon* spp. shown as the number of species which have a spot in that location (38). A total of 42 of 90 species (46.7%) have spots during the day. The posterior midsection of the body was treated separately from the caudal peduncle because the majority of spots occur on the caudal peduncle. In addition to the species with one spot, another five species have two spots which also only occur in the locations where single spots are found.

spots than adults (54 of 66 species) even though juvenile coloration is unknown for 24 species. Nearly half (42.6%) of the diurnal spots found on juveniles are ocellated (Fig. 3). Five of the 54 species having false eyespots as juveniles have two spots. Juvenile spots only occur on five of the ten body sectors and most occur in the dorsal posterior sector (27) or on the caudal peduncle (14). At least one species, *C. ocellatus* is known to have three spots as a very young juvenile. No juvenile is known to exhibit a spot on the nape whereas diurnal false eyespots above the real eye occur on seven species as adults (Fig. 1).

Because related species are likely to be exposed to the same suite of predators, their coloration could be expected to be similar. There are definite relationships in coloration patterns within subgenera. For example, 16 of the 27 species in the subgenus *Chaetodon* have a spot as a juvenile. Eyespots occur on the dorsal posterior of 10 species, another seven species have spots elsewhere on the body, five have none and six are unknown. Sixteen of the 27 species also have eyespots as adults but many of them occur on different parts of the body than do the juvenile spots.

Some species exhibit eyespots as juveniles but not as adults. All nine species in the subgenus *Chaetodontops* have spots as juveniles (seven of nine are located in the dorsal posterior region) but only two species exhibit spots as adults. The subgenus *Tetrachaetodon* is unusual because all four species have spots in the same locations as juveniles and adults.

Accounts of several species reported their distribution as occurring primarily in deep water. Of nine species (*C. aya, C. burgessi, C. falcifer, C. lavocoronatus, C. guezei, C. guyanensis, C. marcellae, C. mitratus* and *C. tinkeri*) known to occur mainly in deep water (Burgess 1978, Steene 1978, Allen 1980), none has diurnal eyespots. Whether or not they have eyespots at night is not known. All nine of those species belong to the subgenera *Roa* and *Prognathodes*.

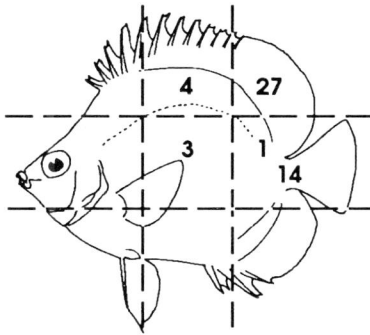

Fig. 3. Location of diurnal eyespots in juvenile *Chaetodon* spp. Of the 66 species for which the juvenile coloration is known, 54 (81.8%) have spots during the day. Among the 54 species with diurnal spots, 23 (42.6%) are ocellated. Numbers indicate the number of species with spots in that location (49). Another five species have two spots which only occur in sectors where single spots are found. Of the five species with two spots, three have a spot on the caudal peduncle and also in the dorsal posterior sector. Very young juveniles of at least one species (*Chaetodon ocellatus*) are known to have three spots.

Nocturnal eyespots

Because direct observations and photographs of *Chaetodon* spp. at night are limited to 32 species, the prevalence of nocturnal spots cannot be completely determined. Nonetheless 11 of those 32 species (34.4%) exhibit false eyespots at night. Species known to exhibit a single dark spot at night are: *C. assarius, C. ocellicaudus, C. rafflesi, C. semeion, C. vagabundus* and *C. zanzibariensis. Chaetodon auripes, C. citrinellus, C. melannotus, C. paucifasciatus* and *C. trifascialis* exhibit two white spots aligned horizontally in the mid and posterior sectors of the body at night.

In addition to the species which display false eyespots at night, another 13 species highlight their diurnal spot at night (*C. auriga, C. bennetti, C. capistratus, C. dolosus, C. fremblii, C. lunula, C. miliaris, C. ocellatus, C. plebeius, C. quadrimaculatus, C. speculum, C. unimaculatus* and *C. weibeli*). Nearly all of these species display a ring around the diurnal spot only at night but *C. quadrimaculatus* darkens the dorsal half of its body to highlight the two white spots which are present during the day. During the day, the ring surrounding the false eyespot on *C. fremblii* and *C. bennetti* is blue but at night it turns white.

There is not enough data on nocturnal patterns to make many comparisons among the subgenera about similarities among related species but some generalities are evident. Of the 27 species in the subgenus *Chaetodon*, three species have nocturnal false eyespots, four species highlight their diurnal spot at night, three species do not have a nocturnal eyespot and the nocturnal coloration of the remaining 18 species is unknown. The nocturnal eyespot pattern found (two white spots) in *C. melannotus* (Steene 1978) is the same as that found in the related species *C. citrinellus* (Allen 1980, personal observation). The related species *C. meyeri* and *C. ornatissimus* (subgenus *Citharoedus*) do not have false eyespots as juveniles, adults or nocturnally but they do highlight their striking lined colors at night.

Discussion

Eye camouflage

Eye camouflage is practical in chaetodontids because diurnal aggression is rare and the eyes are not needed for aggressive displays. The importance of eye obliteration to predation avoidance is also supported by the fact that in many species they occur concomitantly with false eyespots. Deep-bodied species with bands are usually mobile and active just above a complex substratum (Barlow 1972) and while they can turn sharply, they cannot sustain the high swimming speeds that elongate species do (Alexander 1967). These behavioral characteristics of chaetodontids partly explain the need to camouflage eyes from speedy predators.

Eyebars and false eyespots occur together in 45% (41 of 90) of the *Chaetodon* spp. Eyebars conceal the location of the real eye and false eyespots confuse potential predators about the identity and/or orientation of the fish. The fact that 95% (86 of 90) of them have eyebars or black heads suggests that there has been greater selective pressure for eye camouflage than for conspicuousness among butterflyfishes. This large percentage of occurrence is significant because few other reef fishes exhibit camouflaged eyes.

Of the 48 *Chaetodon* spp. in which part of the iris is not covered by the eyebar, the uncovered portion of the eye matches the body color surrounding it in all but three species. When a portion of the iris is not hidden, its color matches the surrounding body color of the fish and therefore tends to blend in and not be noticeable.

Diurnal false eyespots

Eyespots are visual cues directed to some recipient. Because mimetic resemblances provide information about the features of an animal to which other animals respond (Hinton 1977), false eyespots and the concomitant camouflage of actual eyes indicate that predators are attuned to the eyes of butterflyfishes.

The prevalence of diurnal false eyespots among juveniles, even though the coloration of 24 species is not known, suggests that the threat of predation may be greater to juveniles than to adults. For example, all nine species in the subgenus *Chaetodontops* have spots as juveniles but not as adults suggesting a greater threat of predation to the young. The explanation of ocelli on juvenile cichlids and piranhas as a way to avoid cannibalism by their adults (Leccia 1970, Zaret 1977) does not explain these spots on young butterflyfishes. Larvae are planktonic and adults remain on the reef their entire life so they are generally not sympatric. Few adult butterflyfishes are planktivorous and none is piscivorous and juveniles are much too large for the adults too eat.

The interesting fact that none of the nine species known to occur in deep water have spots could suggest that either models or dupes are not abundant in deep water.

That predators can be confused about prey orientation depends on the assumption that the predator anticipates the direction the prey will move if it detects the attack. If butterflyfishes camouflage their real eye and display false eyespots to misdirect attacks of predators, then eyespots should be located in areas of the body that can survive an attack. Since it is likely that predators anticipate the direction of escape and therefore lead their prey when attacking, deception regarding which end of the fish was the front would increase the probability of a miss (Hailman 1981). None of the *Chaetodon* spp. has a ventral posterior spot similar to that found on the *Forcipiger* spp.

Mimicry of a species which does not always kill potential predators depends on trial-and-error learning of the unpalatability of the prey. Therefore for butterflyfishes to teach predators about their noxious qualities, they must be able to repel them and survive an attack. False eyespots among *Chaetodon* spp. are located principally in invulnerable regions of the body. The fact that attacks at these locations can be survived is supported by the successful escapes of two species as presented below. The eyespots on the few species which have a centrally located spot tend to be larger than those found elswhere on the body. These large, centrally located eyespots are thought to discourage rather than misdirect attacks.

The majority of *Chaetodon* spp. are believed to have diurnal false eyespots as juveniles but many disappear with age. However, there are not enough photographs of juveniles to exactly quantify the abundance of eyespots. Juvenile to adult differences in color patterns may indicate a greater threat of predation on juveniles.

Some species have spots in two different parts of their body and sometimes these multiple spots vary in size. Pough (1988) has indicated that if a precise resemblance of a model is not necessary, a mimic can gain protection from a broader spectrum of predators by combining features of two or more species of models. Therefore multiple spots on a species could represent the incorporation of traits of more than one model (see mimicry below).

Nocturnal false eyespots

While it is true that some species do not change their patterns at night, every species I have observed at night and those documented by photographs change at least color intensity through blanching and/or shading. Butterflyfishes forage during the day, spawn at dusk and rest at night. These nocturnal color changes, particularly false eyespots, are important because butterflyfishes are inactive at that time. Since it is unlikely that nocturnal changes in color and/or pattern are useful in conspecific interactions, they are therefore believed to provide visual cues to potential predators.

Besides dusk, butterflyfishes are probably most susceptible to predators at night when resting quietly in holes or down in the reef. The importance of false eyespots as an antipredator device at night is supported by the fact that several species which have an eyespot without a ring in the day, ring the eyespot at night while resting (*Chaetodon bennetti*, *C. miliaris* and *C. speculum*). These fishes are docile at night and appear to be in a state of torpor. Nocturnal false eyespots could serve more to scare away potential nocturnal predators than to confuse them of the orientation of the prey. These species may be trying to mimic an enemy of their potential predators to intimidate them to leave.

Fricke (1973) and Burgess (1978) have stated that nighttime coloration is equal to stress or fright coloration and that nocturnal coloration is often exhibited when the fishes are collected and stressed. I have also noted that most of the over 20 species that I have collected changed to nocturnal coloration when speared.

My observations suggest that nocturnal eyespots appear during the dusk crepuscular period and persist throughout the night. The low ambient light levels at dusk would tend to enhance the confusing effect of eyespots when piscine predators are most active (Hobson 1979). *Chaetodon capistratus* changes into its nocturnal coloration during the dusk crepuscular period in St. Croix (Neudecker & Lobel 1982).

Mimicry

Mimicry theory provides some insight into chaetodontid coloration and offers some tests to help determine the evolutionary significance of poster coloration of butterflyfishes. The evolution of visual mimicry among butterflyfishes depends on the presence of unpalatable or noxious models that advertise perceptible characteristics to potential predators by passive or active displays of brightly colored patterns, mimics which copy the appearance of the model and dupes which are confused by the similarities (Brower 1988, Pough 1988). Models are either attractive or repellent to the dupe (Pasteur 1982, Pough 1988). A model repellent to the dupe is trying to elicit the response of avoidance.

Mimetic resemblances are either precise (the model is an identifiable species or a group of species) or abstract (the model is a general category of organism, e.g. a snake, but cannot be identified as a particular species). This distinction identifies points on a continuum of resemblances. Precise resemblance lies at one extreme and grades into less precise resemblance to a specific model and ultimately into an abstract likeness in which only a general category of model can be identified (Pasteur 1982, Pough 1988). The posterior dorsal eyespot on *Chaetodon epphipium* could be an abstract model of the vertebrate eye.

Most vertebrate mimics are precise and the model is an identifiable species. All well-documented cases of mimicry among fishes involve a precise mimic (Randall & Randall 1960, Leccia 1970, Losey 1972, 1978, McCosker 1977, Zaret 1977). The coral reef fish *Calloplesiops altivelis* darts into a hole when threatened and leaves the posterior part of its body exposed. The fish's dorsal fin bears an ocellus and when dorsal, caudal and anal fins are expanded, the posterior end of the fish resembles the sympatric moray eel, *Gymnothorax meleagris* (McCosker 1977). Because the mimic was a precise copy of an identified model, McCosker (1977) termed the mimicry Batesian. The mimic is edible but it copies the look of a dangerous predator. The cases of juvenile fishes avoiding cannibalism by adults and their mimics (Leccia 1970, Zaret 1977) are also cases of Batesian mimicry.

Batesian mimics have no deterrent, their sole means of surviving an attack is by resembling their model and strict Batesian mimicry implies precise resemblances (Huheey 1984, 1988). Rothschild (1972) has called this 'a sheep in wolf's clothing'. Under Batesian mimicry, the mimic benefits but neither model nor potential predator benefits. The requisite conditions for Batesian and Mullerian mimicry are listed in Cott (1957), Rettenmeyer (1970) and Huheey (1984), and reviewed by Huheey (1988).

Hailman's (1981) evidence of mimicry, which he called symmetry-deception, was based on a visual match of the features of a mimic and model, i.e. an abstract generalized model of a 'fish-like creature'. This would be the first case of an abstract resemblance to a model in fishes since the model is thought to resemble the generalized morphology of a fish rather than a particular species. It would also be atypical because while Batesian mimicry involves deception of predators by prey it usually requires a very close resemblance to the model (Cott 1957, Huheey 1984, 1988).

Mullerian mimicry is a system of mutual warning coloration arising from the convergence of two or more warningly colored species which mimic a relatively inedible model; it is beneficial to the model, mimic and potential predator. Animals involved in Mullerian mimicry tend to have open, slow movements and are noted for their conspicuousness. Selection leads to convergence to the most common or average pattern or to the pattern associated with the most noxious characteristics (Huheey 1984, 1988, Pough 1988). Because predators of warningly colored species are encouraged to generalize, since they benefit by recognizing similar patterns as indicating unpalatability (Huheey 1988), the general pattern of eyespots as an advertisement of noxious traits in butterflyfishes could have been favored.

Mullerian mimicry does not have to be exact enough to cause a misidentification by a potential predator, just close enough to remind it of a past experience with a noxious prey (Huheey 1988). The precision of protective mimicry may be inversely correlated with the consequences to a dupe for mistaking a model for its mimic. When the consequences are minor, mimicry tends to be precise (concrete) but when the model is deadly, imprecise (abstract) mimicry is possible (Pough 1988). In Mullerian systems, variation in color pattern is common and even a poor resemblance can offer some protection (Pough 1988). In fact, the resemblance among mimics is often not close; if it is, it is probably not Mullerian mimicry (Huheey 1988).

The fact that so many *Chaetodon* spp. have similar false eyespots located on the same portion of the body suggests Mullerian mimicry. Several of the anatomical and behavioral traits are consistent with those which typify Mullerian mimicry. They all have deep bodies and spines and exhibit warning coloration. They have the open and slow movements and related species exhibit similar color patterns, particularly eyespots. While no specific models have yet been identified, models tend to converge to a generalized pattern. An exact resemblance is not prerequisite because neither the model nor the mimic depends on deceiving the potential predator, only in reminding it of noxious qualities (Huheey 1988, Pough 1988).

Because butterflyfishes do not exhibit conspicuous changes in morphology from juveniles to adults, size limitation could restrict mimetic resemblances (Pough 1988). The changes in color patterns between juvenile and adult butterflyfishes, such as the presence of ocelli on juveniles but not adults of a species, might be explained by the mimicry of a size-limited model.

The aposematic hypothesis

Because piscivorous predators usually swallow their prey whole, spininess is an important anatomical adaptation against predators (Hobson 1979). These predators must also swallow such prey headfirst to avoid locking the spines erect and lodging them in the esophagus; failure to do so has killed many piscine predators (Hobson 1979). Motta (personal communication) has seen a trumpetfish reject an acanthurid which was lodged in its mouth with erect dorsal spines and he has also observed a piscivore reject a butterflyfish under similar circumstances.

The aposematic hypothesis is supported by the absence of chaetodontids in the stomachs of reef predators (Ehrlich 1975). Chaetodontid larvae are taken in abundance by large pelagic predators such as tuna and dolphin (Burgess 1978) but adults are conspicuously absent from the stomachs of the piscivorous predators found on reefs. It has been reported that moray eels, scorpion fishes, snappers and groupers are predators of adult chaetodontids (Burgess 1978) but the available data on the feeding patterns of these predators (Hiatt & Strasburg 1960, Randall 1967, Hobson 1974, 1979) do not support this contention. I know of only two records of adults taken as prey and one of the predators was a shark. However related diurnal teleosts such as acanthurids, labrids, pomacentrids and scarids, some of which are monochromatic and others highly colored, are all taken by piscivorous reef fishes including barracuda, jacks and eels (Hiatt & Strasburg 1960, Randall 1967, Hobson 1974, 1979).

Butterflyfishes are conspicuous, abundant and clearly available to piscivorous fishes but are apparently not taken by them. Since predation on adult chaetodontids appears to be a rare event and they seem to be defended by noxious anatomical features such as spines, poster coloration is thought to be aposematic. I believe that the presence of chaetodontid larvae in pelagic piscivores and the absence of juveniles and adults from piscivorous reef species suggests that larger fish attain a refuge in size after which they are not susceptible to predation unless debilitated.

When the prey has a defensive mechanism which makes it unpalatable, the stimulus presented by a large individual is likely to be more repellent to the predator than that presented by a small individual (Pough 1988). This could explain why diurnal spots tend to be two to three time larger than the fishes' real eyes. False eyespots may be a kind of superstimulus and mimic a much larger individual.

One of the central contentions of this paper is that butterflyfish coloration advertises that predators can be choked if they try to swallow these spiny, unpalatable fishes. However it is also possible that they could possess some degree of toxicity even though there is some evidence to suggest that they are palatable (Reighard 1908). The spines, skin and flesh of butterflyfish need to be biochemically analyzed for toxins.

Prey escape and predator learning

It has not been demonstrated that predators are misled by these false eyespots (Wickler 1968) or that piscivores can learn that butterflyfishes are unpalatable. Field observations with photographic documentation indicate that *Chaetodon* spp. can survive attacks in the anterior and posterior dorsal sectors. On June 9, 1976, a pair of adult *Chaetodon citrinellus* (about 10 cm SL) was photographed off the northeastern coast of Guam. One of the pair had lost 10% of the posterior dorsal region of its body (Fig. 4).

At night, *C. citrinellus* blanches to expose two lateral eyespots (Allen 1980, personal observation). The location of the attack was at the posterior eyespot, suggesting that the potential predator was misled about the orientation of the fish and that the attack occurred at night. The wound healed quickly and the fish was observed and photographed with its mate nine times over a 10 month period between June 9, 1976 and April 7, 1977. Because the pair remained together in the same home range area and behaved normally, the damaged individual was believed to be reproductively active.

Chaetodon quadrimaculatus exhibits the same pattern of spots day and night that *C. citrinellus* displays only at night. Figure 5 shows a *C. quadrimaculatus* which also apparently survived an attack in the same part of the body. Figures 4 and 5 suggest that butterflyfishes can mislead predators and survive attacks from them. Other photographs of 'deformed' butterflyfishes missing two or three anterior dorsal fin rays (Burgess 1978) may also be the result of attacks on fishes poised in their defensive posture. Consequently, these photographs can be interpreted as evidence that potential predators learn that butterflyfishes are noxious through trial-and-error. Additional support comes from Motta's observation of a predatory fish spitting out a butterflyfish caught in its mouth with erect fin rays. Predator learning through negative experiences is a central part of aposematic coloration.

Field testing

To establish the presence of mimicry systems among butterflyfishes, models and dupes need to be identified. The literature on experimental tests of mimicry, cited above, is valuable in constructing adequate field experiments. There is currently no

154

Fig. 4. A *Chaetodon citrinellus* thought to have escaped predation in Guam. Even though it lost 10% of its body, the fish recovered well and was observed with its mate for 10 months in 1976 and 1977.

Fig. 5. A *Chaetodon quadrimaculatus* thought to have escaped predation in Hawaii. This species exhibits false eyespots day and night in the same location where *C. citrinellus* only exhibits them at night. (Photo. by Phil Motta).

available data on the presence or absence of poisons in butterflyfishes. They need to be analyzed for toxins in the spines, skin and flesh. Poisons in addition to spines would support the warning coloration and Mullerian mimicry hypothesis.

Various species of reef fishes, including butterflyfishes, could be enclosed in large canopies over the reef and predators introduced to see if the chaetodontids are avoided. It needs to be tested whether or not the color patterns themselves deter predators. Whether or not butterflyfishes are edible could be tested by starving predators before introduction. The same type of feeding trials should be conducted using chaetodontids with their coloration removed. The significance of eyespots could be tested by placing food in models with and without eyespots to see if eyespots can fool potential predators. Some of these tests could also be run in the laboratory and field using octopuses, sharks, eels, scorpaenids and other fishes as potential predators.

Conclusions

Butterflyfishes are deep-bodied, bony fishes which erect formidable fin spines when threatened. They communicate both inter- and intraspecifically with their coloration and changes to it. The fact that 95% of the species of the genus *Chaetodon* have eyemasks, eye camouflage or black faces and the prevalence of false eyespots demonstrates that eyes are important signals for predators which must be hidden. This conclusion is supported by the fact that 45.5% of the species have false eyespots along with eye camouflage to confuse predators about their direction of movement and the location of their real eyes.

False eyespots are visual cues important in minimizing predation. In cases in which eyespots are centrally located on the flank of fishes (e.g. the two-eye patterns of *C. citrinellus*, *C. melannotus* and *C. trifascialis*) and appear or are accentuated at night, they are thought to function to intimidate and discourage potential predators mainly by appearing to be an enemy of the potential predator. When eyespots are ever-present (even though shading is dynamic) and located in the dorsal posterior, the nape, or on the caudal peduncle, they probably serve more to misdirect attacks than to discourage them. It is important to note that these two functions are not mutually exclusive and could act together.

False eyespots are located primarily in areas of the body which allow survival following an attack. No single explanation of the role of brilliant coloration of chaetodontids is sufficient or comprehensive enough to account for all circumstances. This paper has outlined the functional significance of butterflyfish coloration at different times. Part of this functional plasticity results from the fact that butterflyfish colors and patterns are not fixed but can be hormonally regulated to vary pigmentation and contrast depending on the social milieu.

For chaetodontids, the aggression release hypothesis (Lorenz 1962, 1966) is rejected because poster coloration is not unchangeable and few species are territorial. The species recognition hypothesis (Zumpe 1965) is rejected from the results of field experiment with conspecific and heterospecific models. The disruptive coloration hypothesis (Longley 1917) is not supported as a general explanation of poster coloration because butterflyfishes are highly visible. Eyespots are visual cues to be seen not to conceal the fish. However disruptive coloration does explain the prevalence of eyebars among these fishes.

The aposematic hypothesis (Gosline 1965) is supported by chaetodontid morphology, behavior, their absence from the thousands of stomachs of piscivorous predators that have been examined and the prevalence of false eyespots. Poster coloration in chaetodontids has evolved partly in response to selective pressures of predation. In general, this coloration is thought to be aposematic which advertises a dangerous, low-quality meal. The prevalence of false eyespots among *Chaetodon* spp. could indicate the presence of Mullerian mimicry. Eye camouflage and false eyespots represent further refinements in minimizing the threat of predators. False eyespots in conjunction with camouflaged eyes can misdirect attacks and repel potential predators, and both effects are enhanced during low light levels when piscivorous predators are most active and butterflyfishes are most vulnerable.

Acknowledgements

The help of George Barlow, Bill Hamiliton and Phil Lobel is greatly appreciated. Reviews by Jack Hailman and Phil Motta helped to significantly improve this paper.

References cited

Allen, G.R. 1980. Butterfly and angelfishes of the world, Vol. 2. Wiley-Interspace, New York. 352 pp.
Alexander, R. 1967. Functional design in fishes. Hutchinson, London. 160 pp.
Barlow, G.W. 1972. The attitude of fish eye-lines in relation to body shape and to stripes and bars. Copiea 1972: 4–12.
Brower, L.P. 1988. Preface: mimicry and the evolutionary process. Amer. Nat. Suppl. 131: 1–3.
Burgess, W.E. 1978. Butterflyfishes of the world. T.F.H. Publications, Neptune City. 832 pp.
Cott, H.B. 1957. Adaptive coloration in animals. Methuen Press. London. 508 pp. (Reprinted with minor corrections; first published 1940).
Edmunds, M.E. 1974. Defence in animals: a survey of anti-predator defences. Longman, Burnt Mill. 357 pp.
Ehrlich, P.R. 1975. The population biology of coral reef fishes. Ann. Rev. Ecol. Syst. 6: 211–247.
Ehrlich, P.R., F.H. Talbot, P.C. Russell & G.R. Anderson. 1977. The behavior of chaetodontid fishes with special reference to Lorenz' 'poster colouration' hypothesis. J. Zool. Lond. 183: 312–228.
Fricke, H.W. 1966. Atrappenversuche mit einigen plukatfarbigen Korallenfishen in Roten Meer. Z. Tierpsychol. 23: 4–7.
Fricke, H.W. 1973. Der Einfluss des Lichtes auf Korperfarbund und dammerungsverhalten des Korallenfischen *Chaetodon melannotus*. Mar. Biol. 22: 251–262.
Gawlik, R.J. 1984. Avoidance learning and memory in the largemouth bass (*Micropterus salmoides*), fed *Bufo* tadpoles. Masters' thesis, University of Eastern Illinois, Charleston. 45 pp.
Gosline, W.A. 1965. Thoughts on systematic work in outlaying areas. Syst. Zool. 14: 59–61.
Guilford, T. 1968. The evolution of conspicuous coloration. Amer. Nat. (Suppl.) 131: 7–21.
Hailman, J.P. 1977. Optical signals. Indiana University Press, Bloomington. 362 pp.
Hailman, J.P. 1981. A test of symmetry-deception in a chaetodontid fish. Anim. Behav. 29: 1266–1267.
Hamilton, W.J. III & R.M. Peterman. 1971. Countershading in the colourful reef fish *Chaetodon lunula:* concealment, communication, or both? Anim. Behav. 19: 357–364.
Hiatt, R.W. & D.W. Strasburg. 1960. Ecological relationships of the fish fauna on coral reefs on the Marshall Island. Ecol. Monogr. 30: 65–127.
Hinton, H.F. 1977. Mimicry provides information about the perceptual capacities of predators. Folia Entomol. Mex. 37: 19–29.
Hobson, E.S. 1979. Interactions between piscivorous fishes and their prey. pp. 231–242. *In:* H.E. Clepper (ed.) Predator-Prey Systems in Fisheries Management, Sport Fishing Inst., Washington.
Hobson, E.S. 1974. Feeding relationships of teleostean fishes on coral reefs in Kona, Hawaii. U.S. Fish. Bull. 77: 915–1031.
Hobson, E.S. & E.H. Chave. 1972. Hawaiian reef animals. University Hawaii Press, Honolulu. 135 pp.
Huheey, J.E. 1984. Warning coloration and mimicry. pp. 257–297. *In:* W.J. Bell & R.T. Cadre (ed.) Chemical Ecology of Insects, Chapman & Hall, London. 524 pp.
Huheey, J.E. 1988. Mathematical models of mimicry. Amer. Nat. (Suppl.) 131: 22–41.
Kruse, K.C. & B.M. Stone. 1984. Largemouth bass (*Micropterus salmoides*) learn to avoid feeding on toad *Bufo* tadpoles. Anim. Behav. 32: 1035–1044.
Leccia, F.M. 1970. Estudios preliminares sobre la ecologia de los peces de los Ilanos de Venezuela. Acta Biol., Venezuela 7: 71–102.
Levine, J.S., P.S. Lobel & E.F. MacNichol. 1980. Visual communication in fishes. pp. 447–476. *In:* M.A. Ali (ed.) Environmental Physiology of Fishes, NATO Advanced. Studies Inst. Ser. A, Life Science V35, Plenum Press, New York.
Longley, W.H. 1917. Studies upon the biological significance of animal coloration. I: The colors and color changes of West Indian reef-fishes. Z. exp. Zool. 1: 533–399.
Lorenz, K. 1962. The function of colour in coral reef fishes. Proc. Roy. Inst. Gt. Brit. 39: 282–296.
Lorenz, K. 1966. On aggression. Harcourt, Brace & World, Inc., New York. 306 pp.
Losey, G.S. 1972. Predation protection in the poison-fang blenny, *Meiacanthus atrodorsalis*, and its mimics, *Ecsenium bicolor* and *Runula laudandus* (Blenniidae). Pac. Sci. 26: 129–139.
Losey, G.S. 1978. The symbiotic behavior of fishes. pp. 205–235. *In:* D.I. Mostofsky (ed.) The Behavior of Fish and Other Aquatic Animals, Academic Press, New York.
McCosker, J.E. 1977. Fright posture of the plesiopid fish *Calloplesiops altivelis:* an example of Batesian mimicry. Science 197: 400–401.
Meyers, R.F. 1980. *Chaetodon flavocoronatus*, a new species of butterfly fish from Guam. Micronesica 16: 297–303.
Motta, P.J. 1984. Response by potential prey to coral reef predators. Anim. Behav. 31: 1257–1259.
Myrberg, A.A. & R.E. Thresher. 1974. Interspecific aggression and its relevance to the concept of territoriality in fishes. Amer. Zool. 14: 81–96.
Neudecker, S. 1979. Effects of grazing and browsing fishes on the zonation of corals in Guam. Ecology 60: 666–672.

Neudecker, S. & P.S. Lobel. 1982. Mating systems of chaetodontid and pomacanthid fishes at St. Croix. Z. Tierpsychol. 59: 299–318.

Pasteur, G. 1982. A classificatory review of mimicry systems. Ann. Rev. Ecol. Syst. 13: 169–199.

Pough, F.H. 1988. Mimicry of vertebrates: are the rules different? Amer. Nat. (Suppl.) 131: 67–102.

Randall, J. 1967. Food habits of reef fishes of the West Indies. Stud. Trop. Oceanogr. 5: 665–847.

Randall, J.E. & H.A. Randall. 1960. Examples of mimicry and protective resemblance in tropical marine fishes. Bull. Mar. Sci. Gulf Caribb. 10: 444–480.

Rasa, O.A.E. 1969. Territoriality and the establishment of dominance by means of visual cues in *Pomacentrus jenkinsi* Pisces: Pomacentridae. Z. Tierpsychol. 26: 825–845.

Reese, E.S. 1973. Duration of residence by coral reef fishes on 'home' reefs. Copiea 1: 145–149.

Reese, E.S. 1975. A comparative field study of the social behavior and related ecology of reef fishes of the family Chaetodontidae. Z. Tierpsychol. 37: 37–61.

Reighard, J. 1908. An experimental study of warning coloration in coral reef fishes. Pap. Tortugas Lab. 2: 257–325.

Rettenmeyer, C.W. 1970. Insect mimicry. Ann. Rev. Entomol. 15: 43–74.

Rothschild, M. 1964. An extension of Dr. Lincoln Brower's theory on bird predation and food specificity, together with some observations on bird memory in relation to aposematic colour patterns. Entomologist 1964: 73–78.

Rothschild, M. 1972. Colour and poisons in insect protection. New Sci. 54: 318–320.

Steene, R.C. 1978. Butterfly and angelfishes of the world, Vol. 1. John Wiley & Sons, New York. 144 pp.

Thresher, R.E. 1977. Eye ornamentation of Carribbean reef fishes. Z. Tierpsychol. 43: 152–158.

Waldbauer, G.P. 1988. Asynchrony between batesian mimics and their models. Amer. Nat. (Suppl.) 133: 103–121.

Wickler, W. 1968. Mimicry in plants and animals. McGraw-Hill, New York. 255 pp.

Zaret, T.M. 1977. Inhibition of cannibalism in *Cichla ocellaris* and hypothesis of predator mimicry among South American fishes. Evolution 31: 421–437.

Zumpe, D. 1965. Laboratory observations on the aggressive behavior of some butterfly fishes Chaetodontidae. Z. Tierpsychol. 22: 226–236.

Chaetodon bennetti from the Fanning Atoll, Line Islands, Pacific Ocean. Photo by P.S. Lobel.

Dentition patterns among Pacific and Western Atlantic butterflyfishes (Perciformes, Chaetodontidae): relationship to feeding ecology and evolutionary history

Philip J. Motta
Department of Biology, University of South Florida, Tampa, FL 33620, U.S.A.

Received 12.4.1988   Accepted 12.8.1988

Key words: Jaws, Evolution, Corals, Teeth, Cladistics, Specialists

Synopsis

The jaw dentition of fifteen species of Pacific and Western Atlantic chaetodontid butterflyfishes was examined in light of their feeding habits and phylogenetic relationships. The ancestral tooth pattern is typical of many of the butterflyfishes, and variations on this basic pattern involve changes in the arrangement, length and number of teeth, and tooth shape to a lesser extent. Many of the more derived conditions can be explained by simple changes in relative jaw shape and size. Despite what appears to be adequate time for evolutionary changes to occur between the Pacific and Western Atlantic faunas, many species retain the generalized tooth arrangement permitting efficient exploitation in a very generalized manner. However, Pacific species as a whole show more specialized morphologies for hard coral feeding than do Western Atlantic species. Cases of parallel and divergent evolution are identified between and among the two faunas. Most morphological change associated with feeding in butterflyfishes is confined to the anterior region of the head, and particularly a few key elements. Suggestions for future morphological studies on the chaetodontids are outlined.

Introduction

Feeding is an important biological role of the vertebrate head, and as such we expect to find morphological differences among species, particularly among species with divergent foraging strategies. Because the dentition and jaw structure of fishes appears to be evolutionary labile many studies have found good correlation between these parameters and foraging (Fryer & Iles 1972, Emery 1973, Greenwood 1974). Many of these differences are emphasized in sympatric populations of fishes that have undergone dramatic adaptive radiations, such as the African cichlid fishes.

Despite the fact that butterflyfish jaw dentitions look superficially alike, Hobson (1974) and Burgess (1978) described some differences, and I (Motta 1984, 1985, 1987, 1988) have elucidated major differences in shape, size, arrangement, number and even biochemical composition of their teeth that are related to how they forage.

In many cases we wish to interpret these differences or lack thereof, in an evolutionary context. While it is interesting to speculate about evolutionary changes that has occured in the feeding apparatus within certain taxa, it is most instructive to have an independently derived classification of the organisms upon which data can be superimposed. Blum's (1988, 1989) butterflyfish cladogram presents such a unique opportunity. With morphological data on fifteen species of Western Atlantic and Pacific butterflyfishes I can interpret my data on dentition patterns and jaw structure in an evolutionary context.

The chaetodontid butterflyfishes are a diverse group of circumtropical reef fishes (114 species, Burgess 1978) that derive their name from their bristle-like teeth (chaite = bristle, odon = teeth). The family is most speciose in the Indo-West Pacific, with diminishing species numbers with increasing distance from the area (Burgess 1978). In the Hawaiian Islands where this study was partly conducted there are approximately 20 species (Gosline & Brock 1960). In the Western Atlantic, particularly the Caribbean, there are seven species, five of which are primarily shallow water inhabitants (Hubbs 1963, Randall 1968). The biomass of butterflyfishes is also lower in the Caribbean-Eastern Pacific fauna as compared to the Indo-West Pacific fauna (Findley & Findley 1989).

Of all the Mid-Pacific forms already studied by me, only one species bridges the East Pacific barrier and is found on the west coast of North and Central America (*Forcipiger flavissimus*, Rosenblatt 1967, Burgess 1978). The Isthmus of Panama has separated the Pacific fauna from the Western Atlantic butterflyfishes for approximately two to five million years. None of the five Caribbean species studied here is found in the Pacific. Given this relatively long period of geographical separation and divergence of diets, this group provides an ideal opportunity to examine patterns in dentition that relate to foraging strategies and/or phylogenetic proximity among species.

Materials and methods

Feeding data

Diets and feeding habits of all the species are based on observations and gut content analyses by others (see Motta 1985, 1988, and below) and by field observations by myself. My observations are on adult specimens using SCUBA at Carrie Bow Cay and adjacent islands, Belize barrier reef; Port Royal Cays and Discovery Bay, Jamaica; south-west and north shore, St. Thomas; and Salt River canyon, St. Croix, U.S. Virgin Islands.

Species studied

Ten species of adult butterflyfishes were studied from the Hawaiian Islands and five from the Caribbean. These represent some of the most common and abundant species at the two locations. The diets and feeding behaviors of the Pacific species are detailed in Motta (1985, 1988) but are in brief: *Chaetodon miliaris* is an opportunistic zooplanktivore that feeds primarily on calanoid copepods; *C. trifascialis* is an obligate hard coral browser that is exclusively associated with *Acropora* corals (and therefore not very abundant in the Hawaiian archipelago); *C. auriga* is a benthic omnivore that tears off pieces of noncoralline and coralline invertebrates, particularly alcyonarians, polychaete worms, sea anemones, scleractinians and algae; *Chaetodon trifasciatus* browses on hard corals as does *C. ornatissimus* and *C. multicinctus*; *C. unimaculatus* is a facultative soft and hard coral grazer; *C. quadrimaculatus* is an omnivore that browses on algae, anthozoans, polychaetes and hydroids; *Forcipiger longirostris* is an inertial suction feeder on small invertebrates, mostly shrimps, feeding in crevices and between coral branches; and *F. flavissimus* is a predator grabbing and tearing pieces of larger, benthic noncoralline invertebrates.

These species are all distributed in the tropical Indopacific as far east as the Hawaiian Islands, except for the two long-nosed forms, *Forcipiger longirostris* the distribution of which is still not resolved but is found as far east in the Pacific as the Tuamotus, and *F. flavissimus* which is found as far east as the west coast of Mexico (Burgess 1978, Thomson et al. 1979, J. Briggs personal communication). *Chaetodon miliaris* is endemic to the Hawaiian Islands (Burgess 1978).

There is not as much data on the diets and feeding habits of the Caribbean species. *Chaetodon capistratus* the most common Caribbean butterflyfish, is a browser tearing one polyp loose per bite, primarily exposed (expanded) polyps. It uses slight rapid forward lunges and occasionaly lateral jerks of the head to capture its prey. Prey preference varies with location but includes gorgonians, scleractinians, zoantharians, polychaetes, anemones, and other small amounts of animal matter (Randall 1967, Birkeland & Neudecker 1981, Gore 1984, Lasker 1985, Motta personal observation).

Chaetodon striatus uses lateral jerks of the head and slight forward lunges during its prey capture. Randall (1967) found polychaete worms and anthozoans to predominate the gut contents, accompanied by a small amount of unidentified crustaceans, and mollusc eggs. Kaufman & Ebersole (1984) classed it as an invertebrate specialist. Therefore, this species apparently browses mostly on sessile non-scleractinian invertebrates, mostly by grabbing and tearing its prey.

The data on *C. sedentarius* is likewise depauperate. Based on my observations, and Randall's (1967) gut content data on three specimens, it appears to be very generalized in diet. It mostly grabs and tears small, sessile, non-scleractinian invertebrates, but also takes scleractinian and gorgonian polyps, algae, small motile invertebrates such as shrimp, and may often feed on plankton. It too uses slight forward lunges and lateral swipes of the head.

Chaetodon ocellatus uses lateral swipes a great deal, as well as forward lunges. Aiken (1975) reports that it feeds mostly on polychaetes, echinoderm tube feet, amphipods, algae, unidentified crustaceans and unidentified eggs. It also appears to primarily feed by grabbing and tearing its sessile prey.

Chaetodon aculeatus is a highly selective predator grabbing and tearing polychaetes, crustaceans, and eggs, oftentimes with very rapid forward lunges. However, it also browses tentacles from tubeworms, and pedicellariae and tubefeet from echinoids (Randall 1967, Birkeland & Neudecker 1981, personal observation). Similar to *C. auriga* and *F. flavissimus* it spends considerable time searching for its prey in rubble, and in-and-around ledges and holes.

These five species are distributed throughout the Western Atlantic Ocean from South America partially up the east coast of North America, and are generally found throughout the Caribbean. They are not found in the Pacific Ocean and do not bridge the Isthmus of Panama (Burgess 1978, J. Briggs personal communication).

Dentitions

The tooth bearing premaxillae and dentaries were removed and cleaned briefly in 5% sodium hypochlorite solution in an ultrasonic cleaner. These bones and the teeth of the ten Pacific species are detailed in Motta (1985, 1988) and butterflyfish tooth structure is given in Motta (1984). The dentaries of all species were illustrated with a camera lucida. Tooth rows on five to ten individuals of each species were counted in the symphysial region of the upper and lower jaw. In species without discrete tooth rows visible, the jaws were cleaned until all the teeth were removed by dissolving their collagenous attachment, and the tooth pedicel rows counted on the jaw.

Cladistics

The cladogram is that of Blum (1988, 1989) and includes only those species investigated here. Blum's cladogram encompasses many more species than presented here, and is based on a parsimony analysis of 34 variable features of osteology and soft anatomy, coded as 50 binary and two multistate unordered characters. These include ten post cranial, seven branchial, and 17 skull characters. Pomacanthids are used as the first outgroup, and the second outgroup contains *Drepane*, ephippids, scatophagids, and the Acanthuroidei.

Results

The dentitions of the ten Pacific species are detailed and illustrated in Motta (1985, 1988). The dentaries of all species, with the tooth arrangement, is illustrated in Figure 1. The premaxillae of these species have similar tooth arrangements to the dentaries. The dentition patterns of the 15 species examined can be grouped into the following

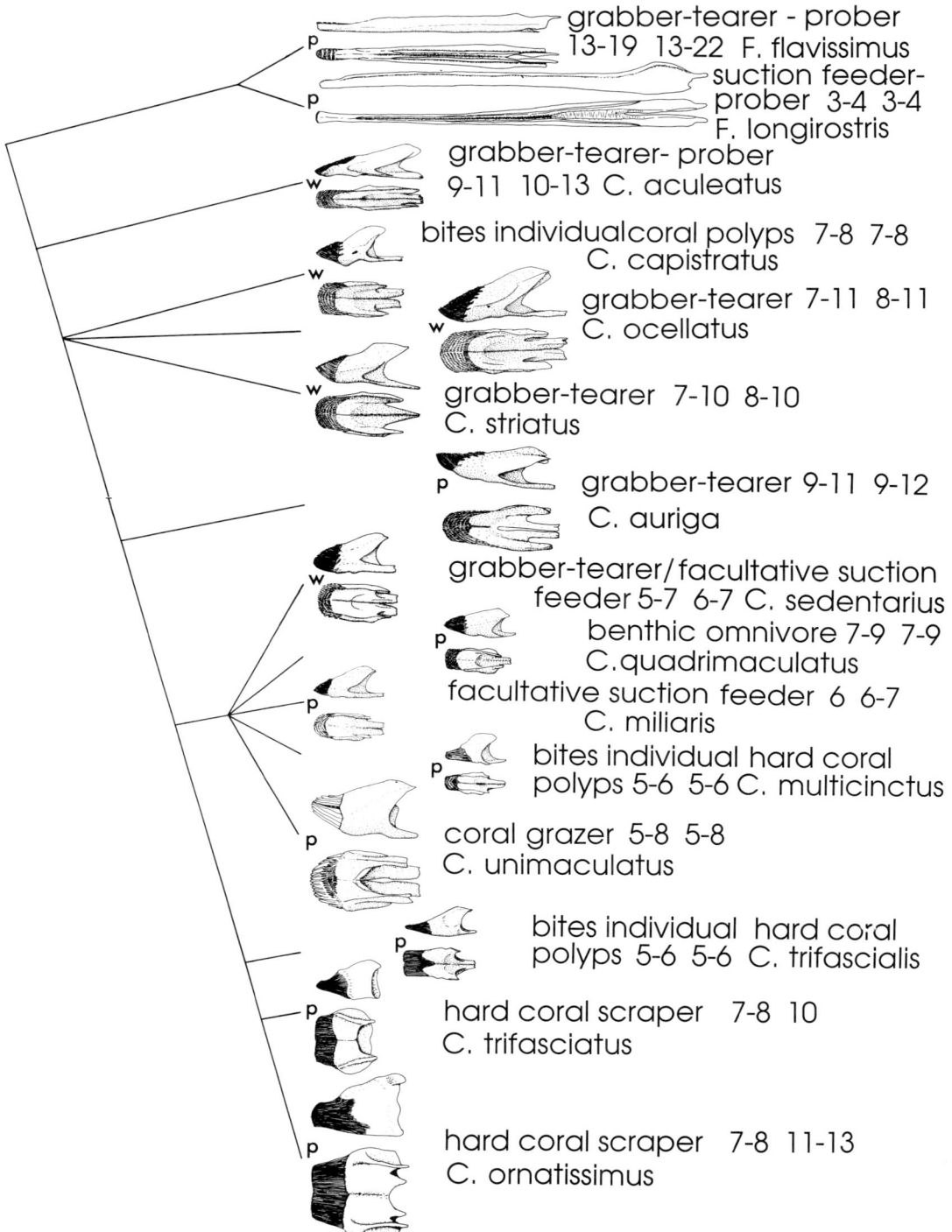

Fig. 1. Left lateral and dorsal view of the dentary of fifteen species of Pacific and Western Atlantic butterflyfishes arranged in a cladogram representing the relative degrees of relatedness. The extent and distribution of the jaw teeth are indicated as well as the summarized feeding behavior for each species. Number of symphysial tooth rows in the premaxilla (left) and dentary (right) is indicated for each species beside the feeding behavior (p – Pacific species, w – Western Atlantic species).

general categories:

1. Teeth in distinct rows with the more lingual teeth being shorter, finer, and more villiform; and more labial teeth more spatulate such that a gradation exists. Teeth lie on the ascending process of the dentary and the descending process of the premaxilla, so that the teeth encircle the mouth. Included are: *C. aculeatus, C. auriga, C. capistratus, C. quadrimaculatus, C. miliaris, C. ocellatus, C. sedentarius, C. striatus,* and *C. unimaculatus.*

2. Teeth in distinct rows with the more lingual teeth being shorter, finer, and more villiform, and more labial teeth more spatulate such that a gradation exists. Teeth do not lie on the ascending process of the dentary and descending process of the premaxilla, otherwise, teeth do not encircle the mouth. *C. multicinctus* is included in this category.

3. Teeth lie in very few distinct rows, mostly on the anterior edges of the jaws. The teeth are all short, villiform, and oriented approximately dorsoventrally, however, lingual teeth are slightly shorter than more labial ones. This includes *F. longirostris.*

4. Teeth lie in numerous distinct rows, groups of rows forming externally visible bands. The short, villiform teeth encircle the mouth, lying on the ascending process of the dentary and the descending process of the premaxilla, and are mostly oriented in a dorsoventral direction. Lingual teeth are only slightly shorter than more labial ones which are slightly more spatulate and longer. This includes *F. flavissimus.*

5. Teeth are mostly massed towards the anterior, spatulate, and approximately the same length, however, there are a few more lingual teeth that are shorter and more villiform. Examination of the tooth pedestals with the teeth removed reveals, as in all the other species examined, that the teeth actually lie in distinct rows grouped into bands. However, because the tooth tips lie in approximately the same transverse plane, they appear as an anteriorly massed pad. Teeth do not lie on the ascending process of the dentary or on the descending process of the premaxilla, otherwise, they do not encircle the mouth. This includes *C. ornatissimus, C. trifascialis,* and *C. trifasciatus.*

The species within the group-one category are, however, not entirely similar. Some of the species have very few rows of teeth, *C. unimaculatus* and *C. miliaris* for example, and others have more tooth rows, *C. aculeatus* and *C. auriga.* The peripheral teeth of *C. unimaculatus*, a grazer that damages the coralline portion of the coral skeleton, are extremely robust. The orientation of the teeth also varies among these species.

In the series *C. auriga, C. aculeatus,* and *F. flavissimus* the tooth pedestals become increasingly dorsoventrally directed as the dentigerous or tooth bearing portion of the jaw becomes more longitudinally oriented. Furthermore, the teeth lie further up the ascending process of the dentary (Fig. 1), and further down the descending process of the premaxilla. In essence, these jaws become increasingly depressed, the ascending and descending processes less pronounced, such that the tooth bands become visible when viewed from above as illustrated on the dentaries. Tooth bands are visible on all of the species when the teeth are removed, but it becomes increasingly evident as the teeth become more dorsoventrally oriented.

Choosing four common points on the dentaries of three very disparate jaw types from *C. quadrimaculatus, C. ornatissimus,* and *F. flavissimus* emphasizes the anatomical differences. *Chaetodon quadrimaculatus* has seven to nine discrete rows of teeth in the dentary. These teeth are attached to tooth pedestals (Motta 1984) that lie in rows grouped into bands (Fig. 2). The dentigerous surface of the jaw lies approximately in a transverse plane, with the pedestals projecting in the longitudinal direction. The more dorsal teeth are shorter and finer in this species, whereas the more ventral teeth are longer and thicker. Because the pedestals lie on above the other in the transverse plane, the more dorsal teeth (the short and fine ones) do not extend as anterior as the ventral ones, and distinct tooth rows are visible dorsally.

Chaetodon ornatissimus has 11 to 13 tooth rows on the dentary and the teeth appear as an anteriorly massed pad. The same four reference points on this jaw (Fig. 2) show that the jaw is foreshortened, and the anterior tooth bearing surface is sloped such that the dorsal surface overhangs the ventral surface. The result is that even though the more dorsal

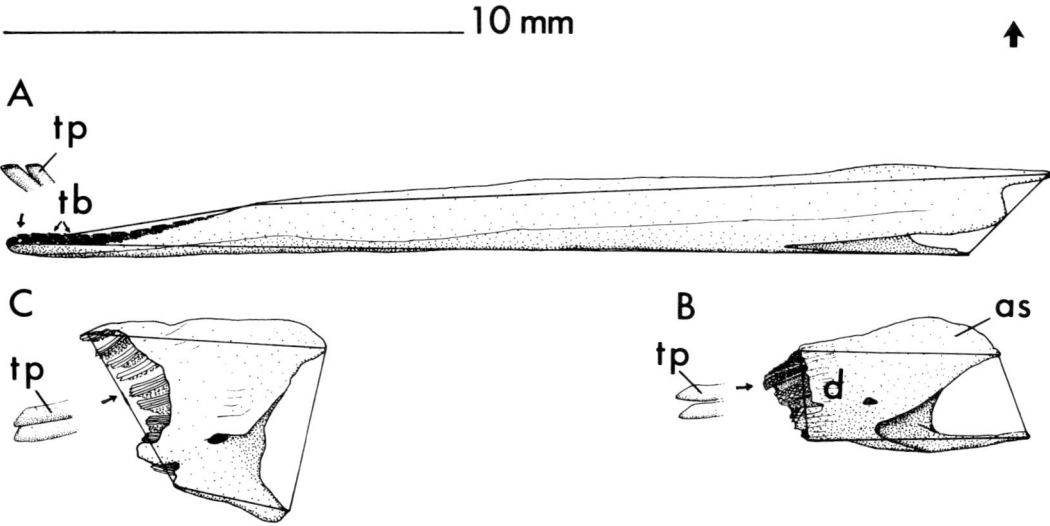

Fig. 2. Dentaries of three representative butterflyfishes with the teeth removed. Four common points on each bone are connected to indicate the relative shape changes in the jaw types. Representative tooth pedestals are enlarged: A – *F. flavissimus;* B – *C. quadrimaculatus;* C – *C. ornatissimus.* Arrow in upper right indicates dorsal (as – ascending process of dentary, d – dentigerous area of jaw, tp – tooth pedestals).

teeth are somewhat shorter and finer, they lie approximately as far anterior as the more recessed ventral teeth and all the teeth then form a pad with all the tooth tips lying at about the same location. Tooth rows are not evident when viewed dorsally.

Forcipiger flavissimus has a lenghtened jaw and with almost a longitudinally oriented tooth bearing surface (Fig. 2). In this case the tooth pedestals are oriented in a more dorsoventral direction, and because the teeth are all of approximately the same length, the 13 to 22 rows of teeth form a dorsoventrally oriented pad of short, villiform teeth. The tooth bands are now visible when viewed dorsally.

Discussion

The structure of butterflyfish teeth has been described in detail elsewhere (Motta 1984, 1985, 1987, 1988). Basically they are comprised of a form of dentine partly covered by an enameloid substance (Orvig 1977, Poole 1971). Each tooth has a shaft with a pulp cavity, and a tooth cap capped with iron (Motta 1984, 1987). Jaw teeth are borne on the premaxilla and dentary in a region referred to as the dentigerous area. Burgess (1978) noticed three distinct patterns of tooth arrangement in butterflyfishes. The most common arrangement is a series of discreet tooth rows. The others are a series of tooth bands (a band being composed of more than one row in close apposition) or a single anterior band. I recognize five categories of tooth pattern in these 15 species. The most common arrangement is a series of discreet tooth rows, showing marked gradation from more villiform, smaller, and shorter teeth in the lingual region, to more spatulate teeth labially. In fact, in all the butterflyfishes examined, the more lingual teeth are smaller and more villiform, although there might be fewer of these teeth in some species.

The functional teeth are all attached to bony pedestals (Motta 1984). The pedestals are in turn arranged in discreet bands in all species although the bands might not be visible externally in some species. In the majority of species with discrete rows of teeth visible, the pedestals lie in a plane that is not quite transverse on the jaw bone, that is, the dentigerous surface is approximately transverse (*C. quadrimaculatus*, Fig. 1). Blum (1988) also found that in most chaetodontids and outgroup taxa, the descending process of the premaxilla is toothed, and the angle between the ascending and descending processes is approximately 90 degrees.

The teeth of the different rows are of different length, such that viewed from the gape one sees discreet rows of teeth and not bands. Because the more dorsal teeth on the dentary are shorter, and their pedestals lie approximately in a transverse plane with the other pedestals, there caps lie more lingual to the other teeth, and so forth for the other rows, such that discrete rows are formed. The same arrangement applies to the premaxilla teeth.

The majority of the species in this type 1 category (see results) have teeth encircling the mouth, that is, teeth lying on most of the descending process of the premaxilla, and on most of the ascending process of the dentary. Other dentition patterns are variations on this arrangement. *Chaetodon multicinctus* varies from this pattern somewhat by not having teeth on most of the ascending and descending processes. In butterflyfishes with primarily a single anterior band of teeth forming a pad, *C. ornatissimus* for example, the dentigerous area of the jaws may be sloped such that the more medial or inner portion lies more anterior than the more outer or distal portion (Fig. 1). There still exist tooth rows and tooth bands, but the shorter inner or proximal teeth lie approximately as far anterior as the longer, more distal teeth because their pedestals lie more anterior than those of the larger teeth. In this way of pad of teeth is formed that has all the teeth with their caps approximately as far anterior as each other, and the tooth rows are not visible externally. This is seen in *C. ornatissimus*, *C. trifasciatus*, and less so in *C. trifascialis*.

In the series *C. ocellatus*, *C. auriga*, *C. aculeatus*, and *F. flavissimus*, the dentigerous area becomes increasingly sloped towards the longitudinal, or frontal plane. *Forcipiger flavissimus* has 13 to 22 rows of teeth in four mid-sagittal bands, the teeth all point approximately dorsoventrally, and the more lingual teeth are still slightly shorter and finer (Motta 1988) (Fig. 1). In this species the tooth bands are visible along with the tooth rows, and the dentigerous area is lengthened.

These arrangements are all variations on a theme, rows of teeth arranged into bands. What varies is the three dimensional arrangement of the teeth, relative length of the teeth, number of teeth, and tooth shape to less extent. Unlike the African cichlid fishes that display a range of tooth types (conical, bicuspid, tricuspid) (Fryer & Iles 1972, Greenwood 1974, Liem & Osse 1975), the butterflyfish tooth type is remarkably constant despite the various specializations.

By lengthening the jaw bones and reorienting the tooth pedestals dorsoventrally during the evolution of the species, *F. flavissimus* for example, has formed numerous rows of recurved teeth that lie in bands (Fig. 1). This species has a relatively large lateral slit in the gape, one that is lined with numerous rows of teeth. This jaw can effectively grasp and hold its worm-like prey. The two long-nosed forms, *C. aculeatus* in the Western Atlantic, and *F. flavissimus* in the Pacific both have similar diets of sessile non-scleractinian invertebrates such as worms, and they both have similar jaw morphologies and dentitions. These two species are ecological equivalents (Pianka 1975) as they fill similar ecological niches in different independently evolved faunas.

Most of these dentition patterns could be accomplished by simple proportional changes in the jaw length and orientation of the tooth pedestals. Using a coordinate system similar to that of Thompson (1961) whereby four coordinates are outlined for each jaw, the relative depression of the dentary and reorientation of the dentigerous area is emphasized (Fig. 2).

Greenwood (1974, 1984) found that what differences there are in jaw shape among the Lake Victoria *Haplochromis* cichlid fishes are attributed to allometric growth changes. The generalized oral dentition among haplochromines in Lake Victoria and elsewhere comprise an outer row of unequally bicuspid teeth backed by two or three rows of smaller, tricuspid teeth. Departures from this pattern are principally by changed relative proportions in shape, in number, in differentition of cusp pattern, and in the pattern of tooth distribution in the jaws. None of these changes seem to involve macromutations or other extraordinary evolutionary events (Liem & Osse 1975). Similar to the coral scraping butterflyfishes *C. ornatissimus* and *C. trifasciatus*, there is a trend among the epiphytic algal-scrapers of Lake Victoria for increase in the inner tooth rows and a tendency for the inner teeth to match the outer teeth (Greenwood 1984).

The most plausible vicariance model of Caribbean and Eastern Pacific fish faunas implies that butterflyfishes of the two regions were at once congruent. This original Eastern-Pacific-Caribbean track was fragmented (Rosen 1975). The butterflyfish fauna of the Eastern Pacific and Caribbean were separated at latest, during the rise of the Isthmus of Panama some two to five million years ago (Woodring 1966, Vawter et al. 1980, Futuyma 1986). The new world landmass constitutes a barrier that is virtually complete as witnessed by the fact that there are probably less than a dozen shore fishes common to the tropical waters on both sides of the isthmus (Briggs 1961). Damselfishes on both sides of the isthmus have been shown to have considerable genetic differentiation despite little morphological change (Gorman et al. 1976, Gorman & Kim 1977), and the presence of so called geminate pairs of fishes, Atlantic and Pacific fishes which are morphologically close but different species, indicates the degree of genetic divergence possible. However, distinct morphological changes can occur in a much shorter time period; the explosive speciation of the African lake cichlid fishes took place in the geologically short time span of 750 000 years (Greenwood 1984). Greenwood (1974, 1984), Liem & Osse (1975), and Liem (1970) have argued that relatively simple allometric changes such as these might be relatively rapid in the evolutionary time frame. In part, the evolutionary success of the haplochromine cichlid fishes in the African lakes could be due to the relatively simple and rapid adaptive modifications of their cranial anatomy (Greenwood 1974).

As most of these Pacific butterflyfishes, with the exception of *F. flavissimus*, are Indopacific in distribution and not found in the eastern Pacific, their time of separation from the Western Atlantic fauna is most likely more than two to five million years. Blum (1988, 1989) has provided a hypothesis of phylogenetic relationship for these 15 species (Fig. 3). This permits discussion of the evolutionary changes in the dentition as discrete events in the phylogeny of chaetodontids.

Based on outgroup comparisons (Blum 1988, personal communication) the ancestral butterflyfish dentition might be best exemplified by *C. capistratus*, *C. ocellatus*, or *C. striatus*. This type of dentition is characterized by setiform teeth; tooth caps unicuspid and slightly spatulate; four to five fold differences between the most labial and lingual teeth; teeth at about 30 degrees from the horizontal; three bands of tooth rows along the medial symphysis with three overlapping rows in the most labial band. The ancestral condition is exemplified by the majority of the species examined here: *C. capistratus*, *C. ocellatus*, *C. striatus*, *C. sedentarius*, *C. quadrimaculatus*, and *C. miliaris* (category 1 of results).

If this is the case, parallel evolution in dentition patterns and jaw form has occured among *F. flavissimus*, *C. aculeatus*, and *C. auriga* with their apomorphic elongated jaw bones, reorientation of the tooth rows in a more dorsoventral direction, and numerous rows of teeth that encircle the mouth. This dentition and jaw structure is suited for probing in crevices, and grasping and tearing their prey (Motta 1985, 1988). I previously indicated that this implies convergent evolution (Motta 1988), but it is better described as parallel evolution. Ideally, convergent evolution involves cases in which similar phenotypes have evolved by different developmental pathways (Futuyma 1986), or it can be defined as the production of a set of similar phenotypic characteristics in phylogenetically unrelated organisms subject to similar abiotic or biotic agents on natural selection (Cody & Mooney 1978). On the other hand, parallel evolution refers to independent developmental modifications of the same kind. Whereas, related species have similar developmental programs, parallelism is frequent among closely related species (Futuyma 1986).

Similarly, the dentition and jaws of *C. multicinctus* and *C. trifascialis* show parallelism. These species are obligate hard coral browsers that nip one polyp per bite, and both have small jaws with teeth that are massed towards the anterior (Motta 1985, 1988). Most of the teeth are similar in length, but more so in *C. trifascialis*, and there are no teeth on

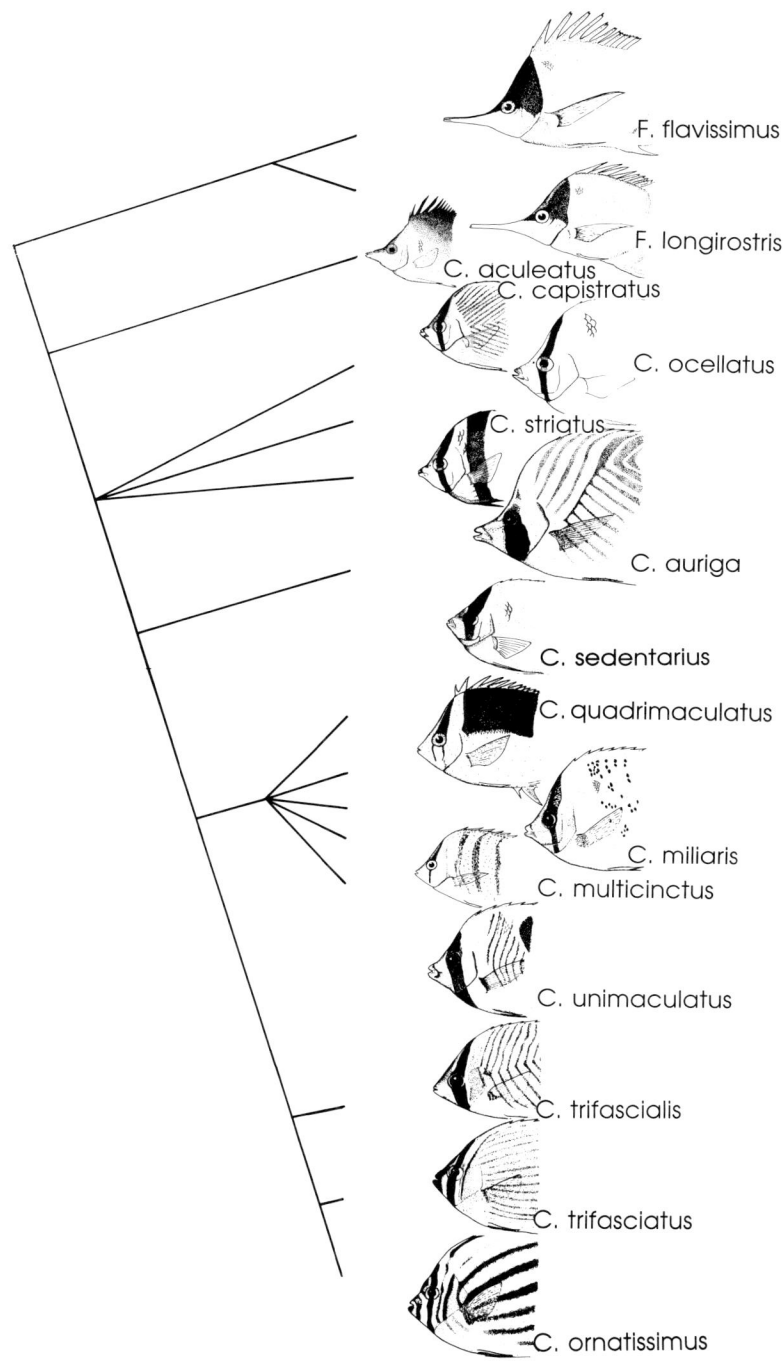

Fig. 3. The relative degrees of relatedness of some Hawaiian and Western Atlantic butterflyfishes based on thirty four morphological characters (see text for explanation). The heads are illustrated with the mouth in the relaxed position. *Forcipiger flavissimus* and *F. longirostris* are sister species. The group comprised of *C. sedentarius, C. quadrimaculatus, C. miliaris, C. multicinctus,* and *C. unimaculatus* are monophyletic. Likewise, *C. trifascialis, C. trifasciatus,* and *C. ornatissimus* are believed to be monphyletic. *C. trifasciatus* and *C. ornatissimus* are almost certainly sister subgenera (S. Blum personal communication).

most of the ascending process of the dentary and the descending process of the premaxilla.

The robust peripheral teeth of the coral grazing *C. unimaculatus* (Motta 1985, 1988) is an autapomorphic character that apparently is not shared with any other butterflyfish. The dentition pattern of teeth massed towards the anterior to form a pad with fewer teeth on the ascending process, such as on *C. trifascialis*, *C. trifasciatus*, and *C. ornatissimus* (Fig. 1) is synapomorphic. However, the sister species (Blum 1988, personal communication) *C. trifasciatus* and *C. ornatissimus* have a highly derived synapomorphic shovel-like mouth (Motta 1985, 1988) that is suited for combing corals to remove soft tissue.

The sister species *F. flavissimus* and *F. longirostris* have undergone divergent evolution in their dentition, with the latter being specialized for high speed inertial suction feeding on small, suspended invertebrates (Motta 1988).

Among the species there is also a range in number of tooth rows in addition to the shape, size, and distributional differences of the teeth. The most specialized suction feeder *F. longirostris* has a greatly reduced dentition typical of plankton feeding fishes (Gregory 1933, Suyehiro 1942, Lagler et al. 1962, Davis & Birdsong 1973, Emery 1973, Alexander 1974) although the reasons for this are not entirely clear. Both *C. miliaris* and *C. sedentarius* are very generalized feeders taking a wide range of prey types, and both are facultative inertial suction feeders on plankton, *C. miliaris* more so. Both of these species have about five to seven rows of teeth in each jaw.

The two obligate coral browsers *C. multicinctus* and *C. trifascialis* that nip individual hard coral polyps also have few tooth rows which is correlated with the very small jaws (Motta 1988) that form pincher-like mouths. *Chaetodon capistratus* has a broader diet, removing individual stony and soft coral polyps along with a variety of other prey, and its dentition is more pleisiomorphic compared to the very specialized Pacific hard coral feeders. On the other extreme, the species that browse on sedentary, non-scleractinian invertebrates by grabbing and tearing them loose, *F. flavissimus*, *C. aculeatus*, *C. ocellatus*, *C. striatus*, and *C. auriga*, have an increase in tooth row number, forming numerous distinct rows of teeth that are often more dorsoventrally oriented. These teeth encircle the mouth and allow them to hold onto their vagile prey (Motta 1985, 1988). Elongation of the jaws and dentigerous area may accompany the increase in tooth row number.

Butterflyfish diversity diminishes from the Indo-Pacific, with 30 species in the Marshall and Marianas Islands for instance (Schultz et al. 1953), 20 in the Hawaiian Islands (Gosline & Brock 1960), and seven in the Western Atlantic (Hubbs 1963, Randall 1968). There is a greater diversity of dentition and jaw morphologies among the Pacific species examined than among the Western Atlantic species, but this could be an artifact of the greater number of species in the Pacific. If a random sample had only examined five Pacific species, *C. auriga*, *C. quadrimaculatus*, *C. miliaris*, *C. multicinctus*, and *C. unimaculatus* for example, then the morphological diversity would have been little greater than that exemplified by the Western Atlantic species.

The five common Western Atlantic species investigated here have been isolated from the ten Pacific species for a considerable period of time, long enough for distinct morphological changes to occur. If the selection pressures associated with feeding are similar in the Pacific and the Western Atlantic butterflyfish faunas, then I would expect ecomorphological convergence between them, similar to that found in other vertebrates, for example birds from different continents (Karr & James 1975, Cody 1975, Cody & Mooney 1978), and desert lizards (Pianka 1975, Cody & Mooney 1978). However, I would expect to find differences under varying selection pressures, for example, alcyonacian soft corals are more represented in the diet of some of the Caribbean species examined, whereas hard scleractinian corals are proportionaly more important in the diet of some of the Pacific species.

There appears to be greater morphological specialization for feeding on hard corals in the Pacific, with the nipping jaws of *C. multicinctus* and *C. trifascialis*, the scraping jaws of *C. ornatissimus* and *C. trifasciatus*, and the robust jaws and teeth of the coral grazing *C. unimaculatus*. However, feeding on sessile non-coralline invertebrates such as tubeworms, pieces of echinoids, and the like, appears to pose similar evolutionary pressures in both faunas, and beak-like jaws for probing in crevices and numerous rows of short, dorso-ventrally oriented teeth have evolved to a greater and lesser extent in *C. ocellatus, C. auriga, C. aculeatus,* and *F. flavissimus*. However, despite at least two to five million years of separation, many of the species retain an ancestral dentition that is extremely versatile, permitting efficient exploitation of the benthos and the plankton. In these cases it appears that behavior determines dietary preference more than feeding morphology.

Butterflyfishes are remarkably consistent in body form. The morphological variation that exists is mostly in cranial anatomy, particularly the anterior regions of the feeding apparatus. Greenwood (1974, 1981) similarly found that most of the adaptive radiation in the Lake Victoria haplochromine cichlids occurs chiefly in the cranial anatomy and is developed through simple morphological transformations involving differential growth patterns in the various skull regions, coupled with similar changes in the jaws and suspensorium. Relatively simple changes in dentition, accompanied by other changes in a few key elements of the head such as the premaxilla and dentary (Motta 1985, 1988) have resulted in a diversity of feeding types among the butterflyfishes.

Future morphological research on butterflyfish feeding adaptations should expand the investigation on comparative osteology and myology of other species to reveal if and how the neurocranium as a whole has responded to these evolutionary changes, and further interspecific comparisons should be made on the cranial feeding adaptations. Further studies are needed to reveal whether my studies on 15 species have exhausted the major morphotypes in the group, or if there is additional diversity in the feeding structures. Another area that needs to be thoroughly explored, is the ecomorphological hypothesis that butterflyfish feeding and morphology are intimately related. Using various morphological parameters that appear to be directly related to feeding (Motta 1985, 1988), a multispecies study using principal component analysis and cannonical correlation should be made to investigate the relationship between feeding morphology and feeding ecology (Motta 1988).

Acknowledgements

I thank Stanley Blum, Peter Meylan and two anonymous reviewers of this journal for commenting on this manuscript. Stanley Blum very kindly provided the cladogram and discussion on butterflyfish relatedness. Joseph and Claude Berry, Louis and Louis Anthony Blanchard, and Frank Gumbs and other helpfull fishermen of Frenchtown, St. Thomas provided numerous specimens from their catches. The research was funded in part by faculty professional development grants from the University of Montana and the University of The Virgin Islands.

References cited

Aiken, K.A. 1975. The biology, ecology and bionomics of Caribbean reef fishes: Chaetodontidae (Butterfly and Angelfishes). Res Rept. Zool. Dept., Univ. West Indies (3). 57 pp.

Alexander, R.McN. 1974. Functional design in fishes. Hutchinson University Library, London. 160 pp.

Birkeland, C. & S. Neudecker. 1981. Foraging behavior of two Caribbean chaetodontids *Chaetodon capistratus* and *C. aculeatus*. Copeia 1981: 169–178.

Blum, S.D. 1988. Osteology and phylogeny of the Chaetodontidae (Pisces: Perciformes). Ph.D. Dissertation, University of Hawaii, Honolulu. 300 pp.

Blum, S.D. 1989. Biogeography of the Chaetodontidae: an analysis of allopatry among closely related species. Env. Biol. Fish. 25: 9–31.

Briggs, J.C. 1961. The East Pacific barrier and the distribution of marine shore fishes. Evolution 15: 545–554.

Burgess, W.E. 1978. Butterflyfishes of the world. T.F.H. Publications, Neptune City. 832 pp.

Cody, M.L. 1975. Towards a theory of continental species diversities: Bird distribution over Mediterranean habitat gradients. pp. 214–257. *In:* M.L. Cody & J.M. Diamond (ed.) Ecology and Evolution of Communities, Belknap, Cambridge.

Cody, M.L. & H.A. Mooney. 1978. Convergence versus nonconvergence in Mediterranean-climate ecosystems. Ann. Rev. Ecol. Syst. 9: 265–321.

Davis, W.P. & R.S. Birdsong. 1973. Coral reef fishes which forage in the water column. Helgol. Wiss. Meeresunters. 24: 292–306.

Emery, A.R. 1973. Comparative ecology and functional osteology of fourteen species of damselfish (Pisces: Pomacentridae) at Alligator Reef, Florida Keys. Bull. Mar. Sci. 23: 649–770.

Findley, J.S. & M.T. Findley. 1989. Circumtropical patterns in butterflyfish communities. Env. Biol. Fish. 25: 33–46.

Fryer, G. & T.D. Iles. 1972. Cichlid fishes of the great lakes of Africa: their biology and evolution. Oliver & Boyd, Edinburgh. 641 pp.

Futuyma, D.J. 1986. Evolutionary biology. Sinauer Associates, Sunderland. 600 pp.

Gore, M.A. 1984. Factors affecting the feeding behavior of a coral reed fish, *Chaetodon capistratus*. Bull. Mar. Sci. 35: 211–220.

Gorman, G.C. & Y.J. Kim. 1977. Genotypic evolution in the face of phenotypic conservativeness: *Abudefduf* (Pomacentridae) from the Atlantic and Pacific sides of Panama. Copeia 1977: 694–697.

Gorman, G.C., Y.J. Kim & R. Rubinoff. 1976. Genetic relationships of the three species of *Bathygobius* from the Atlantic and Pacific sides of Panama. Copeia 1976: 361–364.

Gosline, W.A. & V.E. Brock. 1960. Handbook of Hawaiian fishes. University of Hawaii Press, Honolulu. 372 pp.

Greenwood, P.H. 1974. The cichlid fishes of Lake Victoria, East Africa: The biology and evolution of a species flock. Bull. Brit. Mus. (Nat. Hist.) Zool. Suppl. 6: 1–34.

Greenwood, P.H. 1981. Species-flocks and explosive evolution. pp. 819–839. *In:* P.H. Greenwood (ed.) The Haplochromine Fishes of The East African Lakes, Cornell University Press, New York.

Greenwood, P.H. 1984. African cichlids and evolutionary theories. pp. 141–154. *In:* A.A. Echelle & I. Kornfield (ed.) Evolution of Fish Species Flocks, University of Maine Press, Orono.

Gregory, W.K. 1933. Fish skulls; a study of the evolution of the natural mechanisms. Trans. Amer. Philos. Soc. 23: 75–479.

Hobson, E.S. 1974. Feeding relationships of teleostean fishes on coral reefs in Kona, Hawaii. U.S. Fish. Bull. 72: 915–1031.

Hubbs, C.L. 1963. *Chaetodon aya* and related deep-living butterflyfishes: their variation, distribution and synonymy. Bull. Mar. Sci. 13: 133–192.

Karr, J.R. & F.C. James. 1975. Ecomorphological configurations and convergent evolution in species and communities. pp. 258–291. *In:* M.L. Cody & J.M. Diamond (ed.) Ecology and Evolution of Communities, Belknap, Cambridge.

Kaufman, L.S. & J.P. Ebersole. 1984. Microtopography and the organization of two assemblages of coral reef fishes in the West Indies. J. Exp. Mar. Biol. Ecol. 78: 253–268.

Lagler, K.F., J.E. Bardach & R.R. Miller. 1962. Ichthyology. John Wiley & Sons, Inc., New York. 497 pp.

Lasker, H.R. 1985. Prey preferences and browsing pressure of the butterflyfish *Chaetodon capistratus* on Caribbean gorgonians. Mar. Ecol. Prog. Ser. 21: 213–220.

Liem, K.F. 1970. Comparative functional anatomy of the Nandidae (Pisces: Teleostei). Fieldiana Zool. 56: 7–166.

Liem, K.F. & J.W.M. Osse. 1975. Biological versatility, evolution and food resource partitioning in African cichlid fishes. Amer. Zool. 15: 427–454.

Motta, P.J. 1984. Tooth attachment, replacement, and growth in the butterflyfish *Chaetodon miliaris* (Chaetodontidae, Perciformes). Can. J. Zool. 62: 183–189.

Motta, P.J. 1985. Functional morphology of the head of Hawaiian and Mid-Pacific butterflyfishes (Perciformes, Chaetodontidae). Env. Biol. Fish. 13: 253–276.

Motta, P.J. 1987. A quantitative analysis of ferric iron in butterflyfish teeth (Chaetodontidae, Perciformes) and the relationship to feeding ecology. Can. J. Zool. 65: 106–112.

Motta, P.J. 1988. Functional morphology of the feeding apparatus of ten species of Pacific butterflyfishes (Perciformes, Chaetodontidae): an ecomorphological approach. Env. Biol. Fish. 22: 39–67.

Orvig, J. 1977. A survey of odontodes ('dermal teeth') from developmental, structural, functional, and phyletic points of view. pp. 53–75. *In:* S.M. Andrews, R.S. Miles & A.D. Walker (ed.) Problems in Vertebrate Evolution, Linn. Soc. Symp. Ser. (4).

Pianka, E.R. 1975. Niche relations of desert lizards. pp. 292–314. *In:* M.L. Cody & J.M. Diamond (ed.) Ecology and Evolution of Communities, Belknap, Cambridge.

Poole, D.F.G. 1971. An introduction to the phylogeny of calcified tissues. pp. 65–79. *In:* A.A. Dahlberg (ed.) Dental Morphology and Evolution, University of Chicago Press, Chicago.

Randall, J.E. 1967. Food habits of reef fishes of the West Indies. Stud. Trop. Ocean. 5: 665–847.

Randall, J.E. 1968. Caribbean reef fishes. T.F.H. Publications, Neptune City. 318 pp.

Rosen, D.E. 1975. A vicariance model of Caribbean biogeography. Syst. Zool. 24: 431–464.

Rosenblatt, R.H. 1967. The zoogeographic relationships of the inshore fishes of tropical America. Stud. Trop. Oceangr. 5: 579–592.

Schultz, L.P., E.S. Herald, E.A. Lachner, A.D. Welander & L.P. Woods. 1953. Fishes of the Marshall and Marianas Islands, Vol. 1, Smithsonian Inst. Bull. 202, Washington. 685 pp.

Suyehiro, Y. 1942. A study on the digestive system and feeding habits of fish. Jap. J. Zool. 10: 1–301.

Thompson, D.A.W. 1961. On growth and form (abridged edition). J.T. Bonner (ed.). Cambridge University Press, London. 364 pp.

Thomson, D.A., L.T. Findley & A.N. Kerstitch. 1979. Reef fishes of the Sea of Cortez. John Wiley and Sons, New York. 302 pp.

Vawter, A.T., R. Rosenblatt & G.C. Gorman. 1980. Genetic divergence among fishes of the Eastern Pacific and the Caribbean: support for the molecular clock. Evolution 34: 705–711.

Woodring, P.W. 1966. The Panama land bridge as a sea barrier. Proc. Amer. Phil. Soc. 110: 425–434.

Prey selection by coral-feeding butterflyfishes: strategies to maximize the profit

Timothy C. Tricas
Department of Zoology, University of Hawaii at Manoa, Honolulu, Hawaii 96822, U.S.A.
Present address: Department of Otolaryngology, Washington University School of Medicine, 517 S. Euclid Ave., Box 8115, St. Louis. MO 63110 U.S.A., and Marine Biological Laboratory, Woods Hole, MA 02543, U.S.A.

Received 7.4.1988   Accepted 15.11.1988

Key words: Chaetodontidae, Corallivore, Coral reef fish, Energy maximizer, Feeding, Optimal foraging

Synopsis

Factors that structure preferences among food corals were examined for the obligate coral-feeding butterflyfish *Chaetodon multicinctus*. In the field, fish show a simple repetitious pattern of foraging composed of (1) pre-encounter search for coral colonies, and (2) post-encounter inspection/orientation, bite, and consumption of polyps. Rose coral, *Pocillopora meandrina,* and the massive coral, *Porites lobata*, were taken in higher proportions than their percentage substrate cover, while finger coral, *Porites compressa*, was taken in lower proportion. Paired presentations of coral colonies in the lab gave similar results: *Poc. meandrina* was preferred over *Por. lobata* which was preferred over *Por. compressa. Poc. meandrina* tissue had the highest energy content, lowest handling time, and highest profitability. Energy content did not differ among *Porites tissues,* but handling time was greater and more inspective eye movements were made while foraging on the branched finger coral, *Por. compressa*. Experimental manipulation of coral colony morphology indicate preferences among *Porites* are most likely structured by handling costs. Predictions of a simple prey-choice foraging model are supported in the *C. multicinctus* system if abundance of the branched coral *Por. compressa* is estimated as that available to fishes rather than percentage substrate cover. The relative size and abundance of stinging nematocysts are also consistent with observed foraging patterns in the field, but await immunological confirmation. Coral-feeding butterflyfishes offer unique opportunities to test models of foraging ecology in reef fishes, and the direction of future studies is suggested.

Introduction

Foraging behavior is an important aspect of individual fitness because it accrues metabolic substrates for maintenance, growth, and reproduction. When in short supply, food can limit one or more of these physiological processes. Under more favorable conditions, differential selection of items from a potential diet set may be advantageous when quality or accessibility vary among prey types.

While the selection of food items by animals has been extensively modelled and tested (recently reviewed by Krebs & McCleery 1984, Stephens & Krebs 1986), surprisingly little field and experimental work exists on reef fishes. This paucity is due in part to difficulties in prolonged observations of individuals in the marine environment, identifying prey species and unobtrusively quantifying their abundance, and performing experimental manipulations of prey to test specific models. Coral-feeding butterflyfishes of the family Chaetodontidae offer perhaps the best features available among marine teleosts to conduct long-term field studies of foraging. They are abundant on most shallow coral reefs in tropical and subtropical oceans, feed conspicuously on corals, and can be observed at close range for long periods. Individuals are conspicuous and usually identifiable by unique features in their color patterns, and are site attached in territories or stable home ranges (Reese 1975, Sutton 1985, Tricas 1985, 1989). Corals are non-cryptic, relatively easy to identify, and are a stable food resource. The combined features of butterflyfishes and their food corals make it relatively easy to quantify feeding rates on specific prey items as well as associated behaviors such as search and handling time, defense of resources, and foraging paths. In addition, chaetodontids generally adapt well to aquaria for studies where more control over experiments is necessary (e.g. Zumpe 1965, Reese 1977, Gore 1984).

There is good evidence that the fitness of butterflyfishes is enhanced by maximizing food intake. Growth is extremely rapid during the first year of life, and fecundity increases exponentially with body size in mature females (Ralston 1976a, b, 1981, Tricas 1986). For some species, male body length is positively correlated with feeding territory area and the size of his female mate (Tricas 1989). Food supply can control egg production in fishes (Tyler & Dunn 1976, Wooton 1977, Hirschfield 1980) and food shortages may induce oocyte resorption and atresia (Scott 1962, Hunter & Macewicz 1985) which is known to occur in chaetodontids (Tricas & Hiramoto 1989). Thus, it is likely that strong selection exists for foraging on food items that maximize growth and reproductive output.

Although measurable features of food quality such as prey size (Werner & Hall 1974) or protein content (Milton 1979, Owen-Smith & Novellie 1982) can influence fitness, energy content is most conveniently measured and most commonly modelled because it usually covaries with measures of organic content. For animals with energy maximization strategies, the optimal diet set consists of those items that maximize the net rate of energy intake. Less profitable items should be excluded from the diet if their inclusion would decrease the net rate of return (Pyke et al. 1977, Krebs 1978). The selection of profitable prey, however, may be moderated by toxic compounds that affect feeding patterns in fishes (Lobel & Ogden 1981, Targett et al. 1986). For corallivorous fishes, nematocyst defense by food corals must be considered as a potential influence on food selection as suggested for invertebrate corallivores (Barnes et al. 1970, Glynn & Krupp 1986).

Many studies demonstrate that butterflyfishes show distinct preferences for particular coral species (e.g. Reese 1977, Birkeland & Neudecker 1981, Gore 1984, Tricas 1985, Hourigan et al. 1988) but the proximate factors that structure these patterns are not understood. An a priori model can be developed for the Hawaiian obligate corallivore *Chaetodon multicinctus* based on the natural history of this butterflyfish and the biology of its food corals. This species feeds on the polyps of the corals *Pocillopora meandrina, Porites lobata,* and *Porites compressa* that occur naturally within feeding territories (Hobson 1974, Reese 1975, Tricas 1985, Hourigan et al. 1988, Motta 1988). Both males and females spend over 90% of their daily time budget feeding and behave as energy maximizers constrained primarily by the time available for feeding (Tricas 1989). This paper tests the prediction that corals with the highest energetic profit (defined as energy intake per unit foraging time) are taken preferentially over species of lower profit. Feeding patterns of *C. multicinctus* are examined in the field and compared with food preferences observed in the laboratory. The energy content, gross morphology, polyp density, and nematocyst defense are examined for food corals in reference to how each may influence the feeding patterns of *C. multicinctus* in their coral reef environment.

Methods

Study site and corals

Field data on food abundance and fish foraging patterns were collected at Puako reef on the northwest shore of the island of Hawaii. The primary study area is located at the base of the shallow reef flat and gradually slopes seaward to depths of about 12 m. This reef habitat consists of extensive fields of the unbranched massive coral, *Por. lobata*, and finger coral, *Por. compressa*. Rose coral, *Poc. meandrina*, is a small, robustly branched species widely scattered in low abundance among *Porites*. In addition, some fish were studied that held feeding territories on the shallow reef flat (<4 m deep) where total coral cover is lower but *Poc. meandrina* is more abundant than on the deep reef.

Field measurements and observations

Coral abundance was estimated within 30 fish territories in the deep coral-rich habitat and five on the shallow reef flat. A 1 m² quadrat (0.1 m grid) was placed at 20 randomly determined locations within each territory and the coral species under each grid intersection tallied. Percentage coral cover for each species was calculated from its proportion of the total counts for all substrates.

Feeding data for resident pairs in each of the 35 territories were recorded between 1000 and 1400 h. Divers followed at distances (1–2 m) that did not disturb the foraging paths of focal fish, and scored on underwater paper the coral species taken in each bite. Each member of a pair was observed during alternate 5 min observation periods for a total of at least 50 min per fish. Qualitative descriptions of foraging behaviors during travel between feeding sites, encounters with coral heads, and handling of corals during feeding bouts were recorded. In addition to direct observations of feeding, gut contents of six fish collected in the afternoon were examined and their volumetric proportions estimated to verify food items in the diet.

Encounter rates of food corals were estimated three times from movements made by fish within territories on the deep reef. Fifty small consecutively-numbered plastic tags were placed along the foraging paths of focal fish at 1 min time intervals. Tag locations were carefully recorded on a 1 m² grid map of the territory, handling and chase times subtracted from total foraging time, and average distance traveled per unit search time deter-

mined. It was assumed that the average horizontal visual field of a foraging fish was a frontal semi-circular field of 25 cm radius. Thus for every 1.0 cm of linear movement a rectangular area of substrate 50 cm² was encountered. Encounter rate (λ) for each coral was calculated as the product of the average area surveyed in the visual field per unit time and the proportion of coral polyp cover in the territory. This can be expressed as,

$$\lambda = 50 \, v \, pd,$$

where v = the average rate of movement during searching, p = proportion of bottom cover for that coral, d = polyp density, and the constant, 50, a conversion factor for substrate area encountered during linear movement.

Energy and water content of coral tissues

The caloric content of the three food corals was determined by microbomb calorimetry (Phillipson 1964). Surface tissue was removed from fresh collected corals with a stream of distilled water shot from a dental water jet (Johannes & Wiebe 1970). Samples were then frozen, lyophylized, pelletized, and bombed. The calorimeter was calibrated using benzoic acid standards. When appropriate, endothermic processes due to combustion of carbonate were adjusted (Paine 1966). Inorganic ash content was determined by ashing pellets for 3 h at 500°C in a muffle furnace. The water content of *Porites* corals was determined by dry weight analysis. Individual coral colonies were drip-dried, placed inside a large dry plastic bag, and surface tissue removed with a high-pressure stream of air. The resultant viscous blastate was transferred from the plastic bag into pre-weighed pans, weighed wet, oven-dried at 40°C, and then reweighed.

Polyp density

The density of coral polyps per unit surface area was estimated for the corals. A circular template (2 cm diameter) was placed haphazardly at 10 locations accessible to fish on the surface of a coral colony and the number of polyps enumerated. Ten colonies were sampled for each species of *Porites* and five for *Poc. meandrina*.

Nematocyst density and size

The relative density and size of stinging nematocysts in polyp tentacles were determined for each coral. Fresh collected colonies were fixed in 10% formalin for two days, rinsed, and whole polyps removed (from colony tips in branching corals) with small needles and forceps. The distal 1 mm of the polyp was then removed, opened by a incision along the oral-aboral axis, and placed flat on a glass slide. The polyp was squashed under a glass cover slip and examined at 400× under a compound microscope. Nematocysts were classified according to Mariscal (1971, 1974). The number of nematocysts in five ocular fields (1 ocular field = approximately 4.2 mm^2) was counted for ten polyps of each coral species. The length (l) and diameter (d) of ten undischarged nematocysts of each species was measured with an ocular micrometer. From these data nematocyst size was estimated by the formula for volume of an ellipse,

$$4/3 \, \pi \, d^2 \, l \, .$$

Laboratory experiments

Fish used in laboratory tests were obtained from shallow coral reefs on the leeward coast of Oahu by divers using handnets and acclimatized for 3 days in aquaria supplied with fresh sea water at Hawaii Institute of Marine Biology. Fish were maintained with fresh colonies of the finely branched coral *Pocillopora damicornis* which occurs on shallow reefs of Kaneohe Bay and is not a normal food for *C. multicinctus*.

Handling time. – Total foraging time is normally defined as that time spent in search and handling of prey. For laboratory study, I defined handling time as that required to locate and ingest polyps after enountering a coral head. This is synonymous with the post-encounter handling component of natural foraging behavior (described in Results) but differs from the more traditional definition of handling time which begins after a prey is captured (e.g. Werner & Hall 1974). Butterflyfish encounter coral colonies sequentially and then must invest additional time to locate polyps on the colony and position their head for a bite. As shown below, this post-encounter inspection/orientation time differs for each coral species and therefore is considered analogous to handling time costs of manipulating prey after capture (e.g. Kislalioglu & Gibson 1976). The beginning of a foraging bout in the tank was defined by the approach to and orientation within one head-length of a coral colony and terminated when the fish departed. Handling time was calculated as the average time interval between bites during foraging bouts.

Relative inspection costs during handling were determined for *Porites* corals. Movement of the black vertical band which passes through the center of the eye made it easy to tally eye movements while fish foraged on colonies. Each of five fish was presented a single coral colony for 10 min and observed from behind a black plastic blind placed against the aquarium glass. For each colony presentation the following were recorded: (1) total foraging bout time, (2) total number of bites, and (3) number of eye movements during foraging bouts. Each test fish was presented each species of *Porites* three times in a randomized sequence.

Coral morphology. – The effect of colony morphology on handling time was tested experimentally. If food preferences were due to handling constraints, the prediction can be made that no preference should exist if handling times are equalized by masking gross coral colony morphology.

Two identical feeders were constructed for paired presentation of live coral fragments (Fig. 1). Sides were made from opaque 6 mm plexiglass and the surface from 2 mm white vinyl. Overall dimensions were 20 × 20 × 5 cm. A square matrix of sixteen 2 cm^2 holes was drilled on the surface, and rows of elastic bands strung so that they crossed under each hole. Small pieces of fresh coral were positioned on the underside of the feeder and held flush with the surface by the bands.

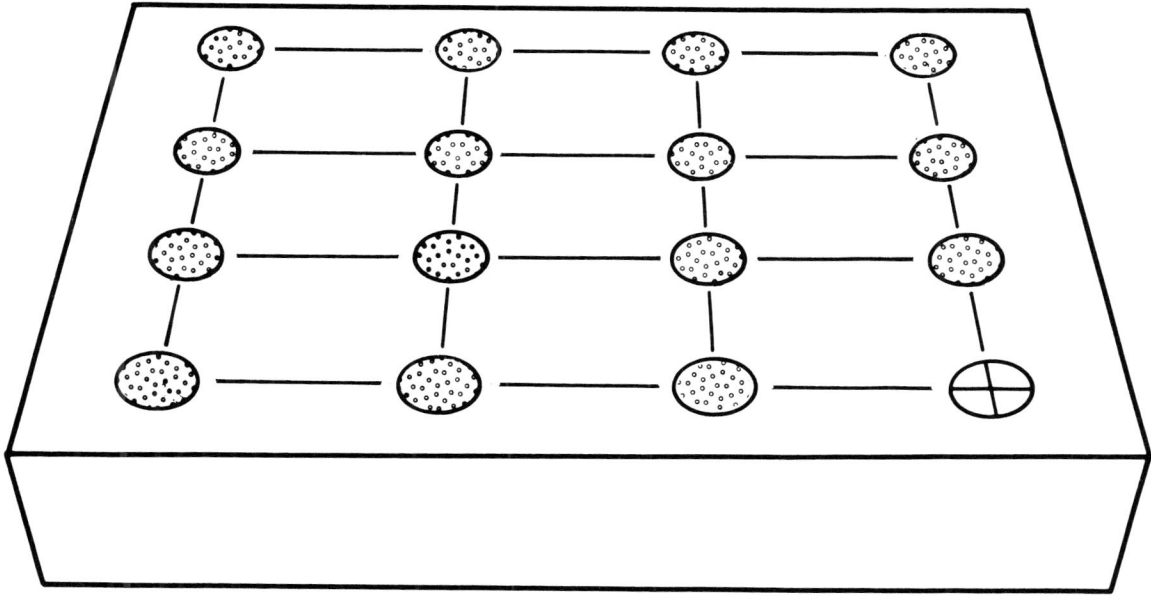

Fig. 1. Feeder used to mask gross morphology of the food corals of *C. multicinctus*. Sixteen fragments from a single coral colony were presented to fish through sixteen 2 cm^2 holes on the surface and held in place by elastic bands that crossed under each hole. A hole without a coral fragment that shows crossed bands is illustrated in right front corner of feeder. Dimensions of feeder are $20 \times 20 \times 5$ cm.

Food preference experiments were performed in a 140 l aquarium. Individuals were presented corals under treatment classes (1) two separate coral colonies, and (2) two corals in separate feeders. In one set of experiments, *Poc. meandrina* was tested with *Por. lobata*. In the second set, all three possible permutations of the two *Porites* species were used in paired treatments: paired conspecific (*Por. lobata* + *Por. lobata* and *Por. compressa* + *Por. compressa*) and heterospecific (*Por. lobata* + *Por. compressa*) combinations. Considerable effort was made to minimize cues that could produce visual or gustatory biases. Coral heads of approximately 20 cm diameter were used in whole colony presentations. Color differences and polyp condition between species were controlled by matching species for color and polyp length, respectively. Coral pieces placed in feeders were taken from the same mother colony as those used in whole colony presentations. The location of paired colonies or feeders in the aquarium (left or right side) was randomized to eliminate position effects. Prior to each test subjects were given a minimum of 5 min to sample each paired colony or feeder for at least 5 feeding bouts. The first presentation treatment (colonies or feeders) was followed immediately by the other. The order of presentation for whole colonies and feeders was randomized for each test sequence and run in triplicate. Experiment durations were 15 min for whole colony and 30 min for feeder presentations. Data were compared between colony and feeder presentations by analysing pooled means of each fish. A paired-comparisons test was used to detect directional changes common to all fish.

Results

Foraging behavior

The foraging of *C. multicinctus* in the field involves a repetitious sequence of search, encounters with colonies, and handling of coral prey (Fig. 2). Search involves travel between individual coral heads or small areas on large *Porites* colonies. As corals are encountered, they are either approached for inspection (the first stage of handling) or passed

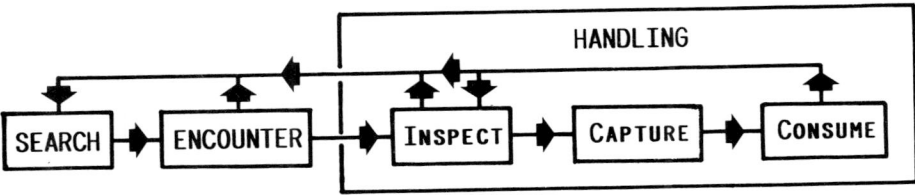

Fig. 2. Flow diagram of foraging patterns among food corals in the field for *C. multicinctus* showing pre-encounter search and post-encounter handling of prey. Handling time includes post-encounter inspection of polyps and orientation of the body for a bite.

for further search behavior. Inspection involves the examination of colonies for polyps and orientation of the body for feeding. When no bites are taken (polyps captured), the fish abandons the colony and returns to search. Captured polyps are consumed (masticated and swallowed) followed by either further inspection of the same colony or departure to search for another colony. This pattern of feeding is only occasionally interrupted by agonistic defense of the feeding territory, brief periods of sheltering, or cleaning by symbiotic cleaner wrasses. In this analysis handling is considered to include time spent during pre-capture inspection and orientation for a bite since it is a function of prey morphology rather than encounter rates with coral colonies.

Feeding patterns in the field

Total coral cover within fish territories in the coral rich habitat ranged from 30 to 91%, and averaged 51%. *Por. lobata* was the most abundant species followed by *Por. compressa* (Table 1). *Poc. meandrina* comprised less that 1% of total bottom cover. Other encrusting corals such as *Montipora verrucosa*, *Leptastrea purpurea*, and *Leptastrea bottae* occurred in patches less than a few centimeters in diameter and were very rare in quadrat samples.

Eighty-two percent of all feeding bites were taken from live corals (Table 1). Fish also consumed small quantities of benthic filamentous algae on dead coral and basalt substrates, and occasionally invertebrates such as amphipod crustaceans and fragments of polychaete worms. Although 18% of the total bites were taken on non-coral substrates, non-coral invertebrate food items comprised less than 5% of the total stomach content volume and algae less than 1% which confirms that the diet of *C. multicinctus* consists largely of scleractinian coral polyps. Undigested filamentous algae and symbiotic coral zooxanthellae were often recognized in the hind gut and indicates that assimilation of all plant material is incomplete.

No individual fish foraged on food items in equal ratios to the proportion of substrate cover (independent G-test: n = 60: df = 3, all test $P<0.001$). Fish fed upon *Por. lobata* in excess of its abundance (Table 1). Although relatively rare in the study area, *Poc. meandrina* was also taken in higher proportion than its abundance. In contrast *Por. compressa* and non-coral substrates (which contained

Table 1. Substrate composition in *C. multicinctus* feeding territories (n = 30) and feeding patterns of resident pairs (n = 60) in the coral rich habitat at Puako. Total observation time = 9025 min. Total number of bites = 93638. B/C = ratio of proportion of bites to proportion of bottom cover. Data expressed as means and standard deviations. *P* determined by Wilcoxon's two-sample test for equal proportions of cover and bites.

Food item	% Total cover	% Total bites	B/C	P
Porites lobata	34.8 ± 12.4	71.9 ± 12.6	2.1	<0.0001
Porites compressa	15.9 ± 14.6	10.1 ± 9.0	0.6	<0.0001
Pocillopora meandrina	0.1 ± 0.1	0.3 ± 0.8	3.0	<0.0001
Hard substrate (non-coral)	49.2 ± 13.5	17.7 ± 10.6	0.4	<0.0001

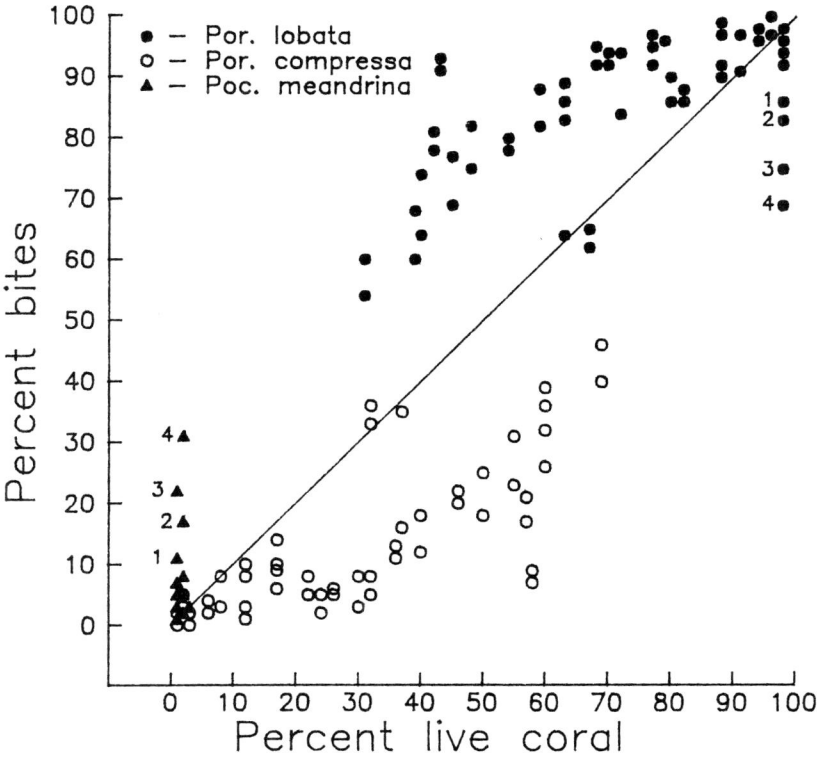

Fig. 3. Relationship between proportion of live coral cover and proportion of feeding bites for individual *C. multicinctus* at Puako. Solid line indicates equal proportions. Examples of the strong preference for *Poc. meandrina* that reduced feeding on *Por. lobata* are shown by numbered points that indicate feeding by 4 fish in territories that contained *Poc. meandrina*.

filamentous algae and invertebrates in much lower densities) were grazed less frequently than expected by their proportion of cover.

Feeding patterns among living corals (exclusive of bites on non-coral substrates) in relation to their relative abundance from the primary study area and the coral-poor reef flat are shown in Figure 3. The relatively uncommon *Poc. meandrina* was taken in greater proportion than expected by its abundance (Wilcoxon's two-sample test: n = 18, $P<0.05$). *Por. lobata* was also taken in a higher proportion than its abundance (Wilcoxon's two-sample test: n = 57, $P<0.05$) except when it cooccurred with the highly preferred *Poc. meandrina*. Finger coral, *Por. compressa*, was taken less than expected (Wilcoxon's two-sample test: n = 53, $P<0.05$). Thus, the feeding relationships observed among live food corals by *C. multicinctus* are consistent with those observed for all food types combined.

Energy and water content of food corals

Caloric densities of coral tissues are shown in Table 2. Mean values for *Por. lobata* and *Por. compressa* were very low on an ash-free dry weight basis and did not differ (t-test: $P<0.3$). The tissues of both species have a water content of about 95%. Thus, energy content on a wet weight basis is also approximately equivalent between the two species of *Porites*. In contrast, *Poc. meandrina* had a mean caloric density approximately 16% higher than the pooled *Porites* data (t-test: $P<0.05$).

Table 2. Caloric densities and water content ($\bar{x} \pm SD$) of coral tissues. n = number colonies assayed.

Species	Calories mg^{-1} AFDW (n)	% water (n)
Pocillopora meandrina	5.29 ± 0.08 (7)	–
Porites lobata	4.63 ± 0.31 (9)	96.1 ± 0.4 (4)
Porites compressa	4.50 ± 0.24 (7)	94.4 ± 0.3 (3)

Nematocyst size, distribution, and abundance

Nematocysts differed qualitatively between the two genera of corals examined. Two types of small nematocysts were observed in *Poc. meandrina* (Table 3). The largest (type 1) and most common had a translucent capsule with the undischarged shaft prominently visible in the interior and was classified as a microbasic mastigophore by Mariscal (1971). The second smaller and less abundant type was not classified although similar in size to cnidocysts described for *Poc. damicornis* by Glynn & Krupp (1986).

Nematocysts were qualitatively similar for both species of *Porites* but much larger than those found in *Poc. meandrina*. Nematocysts of *Por. compressa* have been classified as holotrichous isorhiza (Glynn & Krupp 1986). Threads of some discharged nematocysts were almost 50× their capsule length. A between-species comparison of estimated cell volume shows that nematocysts of *Por. compressa* are larger than those of the preferred *Por. lobata* (t-test: $P<0.001$) although of much lower magnitude than the inter-generic size differences. Nematocyst density in polyp squashes was higher in *Por. compressa* than in *Por. lobata* (t-test of square-root transformed data: $t = 3.14$, $df = 18$, $P<0.01$). The distribution of nematocysts in *Por. compressa* was also more clumped than in *Por. lobata*. This is reflected by a higher coefficient of dispersion (ratio of variance to mean, Sokal & Rohlf 1981) of 1.60 for *Por. compressa* than 1.03 *Por. lobata*.

Polyp densities

Corals of the genus *Porites* have similar calyx morphologies but some minor differences exist. Calyxes were on average smaller in *Por. lobata* than *Por. compressa* and their overall density was higher in the former species (Table 4, t-test: $P<0.005$). In addition, calyxes (and presumably polyps) were more variable in size for *Por. compressa* due to the coral's more complex surface relief. Unlike *Porites*, the surface of *Poc. meandrina* is covered with small verrucae that underlie polyp clusters. The variable size of the verrucae and numbers of associated polyps produce considerable variability in local polyp density although it did not differ from either species of *Porites* (Tukey's multipe comparisons test: $P>0.05$).

Handling costs

In laboratory feeding experiments, fish took 71% more bites per unit foraging time on *Poc. meandrina* than on *Por. lobata* (paired t-test: $P<0.05$), and 47% more bites per unit foraging time on *Por. lobata* than on *Por. compressa* (Table 5, paired t-test: $P<0.05$). The high feeding rate on *Poc. meandrina* was influenced by intense feeding bouts by fish on polyp clusters located on colony verrucae. Assuming a linear inverse relationship between handling time and feeding rate, handling time was 83% greater for *Por. lobata* and 151% greater for *Por. compressa* compared to *Poc. meandrina*. The high handling time for *Por. compressa* is reflected in the greater number of eye

Table 3. Nematocyst dimensions, volume, and density ($\bar{x} \pm SD$) for three food corals measured in a compound microscope ocular field (4.2 mm²). Ten polyps were sampled for each species.

Species	Length (mm)	Diameter (mm)	Volume ($\times 10^{-5}$ mm³)	Density (no. per ocular field)
Porites lobata	0.063 ± 0.003	0.017 ± 0.001	7.36 ± 0.92	4.30 ± 2.1
Porites compressa	0.076 ± 0.001	0.018 ± 0.0	10.21 ± 0.20	6.40 ± 3.2
Pocillopora meandrina				
Type 1	0.040 ± 0.003	0.005 ± 0.0	0.39 ± 0.04	>40
Type 2	0.022 ± 0.003	0.003 ± 0.0	0.06 ± 0.01	–

movements made to inspect polyps while feeding compared to *Por. lobata* (paired t-test: $P<0.01$), and over twice as many eye movements per bite (paired t-test: $P<0.05$). Eye movement data support the hypothesis that foraging on *Por. compressa* has a higher handling cost related to coral inspection that results in fewer bites per unit time compared to *Por. lobata*.

Food preference and the constraint of coral morphology

C. multicinctus shows the same preference for corals in the laboratory as observed in the field. Whole colonies of *Poc. meandrina* were highly preferred over those of *Por. lobata* (Fig. 4) which were preferred over *Por. compressa* (Fig. 5). No preference was found in any conspecific colony or feeder presentation.

Fish discriminated between *Poc. meandrina* and *Por. lobata* when gross coral morphology was masked by the feeders (Fig. 4). The proportion of bites on *Pocillopora* in whole colony presentations (98%) remained very high when the two corals were presented in the feeders (88%). The preference for whole colonies of *Por. lobata* over *Por. compressa*, however, was not replicated when coral fragments were presented in feeders (Fig. 5). For pooled feeding data, fish did not show a preference for either coral when morphology was masked (Wilcoxon signed-ranks test: $P<0.05$) and a decreased proportion of bites on *Por. lobata* in feeder experiments compared with associated presentations of whole colonies (Wilcoxon's two-sample test: $P<0.05$, one-tailed test). Fish did not show selective feeding among *Porites* when the variable of gross morphology was controlled.

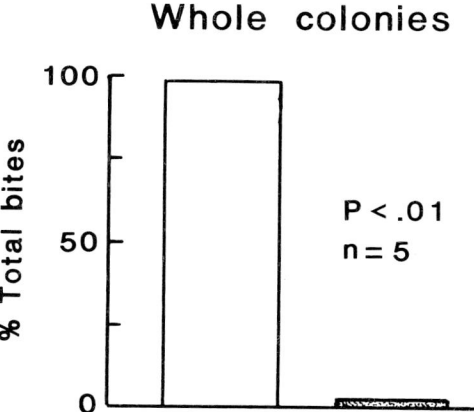

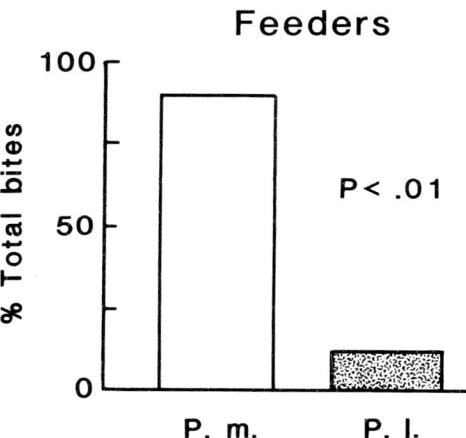

Fig. 4. Feeding preferences of five *C. multicinctus* on the corals *Pocillopora meandrina* (P. m.) and *Porites lobata* (P. l.) in paired presentations of whole coral colonies and paired presentations of coral fragments in experimental feeders that masked coral gross morphology. *P* determined by Wilcoxon's two-sample test of proportions of bites within each treatment.

Table 4. Polyp densities (number per cm^2) of three food corals. Coefficient of dispersion (CD) = variance/mean, n = number of colonies examined.

Species	n	$\bar{x}$ ± SD	CD
Pocillopora meandrina	5	66.4 ± 15.3	3.5
Porites lobata	10	76.7 ± 5.7	0.4
Porites compressa	10	61.0 ± 10.5	1.8

Table 5. Feeding rates and frequency of eye movements ($\bar{x}$ ± SD) of *C. multicinctus* during foraging on coral colonies.

Species	Bites	Eye movements	
	minute^{-1}	minute^{-1}	bite^{-1}
Porites lobata	19.3 ± 7.8	33.3 ± 8.0	2.1 ± 0.9
Porites compressa	13.1 ± 8.1	42.1 ± 11.4	4.6 ± 2.4
Pocillopora meandrina	33.0 ± 8.6	–	–

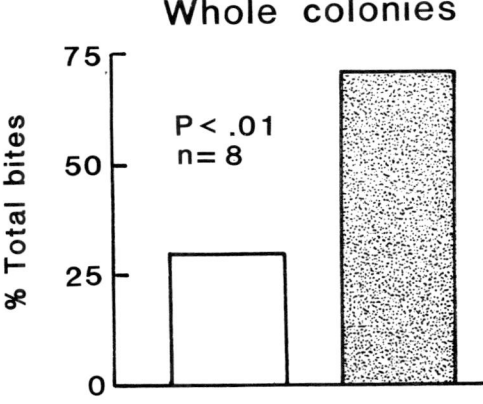

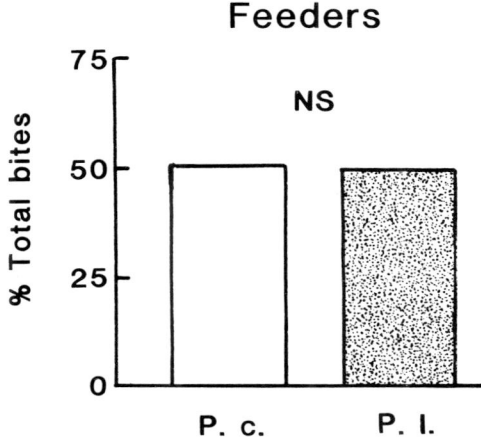

Fig. 5. Feeding preferences of eight *C. multicinctus* on the corals *Porites compressa* (P. c.) and *Porites lobata* (P. l.) in paired presentations of whole coral colonies and paired presentations of coral fragments in experimental feeders that masked coral gross morphology. *P* determined by Wilcoxon's two-sample test of proportions of bites within each treatment.

Discussion

Field studies on corallivorous chaetodontids indicate corals of the genus *Pocillopora* to be a preferred food (Randall 1974, Neudecker 1979, Tricas 1985, Hourigan et al. 1988). Although relatively uncommon on the Puako reef, *Poc. meandrina* was grazed by *C. multicinctus* more frequently than expected both in its proportion of total live coral and total bottom cover. The preference for *Poc. meandrina* by *C. multicinctus* was reported in four

different reef habitats by Hourigan et al. (1988) where this coral species comprised as high as 55% of the coral diet but only 25% of live coral cover.

For many fish, feeding on *Por. lobata* was reduced when *Poc. meandrina* was present in feeding territories. Similarly, the focus on these two corals reduced feeding upon finger coral, *Por. compressa*, which often was the most common coral within feeding territories but rarely taken in proportion to its percentage of bottom cover. Thus the low occurrence of *Por. compressa* in the diet may be due to concentrated feeding on other species rather than active avoidance. In laboratory feeding experiments with corallivorous chaetodontids, corals of the genus *Pocillopora* have been shown to be the preferred food in paired presentations while *Por. compressa* was often the least preferred although the latter species is readily fed upon when presented alone (Reese 1977, Cox 1983, Hourigan et al. 1988).

The preference for Pocillopora *over* Porites

The observed preferences for *Poc. meandrina* by *C. multicinctus* is consistent with its relatively high energy content. Surface tissues of *Poc. meandrina* had a mean caloric density 16% higher than *Porites*. This difference is probably due to fat bodies that comprise approximately 32% of the dry tissue weight of *Poc. meandrina* (Stimson 1987) and the occurrence of all tissue on the surface of the imperforate corallite skeleton. In contrast, *Porites* corals have a perforate skeleton where much of the living tissue and most fat bodies are located beneath the corallite surface (Stimson 1987) and are inaccessible to polyp pickers like *C. multicinctus*. Thus, more tissue (and its energy-rich components) may be available on a per bite basis to foragers on the imperforate *Poc. meandrina*. Hourigan et al. (1988) estimated that 32% more calories per bite are taken by *C. multicinctus* on *Poc. meandrina* than on *Por. lobata*, and found no difference in calories per bite among the two species of *Porites*. The congener, *Poc. damicornis*, which is a highly preferred species by corallivorous chaetodontids (Randall 1974, Reese 1977, Neudecker 1979), also shows a high caloric (Richmond 1982, Glynn & Krupp 1986) and lipid content (Stimson 1987). The acquisition of lipid is important for *C. multicinctus* as seen in pronounced seasonal deposition of body fat (Tricas 1986) and high lipid fraction in reproductively active ovaries (21%) and testes (33%) (Tricas, unpublished data).

Although polyps did not differ in average density among genera, their clustered distribution on *Poc. meandrina* verrucae is probably the primary factor responsible for the low handling time of this coral relative to *Porites*. Fish continued to discriminate *Poc. meandrina* when gross morphology was masked indicating possible cues from surface microstructure. The extreme differences in nematocyst size, however, is also consistent with the preference for species of *Pocillopora* and may also influence feeding patterns and choices among these corals.

The preference for Porites lobata *over* Porites compressa

The preference for *Por. lobata* over *Por. compressa* in the field and in whole colony presentations is probably not a function of energy content of tissues since their caloric densities and water contents are similar. In this case differences in the rate at which fish can harvest polyps also may structure feeding patterns. The branched morphology of *Por. compressa* colonies greatly reduces the number of polyps available to fish compared to *Por. lobata*. More time is required to locate suitable polyps, more inspective eye movements per unit handling time are made, and fewer bites per unit handling time are taken. The indiscriminate feeding among the two species of *Porites* when their gross morphologies were masked supports the existence of a handling time constraint. The differences in gross morphology apparently outweigh microstructural differences between the species such as larger average polyp size for *Por. compressa*. The larger size, higher abundance, and clustered distribution of nematocysts among *Porites* corals are also consistent with observed foraging preferences. Since fish did not show preferences among *Porites* over the 30 min duration of the coral feeder experiments, however, any nematocyst-mediated feeding probably operates over a longer time period. Clearly, more research is needed to determine the role of nematocyst defense in food corals.

The prey choice model

The basic assumptions of the classic prey choice model for energy maximizers (reviewed by Stephens & Krebs 1986) are met by the *C. multicinctus* system, and allow a simple predictive model. Foraging consists of a repetitive pattern of searching among coral heads that are encountered sequentially and in approximate proportion to their abundance. Because fish must forage over wide areas to feed on benthic non-mobile prey, handling and search time are discrete events. Coral tissues contain no large undigestible components (although some portion of the symbiotic zooxanthellae may pass undigested through the gut), thus the assimilable fraction of energy is probably similar among coral species. Each coral has a predictable average energy content and handling time is determined largely by morphological constraints of each prey coral. Energetic cost per unit time of searching and handling are approximately equivalent for different prey. Since handling primarily involves inspection, energetic costs are probably only slightly higher than standard metabolic rates and can be ignored as a differential cost to be subtracted directly from energy intake.

The diet set that will maximize rate of energy intake can be predicted by applying the empirically derived parameters for each coral species (n = 3), in order of ranked profitability until the following equality is maximized,

$$\frac{E_n}{T} = \frac{\sum_{i=1}^{n} \lambda_i E_i}{1 + \sum_{i=1}^{n} \lambda_i h_i},$$

where E = energy gain, h = handling time, T = time foraging, and λ = encounter rate of each coral type during search.

The energy return per bite is highest and handling time lowest for *Poc. meandrina* (Pm) which combine to give this coral the highest profitability (Table 6). *Por. compressa* (Pc) has the highest handling time but approximately equivalent profitability as *Por. lobata* (Pl) because of its higher estimated energy content per bite. By this model only *Poc. meandrina* should be taken,

$$\frac{\lambda_{Pm} E_{Pm}}{1 + \lambda_{Pm} h_{Pm}} = 0.05 < \frac{E_{Pl}}{h_{Pl}} = \frac{E_{Pc}}{h_{Pc}} = 0.01.$$

However, fish can not sustain themselves on the rare *Poc. meandrina* alone and therefore should take both species of *Porites* when encountered. The predicted preference for *Poc. meandrina* by *C. multicinctus* is supported in habitats where the coral occurs in higher abundance (Hourigan et al. 1988) and also when it is supplemented in feeding territories (Tricas 1989) where it is then taken almost exclusively.

The inconsistency between the predicted diet set and the apparent partial preference for *Por. compressa* observed in the field can be explained at least partially by the expression of abundance of *Por. compressa* as percentage total substrate cover. At Puako, beds of *Por. compressa* are a thick matrix of highly twisted, 1–2 cm diameter branches that are broadly spaced. As a result, much of the area covered by this coral does not contain polyps due to large inter-branch gaps and inaccessible covered branches. The 'available abundance' of *Por. compressa* was estimated for some colonies to be less than half that of the total surface area (Tricas, unpublished data) and therefore this coral may actually be taken in equal or greater proportion than its availability to fish. While this consideration is less critical for massive and encrusting corals like *Por. lobata* where virtually all polyps are exposed, coral abundance would be better estimated for food preference studies in terms of that available to the fish rather than simple measures of total substrate cover. This is especially crucial for tests of the prey choice model if branched corals are of the highest profitability ranks because encounter rates (a function of abundance) are used to determine the predicted diet set.

Nutrient constraints and toxic properties of food items may also alter predictions of the simple prey choice model in reef fishes (Lobel & Ogden 1981). The need for rare but essential amino acids or inorganic nutrients unavailable in preferred corals may require searching for less abundant species, or energetically unprofitable and time costly filamentous algae or invertebrates among non-coral substrates. Although untested, foraging on coral tissues may also be limited by the need to add roughage to the diet to facilitate movement of gelatinous coral tissue through their gut. Hobson (1974) reported a similar complement of filamentous algae in the gut of *C. multicinctus* and *Chaetodon ornatissimus* as did Harmelin-Vivien & Bouchon-Navaro (1983) for numerous corallivorous chaetodontids. Nematocyst toxins or metabolic compounds present in live corals may set an upper limit to the number of bites taken per unit time from otherwise energetically profitable corals. In this regard, Glynn & Krupp (1986) reported no adverse affects of crude coral extracts of *Por. compressa* on a freshwater fish. While the differences found for nematocysts among prey corals are consistent with observed foraging patterns, more work is necessary

Table 6. Estimates for parameters that characterize foraging returns for *Chaetodon multicinctus* on food corals. See Appendix for estimate of E.

	Energy gain (E) (cal bite^{-1})	Handling time (h) (sec bite^{-1})	Profitability (E h^{-1})	Encounter rate (λ) (polyps sec^{-1})
Pocillopora	0.10	1.82	0.06	21
Porites lobata	0.04	3.11	0.01	8407
Porites compressa	0.06	4.58	0.01	3055

to establish whether any toxic compounds in corals function as aversive stimuli to coral-feeding butterflyfishes.

Future direction

Coral-feeding butterflyfishes offer many advantages to field studies of foraging ecology, energy allocation, and feeding behavior of fishes. While there currently exists considerable data on the food habits of many species, much work remains to develop and test models of their foraging patterns. Below are suggested topics to address the decision, currency, and constraint assumptions of butterflyfish foraging systems:

(1) *Does maximization of some component of food best characterize the foraging strategies of coral-feeding butterflyfishes?* This important question must be addressed before any rate-maximization model can be tested. Life history data for individual species are needed to establish how reproductive success varies with food intake. If the relationship exists, preferred food items that maximize the rate of intake can be predicted by estimating their relative profitability. In addition to energy content, analyses are needed to assess the relative protein, lipid, and carbohydrate content of coral tissues and how they are allocated in physiological pathways of butterflyfishes. Food choice experiments where energy or component content of food corals are manipulated and compared to controls (for example, by spiking the autotrophic symbiotic zooxanthellae with inorganic nutrients) could provide a direct test of a maximization strategy.

(2) *What features of coral tissues other than energy content influence selection (or avoidance) of food types?* In addition to component and energy content, critical inorganic or rare organic nutrients may govern the selection of relatively non-profitable coral prey. These questions require knowledge of the physiology of growth and assimilation in the fishes and the relative abundance of nutrients in prey corals. In addition, aversion to coral species can be addressed through immunological assays of antibodies to nematocyst toxins and surveys for metabolic compounds such as those known to occur in marine macroalgae (e.g. Targett et al. 1986), soft corals, and gorgonians (Fenical 1982, Gerhart 1984).

(3) *Patch vs. prey models.* In the present study foraging preferences were compared to a classical 'prey choice model' that addressed 'what corals should be included in the diet?'. Alternatively, individual colonies can be regarded as discrete patches of prey and derivatives of the classical 'prey-patch model' used to ask 'how long should a fish feed before moving to another coral head' (sensu Charnov 1976). This question also deals with maximization strategies and would include analyses of polyp depletion, withdrawal, and regeneration rates. Butterflyfish systems offer exciting field and laboratory opportunities to vary the number of polyps per patch, travel time between patches, and patch encounter rates by manipulation of individual coral colonies.

(4) *Feeding ontogeny and periodicity.* Ontogenetic studies of age-specific food preference patterns could be used to address trade-offs between the acquisition of protein for growth in juveniles and lipids for reproduction in adults. Also, unlike adults that are often site attached for long periods (Reese 1973, 1975, Sutton 1985), new recruits to the reef often experience intense predation (Tricas, unpublished data), and may exhibit habitat and food preferences mediated by predation risk (e.g. Mittlebach 1984). Finally, diel, lunar, or seasonal shifts in feeding patterns may follow changes in profitability of food corals such as shown for chaetodontids feeding on gorgonians (Lasker 1985). Closely monitoring changes in foraging behaviors in relation to coral spawning activity would provide valuable insight into the relationships of butterflyfishes and their coral prey.

Acknowledgements

I thank Robin Hori, Tom Hourigan, Phil Motta, Ernst Reese, many Earthwatch volunteers for their field assistance, and Mike Hadfield and Bob Kinzie, for sharing their facilities for coral tissue analyses. This study has benefited greatly from discussions with Bruce Cushing, John Ebersole, Gene Helfman, George Losey, Ernst Reese, Bob Richmond, and John Stimson. Field work was supported by the Center for Field Research and laboratory facilities provided by Hawaii Institute of Marine Biology. This paper is part of a dissertation submitted to the University of Hawaii in partial fulfillment of requirements for the Ph. D. degree in Zoology, and is dedicated to my field companion, Nicole.

References cited

Barnes, D.J., R.W. Brauer & M.R. Jordan. 1970. Locomotory response of *Acanthaster planci* to various species of coral. Nature (Lond.) 228: 342–344.

Birkeland, C. & S. Neudecker. 1981. Foraging behavior of two Caribbean chaetodontids: *Chaetodon capistratus* and *C. aculeatus*. Copeia 1981: 169–178.

Charnov, E.L. 1976. Optimal foraging: the marginal value theorem. Theor. Popul. Biol. 9: 129–136.

Cox, E.F. 1983. Aspects of corallivory by *Chaetodon unimaculatus* in Kaneohe Bay, Oahu. Masters Thesis, University of Hawaii, Honolulu. 60 pp.

Fenical, W. 1982. Natural products chemistry in the marine environment. Science 215: 923–928.

Gerhart, D.J. 1984. Prostaglandin A_2: an agent of chemical defense in the Caribbean gorgonian *Plexaura homomalla*. Mar. Ecol. Prog. Ser. 19: 181–187.

Glynn, P.W. & D.A. Krupp. 1986. Feeding biology of a Hawaiian sea star corallivore, *Culcita novaeguineae* Muller & Troschel. J. Exp. Mar. Biol. Ecol. 96: 75–96.

Gore, M.A. 1984. Factors affecting the feeding behavior of a coral reef fish, *Chaetodon capistratus*. Bull. Mar. Sci. 35: 211–220.

Harmelin-Vivien, M.L. & Y. Bouchon-Navaro. 1983. Feeding diets and significance of coral feeding among chaetodontid fishes in Moorea (French Polynesia). Coral Reefs 2: 119–127.

Hirschfield, M.F. 1980. An experimental analysis of reproductive effort and cost in the Japanese Medaka *Oryzias latipes*. Ecology 61: 282–292.

Hobson, E.S. 1974. Feeding relationships of teleostean fishes on coral reefs in Kona. Hawaii. U.S. Fish. Bull. 72: 915–1031.

Hourigan, T.F. 1987. The behavioral ecology of three species of butterflyfishes (Fam. Chaetodontidae). Ph.D. Dissertation, University of Hawaii, Honolulu, 322 pp.

Hourigan, T.F., T.C. Tricas & E.S. Reese. 1988. Coral reef fishes as indicators of environmental stress in coral reefs. pp. 107–135. *In:* D.F. Soule & G. Kleppel (ed.) Marine Organisms as Indicators, Springer-Verlag, New York.

Hunter, J.R. & B.J. Macewicz. 1985. Rates of atresia in the ovary of captive and wild northern anchovy, *Engraulis mordax*. U.S. Fish. Bull. 83: 119–136.

Johannes, R.E. & W.J. Wiebe. 1970. Method for determination of coral tissue biomass and composition. Limnol. Oceanogr. 15: 822–824.

Kislaliogiu, M. & R.N. Gibson. 1976. Prey 'handling time' and its importance in food selection by the 15-spined stickleback, *Spinachia spinachia* (L.). J. Exp. Mar. Biol. Ecol. 25: 151–158.

Krebs, J.R. 1978. Optimal foraging: decision rules for predators. pp. 23–63. *In:* J.R. Krebs & N.B. Davies (ed.) Behavioural Ecology: an Evolutionary Approach, Blackwell Scientific, Oxford.

Krebs, J.R. & R.H. McCleery. 1984. Optimization in behavioural ecology. pp. 91–121. *In:* J.R. Krebs & N.B. Davies (ed.) Behavioural Ecology: an Evolutionary Approach, Sinauer, Sunderland.

Lasker, H.R. 1985. Prey preferences and browsing pressure of the butterflyfish *Chaetodon capistratus* on Caribbean gorgonians. Mar. Ecol. Prog. Ser. 21: 213–220.

Lobel, P.S. & J.C. Ogden. 1981. Foraging by the herbivorous parrotfish *Sparisoma radians*. Mar. Biol. 64: 173–183.

Mariscal, R.N. 1971. Effect of a disulfide reducing agent on the nematocyst capsules from some coelenterates, with an illustrated key to nematocyst classification. pp. 157–168. *In:* Experimental Coelenterate Biology, University of Hawaii Press, Honolulu.

Mariscal, R.N. 1974. Nematocysts. pp. 129–178. *In:* Coelenterate Biology. Reviews and New Perspectives, Academic Press, New York.

Milton, K. 1979. Factors influencing leaf choice by howler monkeys: a test of some hypotheses of food selection by generalist herbivores. Amer. Nat. 114: 362–378.

Mittlebach, G.C. 1984. Predation and resource partitioning in two sunfishes (Centrarchidae). Ecology 65: 499–513.

Motta, P.J. 1988. Functional morphology of the feeding apparatus of ten species of Pacific butterflyfishes (Perciformes, Chaetodontidae): an ecomorphological approach. Env. Biol. Fish. 22: 39–67.

Neudecker, S. 1979. Effects of grazing and browsing fishes on the zonation of corals in Guam. Ecology 60: 666–672.

Owen-Smith, N. & P. Novellie. 1982. What should a clever ungulate eat? Amer. Nat. 119: 151–178.

Paine, R.T. 1966. Endothermy in bomb calorimetry. Limnol. Oceanogr. 11: 126–129.

Phillipson, J. 1964. A miniature bomb calorimeter for small biological samples. Oikos 15: 130–139.

Pyke, G.H., H.R. Pulliam & E.L. Charnov. 1977. Optimal foraging: a selective review of theory and tests. Q. Rev. Biol. 52: 137–154.

Ralston, S. 1976a. Anomalous growth and reproductive patterns in populations of *Chaetodon miliaris* (Pisces, Chaetodontidae) from Kaneohe Bay, Oahu, Hawaiian Islands. Pac. Sci. 30: 395–403.

Ralston, S. 1976b. Age determination of a tropical reef butterflyfish utilizing daily growth rings of otoliths. U.S. Fish. Bull. 74: 990–994.

Ralston, S. 1981. Aspects of the reproductive biology and feeding ecology of *Chaetodon miliaris*, a Hawaiian endemic butterflyfish. Env. Biol. Fish. 6: 167–176.

Randall, J.E. 1974. The effects of fishes on coral reefs. Proc. Second Internat. Coral Reef Symp. 1: 159–166.

Reese, E.S. 1973. Duration of residence by coral reef fishes on 'home' reefs. Copeia 1973: 145–149.

Reese, E.S. 1975. A comparative field study of the social behavior and related ecology of reef fishes of the family Chaetodontidae. Z. Tierpsychol. 37: 37–61.

Reese, E.S. 1977. Coevolution of corals and coral feeding fishes of the family Chaetodontidae. Proc. Third Internat. Coral Reef Symp. 1: 267–274.

Richmond, R.A. 1982. Energetic considerations in the dispersal of *Pocillopora damicornis* (Linnaeus) planulae. Proc. Fourth Internat. Coral Reef Symp. 2: 153–156.

Scott, D.P. 1962. Effects of food quality on fecundity of rainbow trout, *Salmo gairdneri*. J. Fish. Res. Board Can. 19: 715–731.

Stephens, D.W. & J.R. Krebs. 1986. Foraging theory. Princeton University Press, Princeton. 247 pp.

Sokal, R.R. & F.J. Rohlf. 1981. Biometry. W.H. Freeman and Co., San Francisco. 859 pp.

Sutton, N. 1985. Patterns of spacing in a coral reef fish in two habitats on the Great Barrier Reef. Anim. Behav. 33: 1332–1337.

Targett, N.M., T.E. Targett, N.H. Vrolijk & J.C. Ogden. 1986. The effect of macrophyte secondary metabolites on feeding preferences of the herbivorous parrotfish *Sparisoma radians*. Mar. Biol. 92: 141–148.

Tricas, T.C. 1985. The economics of foraging in coral-feeding butterflyfishes of Hawaii. Proc. Fifth Internat. Coral Reef Symp. 5: 409–414.

Tricas, T.C. 1986. Life history, foraging ecology, and territorial behavior of the Hawaiian butterflyfish, *Chaetodon multicinctus*. Ph.D. Dissertation, University of Hawaii, Honolulu. 247 pp.

Tricas, T.C. 1989. Determinants of feeding territory size in the corallivorous butterflyfish, *Chaetodon multicinctus*. Anim. Behav. (in press).

Tricas, T.C. & J.T. Hiramoto. 1989. Sexual differentiation, gonad development, and spawning seasonality of the Hawaiian butterflyfish, *Chaetodon multicinctus*. Env. Biol. Fish. (in press).

Tyler, A.V. & R.S. Dunn. 1976. Ration, growth, and measures of somatic and organ condition in relation to meal frequency in winter flounder, *Pseudopleuronectes americanus*, with hypotheses regarding population homeostasis. J. Fish. Res. Board Can. 33: 63–75.

Werner, E.E. & D.J. Hall. 1974. Optimal foraging and the size selection of prey by the bluegill sunfish (*Lepomis macrochirus*). Ecology 55: 1042–1052.

Wooton, R.J. 1977. Effect of food limitation during the breeding season on the size, body components and egg production of female sticklebacks (*Gasterosteus aculeatus*). J. Anim. Ecol. 46: 823–834.

Zumpe, D. 1965. Laboratory observations on the aggressive behavior of some butterflyfishes (Chaetodontidae). Z. Tierpsychol. 22: 226–236.

Appendix

Energy gain per bite (E) was estimated for each coral by the combined product of (1) the estimated volume (mm^3) of coral tissue per bite (see below), (2) the density of coral tissue (assumed to be equivalent for all three corals and that of water, $1.0\,mg\,mm^{-3}$), (3) the proportion of dry tissue in a bite (a dimensionless constant estimated as 1 − proportion of water from Table 2, assumed to be 0.05 for *Poc. meandrina*, and tissue ash content assumed to be equivalent for all species), and (4) calories mg $AFDW^{-1}$ from Table 2.

Bites upon *Porites* corals usually remove only polyp tentacles while bites upon *Poc. meandrina* usually take tentacles, the oral disk, and a distal portion of the polyp column (Hourigan 1987, Tricas unpublished data). Mean tentacle length (l) and radius (r) were estimated from photographs of live polyps, and volume for 12 tentacles calculated from the formula for the volume of a cylinder, 12 ($\pi r^2 l$). The volume of an additional 0.5 mm length of polyp column taken in the imperforate *Poc. meandrina* was calculated by the same method and summed with the tentacle volume estimate. Mean tentacle radii (0.1 mm) and lengths (0.6 mm) were equivalent for both *Porites* corals providing an estimate of $0.23\,mm^3$ tissue per bite. Mean *Poc. meandrina* tentacle radius (0.1 mm) and length (0.5 mm) gave a tentacle volume estimate of $0.19\,mm^3$; oral disk/polyp column volume was $0.19\,mm^3$ for a $0.38\,mm^3$ total bite volume estimate. These estimates do not include other coral tissues which may also be ingested (e.g. inter-polyp coenosarc), thus they may underestimate actual energy content.

Chaetodon trifascialis on the reef of Heron Island, Australia. Photo by E.S. Reese.

Temporal and areal feeding behavior of the butterflyfish, *Chaetodon trifascialis*, at Johnston Atoll

Darby K. Irons
Department of Zoology, University of Hawaii at Manoa, Honolulu, HI 96822, U.S.A.

Received 8.4.1988   Accepted 23.9.1988

Key words: Foraging, Ethology, Ecology, Chaetodontid, Territorial, Corallivore, Corals, Reef

Synopsis

The chevron butterflyfish, *Chaetodon trifascialis*, is found throughout the Indo-Pacific. It is a territorial, diurnal, corallivore found in close association with *Acropora* spp. corals. The feeding behavior of 33 individuals was studied over six seasons in three habitats. *Chaetodon trifascialis* spent one third of its active time feeding. However, there was much individual variation. Fish had significantly higher feeding rates during the early afternoon, and there were no significant differences in the feeding rates between the seasons. Feeding rates were significantly different between the three habitats. The *Montipora*-rich habitat had the highest feeding rates ($\bar{x} = 10.74$ bites min^{-1} ± 0.87, all corals combined) and the *Acropora-Montipora* mixed habitat had the lowest feeding rates ($\bar{x} = 4.58$ bites min^{-1} ± 0.63, all corals combined). Females fed significantly more than males. While *C. trifascialis* had been thought to only eat *Acropora* spp. corals, it occasionally fed on *Montipora* spp. and *Pocillopora* sp. corals when *Acropora* spp. were scarce. *Chaetodon trifascialis* exhibited patterns predicted by foraging theory of an energy maximizer. Territory sizes were inversely related to food density and feeding rates were inversely related to intruder rates. This is a promising system for future testing of foraging strategy models.

Introduction

The butterflyfishes of the world (Perciformes, Chaetodontidae) exhibit a wide range of feeding behaviors, from planktivory to corallivory (Hiatt & Strasburg 1960, Talbot 1965, Hobson 1974, Reese 1975, 1977, 1981, Burgess 1978, Birkeland & Neudecker 1981, Harmelin-Vivien 1981, Ralston 1981, Harmelin-Vivien & Bouchon-Navaro 1981, 1983). Though the feeding behaviors of some chaetodontids have recently been studied in detail (Gore 1984, Neudecker 1985, Tricas 1986, Hourigan 1987), there are many species for which there is little information. One such species is *Chaetodon trifascialis* (*Megaprotodon trifascialis*).

Chaetodon trifascialis is found throughout the Indo-Pacific, ranging from the Indian Ocean throughout Polynesia including the Northwest Hawaiian Islands. It does not, however, occur in the high Hawaiian Islands (Burgess 1978, Reese 1981). It is a territorial, diurnal, corallivore found in close association with *Acropora* spp. corals (Reese 1975). This butterflyfish has a specialized, forcep-like jaws which are well suited to removing single coral polyps (Motta 1985, 1988). *Chaetodon trifascialis* has been observed feeding almost exclusively on *Acropora* spp. corals (Reese 1975, 1977, 1981, Masuda et al. 1984). No other chaetodontid is known to be so specialized in terms of its prey choice.

Chaetodon trifascialis is solitary (Reese 1975, 1977) and site attached. Individuals have been observed in the same territory for up to three years in this study and up to seven years at Enewetak Atoll (Reese 1981). It has been postulated that males and females hold adjacent territories (Reese 1973). However, the sexes are monomorphic and can not be distinguished in the field.

Chaetodon trifascialis is inactive at night, hiding in the coral. Fish become active at sunrise and remain active until sunset.

This study posed the following questions about the feeding behavior of *C. trifascialis*: 1. Are there differences in the feeding rates throughout the day or between seasons? 2. Are there differences in the feeding rates of males versus females? 3. Are there differences in the feeding rates of fish in different habitats? 4. Are there preferences for one species of coral or does the percentage of bites on a coral species correspond to its respective abundance? 5. How much time do the fish spend feeding in relation to other activities?

Material and methods

Study area

Johnston Atoll is located approximately 1250 km southwest of the Hawaiian Islands. It is approximately 17 km long and 5 km wide. An estimated 30–40% of the live coral cover in the lagoon area is composed of *Acropora cytherea* and the remaining 60% is mostly *Montipora verrucosa*, *M. patula*, and *M. verrilli* (Irons et al. 1984). Since *M. patula* and *M. verrilli* are virtually impossible to distinguish in the field, I will refer to these two species together as *M. patula/verrilli*.

Coral cover

Data were collected in three separate habitats. The *Acropora*-rich habitat had approximately 90% of the live coral coverage consisting of *Acropora cytherea*. This habitat was located approximately 60 m inside the barrier reef at a depth of 7 m.

The *Acropora-Montipora* mixed habitat had about 75% coverage of *A. cytherea* and about 20% *Montipora* spp. corals. This habitat was located 30 m off the east shore of Johnston Island at a depth of 3 m.

The *Montipora*-rich habitat had less than 1% *A. cytherea* and about 95% *Montipora* spp. corals. This habitat was located in the central lagoon at a depth of 10 m.

The percent coral cover at the *Acropora*-rich and the *Montipora*-rich habitats was calculated by placing five 1 m^2 quadrats randomly along a 100 m transect line. Four transect lines were layed parallel to each other and approximately 20 m apart in each habitat, making a total of twenty 1 m^2 quadrats for each of these two habitats. No transects were done in the *Acropora-Montipora* mixed habitat due to the topography.

Identification

Individual fish were identified from natural variations of their markings (Reese 1973). Photographs of the left and right sides of each fish were taken to assist in the identification of individuals from one sampling period to the next.

Feeding observations

Bites per coral species were counted for ten consecutive 5 min intervals, resulting in a total observation time of 50 min for each fish. A total of 76 50 min feeding periods were recorded for 33 fish. Data were collected in July 1984, January, April, August, October 1985, and January 1986. Four to seven fish in each of the three habitats were observed during each data collection trip. Certain fish, especially those in the *Acropora*-rich and the *Montipora*-rich habitats had as many as five feeding periods recorded, each in a different sampling period. At least two separate feeding periods were recorded for most individuals. Data were collected at all times of the day from sunrise to sunset.

All of the study individuals which remained in April 1986 were collected, except for three individuals which eluded capture. From time to time, individuals disappeared from their territories and were not seen again, especially in the *Acropora-Montipora* mixed and the *Montipora*-rich habitats. Several of the focal individuals disappeared following a major storm which damaged portions of the study areas in February of 1986. The collected fish were sexed, weighed, and measured.

Territory sizes

Each territory was roughly measured by recording its length and width. Territory sizes were estimated by the equation: Territory size = Length × Width.

Statistics

Each 50 min feeding period was tested for the randomness of the 5 min intervals comprising it using the runs test above and below the median (Sokal & Rohlf 1981). The sequence of 5 min intervals in only five of the 76 feeding periods significantly departed from randomness (p<0.05). As a result, each 5 min feeding interval was considered independent of the previous and following intervals. Data from these 5 min intervals were the smallest units of measurement used in the analyses. The data were collapsed by computing a mean of the 10 5 min intervals comprising each 50 min feeding period. These means were then used in the analysis of variances (ANOVAs) used in analyzing the data.

Separate two-way ANOVAs were used to test for differences between the seasons (S = 6) and the hours of the day (D = 11). Nested ANOVAs were used to test for differences between the habitats (H = 3) and the sexes (male, female, and unknown). Individual fish (n = 33) were used as the second factor in each two-way ANOVA and as the nested factor in each nested ANOVA to compensate for the repeated measures on each fish. Each two-way ANOVA and each nested ANOVA had a total of 76 cells. All analyses combined the sexes except for the nested ANOVA of the sexes and fish. Parametric pairwise comparisons were performed using Tukey's studentized range test (SAS 1985) and comparison limits were calculated to aid comparisons (Sokal & Rohlf 1981). All means reported include plus or minus one standard error of the mean.

Results

Are there differences in the feeding rates throughout the day or between seasons?

There was substantial variation in the time spent feeding at the various hours of the day and at the different seasons (Fig. 1, 2). The high variation within and among the individuals made differences and trends in feeding rates throughout the hours of the day and between the seasons difficult to detect (Sokal & Rohlf 1981, Martin & Kraemer 1987).

Results from the two-way ANOVAs indicated significant differences between the hours of the day (p<0.05) and significant differences between fish (p<0.05) for both the feeding rates on all corals combined and the rates on *A. cytherea*. There were no significant differences between the seasons sampled (p>0.35) on both the rates on all corals combined and the rates on *A. cytherea*.

Tukey's studentized range test for the mean feeding rates through the hours of the day (Fig. 3) indicated that the fish fed significantly more (p<0.05) in the early afternoon than in the early morning and later afternoon.

Are there differences in the feeding rates of males versus females?

Females had a significantly higher mean feeding rate ($\bar{x}$ = 9.87 bites min^{-1} ± 0.58, n = 9, all corals combined) than males ($\bar{x}$ = 6.86 bites min^{-1} ± 0.89, n = 8, all corals combined) for both the bites on all corals combined and the bites on *A. cytherea* alone (t-Test, p<0.05). There were no significant differences among females only and among males only (Tukey's studentized range test, p>0.05), except among the male feeding rates on *A. cytherea* alone (Tukey's studentized range test, p<0.05). However those differences were due mainly to two individuals.

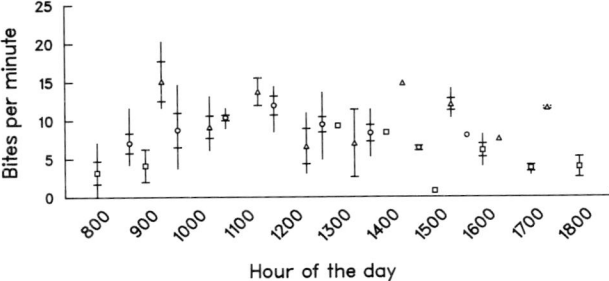

Fig. 1. Mean number of bites per minute ± 1 standard error and ranges on all coral types combined through the hours of the day. Not all habitats are represented for every hour of the day. The mean of each 50 min feeding period was used to calculate the hour means (total n = 76). Points with no standard error or range represent the data from a single fish. ○ = *Acropora*-rich habitat, □ = *Acropora-Montipora* mixed habitat and △ = *Montipora*-rich habitat.

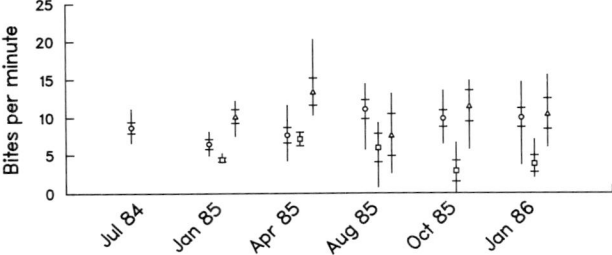

Fig. 2. Mean number of bites per minute ± 1 standard error and ranges on all coral types combined for the seasons sampled. The mean of each 50 min feeding period was used to calculate the season means (total n = 76). ○ = *Acropora*-rich habitat, □ = *Acropora-Montipora* mixed habitat and △ = *Montipora*-rich habitat.

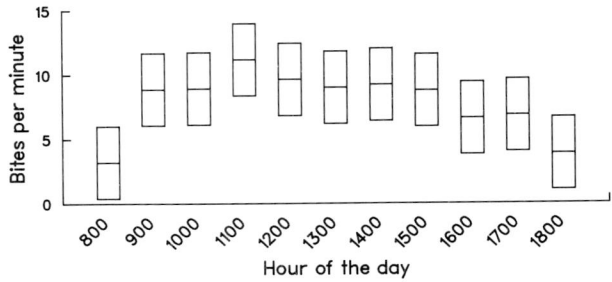

Fig. 3. Comparison limits of the mean feeding rates on all coral types combined through the hours of the day. Each box represents the mean feeding rate and its 95% comparison limits. Boxes which overlap are not significantly different from each other.

Are there differences in the feeding rates of fish in the three study habitats?

The feeding rates on all corals combined were significantly different between the three habitats. However, the mean feeding rate on *A. cytherea* in the *Acropora*-rich habitat ($\bar{x}$ = 9.07 bites min^{-1} ± 0.48, n = 36) and the *Montipora*-rich habitat ($\bar{x}$ = 8.88 bites min^{-1} ± 0.90, n = 22) were not significantly different from each other (Tukey's studentized range test, p>0.05). The feeding rates on *A. cytherea* at both the *Acropora*-rich and the *Montipora*-rich habitats were significantly higher than the feeding rates in the *Acropora-Montipora* mixed habitat ($\bar{x}$ = 4.56 bites min^{-1} ± 0.63, n = 18, Tukey's studentized range test, p<0.05). Unexpectedly, the feeding rates on all corals combined of fish in the *Montipora*-rich habitat ($\bar{x}$ = 10.74 bites min^{-1} ± 0.87) tended to be higher than the feeding rates of fish in the *Acropora*-rich habitat ($\bar{x}$ = 9.07 bites min^{-1} ± 0.48).

Are there preferences for one species of coral or does the percentage of bites on a coral species correspond to its respective abundance?

The percentage of bites on each species of coral was very different than the abundance of the coral (Table 1). Results from the *Montipora*-rich habitat demonstrated *C. trifascialis*' strong preference for *A. cytherea*. Almost 83% of the total bites recorded in the *Montipora*-rich habitat were on *A. cytherea* which comprised less than 0.5% of the live coral in that habitat.

How much time do the fish spend feeding in relation to other activities?

Unlike other butterflyfishes (Tricas 1986, Hourigan 1987), *C. trifascialis* spent less than one-third of its active time feeding (Table 2). The major portion of its time was spent in patrolling (non-feeding activities, which included swimming around its territory) and very little of its time was actually spent interacting (aggressive and non-aggressive behavior) with other fishes.

Individuals in the *Acropora-Montipora* mixed habitat spent much more time patrolling their territories (80%) than did individuals in the other two habitats. Interestingly, fish in the *Montipora*-rich habitat spent less time patrolling (62%) even though their territories were much larger than those of the fish in the other two habitats (Table 2).

Discussion

Chaetodon trifascialis fed almost exclusively on corals of the Family Acroporidae (*Acropora* spp. and *Montipora* spp.). However, they fed on *Pocillopora meandrina* (Family Pocilloporidae) occasionally. Two *C. trifascialis* were also seen feeding on coral mucus in the water column. Suspended coral mucus was shown to contain a significant amount of organic matter and to be enriched with nitrogen when compared to more recently secreted coral mucus or microscopic particulate organic matter (Coles & Strathman 1973). This observation suggests that *C. trifascialis* may be more flexible in its feeding behavior than previously thought.

Feeding rates varied during the day, but fish fed at a significantly higher rate in the early afternoon. Studies on *Acropora acuminata* indicate that lipid production is maximal during the early afternoon (Crossland et al. 1980). If a similar pattern holds for *A. cytherea*, *C. trifascialis* could be taking advantage of this increased lipid production by feeding more during this time of day.

There was no pattern to the differences in the feeding rates between the various seasons sampled. Intuitively, I would not expect any seasonal differences since water temperature at Johnston Atoll only varies within one degree Celsius during the year.

C. trifascialis is similar to other butterflyfishes in that males and females have different feeding rates (Tricas 1986, Hourigan 1987). Males and females do have adjacent territories and a single male has been observed interacting with up to three females. Males also seem to 'visit' females more than females 'visit' males. Chasing intruders and visiting females could prevent males from feeding as much as females. Also, since eggs are considered to be

Table 1. Percent of bites taken on each coral species (calculated as percent of the total bites taken within the respective habitat) and the coral composition of each habitat. Number in the brackets represents the percent of live coral.

Coral	*Acropora*-rich		Mixed		*Montipora*-rich	
Acropora cytherea	100	(91.90)	99.62	(75)	82.68	(0.32)
Montipora patula/verrilli	–	(1.25)	0.38	(8)	4.16	(51.02)
M. verrucosa	–	–	–	(12)	13.11	(44.46)
Pocillopora meandrina	–	–	–	–	0.04	(0.65)

Table 2. Time budgets of *C. trifascialis* in each habitat and relative territory sizes. Percentages represent the mean percent of active time spent performing the particular activity. Number in the brackets represents the standard deviation.

Activity	Combined data	*Acropora*-rich	Mixed	*Montipora*-rich
Feeding	29% (11)	30% (9)	16% (8)	36% (12)
Interactions				
Conspecifics	2% (1)	2% (1)	3% (1)	1% (<1)
Other species	<1% (<1)	<1% (<1)	<1% (<1)	<1% (<1)
Patrolling	68% (13)	68% (11)	80% (10)	62% (13)
Relative territory size (m^2)		small <10	slightly larger <40	very large >400

more energetically costly than sperm (Trivers 1972), females should feed more than males.

C. trifascialis shows a definite preference for *A. cytherea* and correspondingly, territory sizes were inversely proportional to the density of *A. cytherea*. Fish in the habitat with a low density of *A. cytherea* tended to supplement their diet with some other coral species and spent more time feeding than did fish in the other habitats.

The mean feeding rate in the *Acropora-Montipora* mixed habitat was significantly lower than the rates in either of the other two habitats. The *Acropora-Montipora* mixed habitat had many schools of parrotfish, goatfish, jacks, and other butterflyfish species which often invaded individual's territories. The individuals in this habitat spent much more time patrolling their territories than the individuals in the other two habitats (Table 2).

Despite the fact that *C. trifascialis* spent only one-third of its active time feeding and two-thirds of its time patrolling its territory, this species could be classified as an energy maximizer. Foraging theory (Hixon 1980, Schoener 1983, 1987) predicts that if this species is indeed an energy maximizer, it should show the following patterns: 1. As food density increases, territory size decreases. 2. As intruder rate or defense time increases, feeding rate decreases. 3. As intruder rate increases, defense time increases. These are precisely the patterns observed in *C. trifascialis*. However, this system needs more study to test if indeed *C. trifascialis* is an energy maximizer, and what constraints, if any, would apply to this fish as an energy maximizer (Hixon 1980, 1982).

This system seems to be suitable for testing some of the foraging models proposed by Hixon (1980) and Schoener (1983, 1987). Future studies could provide insight into this system and the refinement of foraging strategy models and their predictions.

Acknowledgements

I would like to thank Ernst Reese, Philip Motta, James Archie, James Parrish, Marvin Lutnesky, William Tyler, Anderson Dee, Steven Poet, and Randall Kosaki for invaluable input and help. I would especially like to thank the people of Johnston Atoll for their support and encouragement. This research was funded by the U.S. Army under contract number DACA83-84-C-0019. This paper is part of a thesis to be submitted to the University of Hawaii in partial fulfillment of the requirements for the M.S. degree in Zoology.

References cited

Birkeland, C. & S. Neudecker. 1981. Foraging behavior of two Caribbean chaetodontids: *Chaetodon capistratus* and *C. aculeatus*. Copeia 1981: 169–178.

Burgess, W.E. 1978. Butterflyfishes of the world. T.F.H. Publications, Neptune City. 832 pp.

Coles, S.L. & R. Strathmann. 1973. Observations on coral mucus 'flocs' and their potential trophic significance. Limnol. Oceanogr. 18: 673–678.

Crossland, C.J., D.J. Barnes & M.A. Borowitzka. 1980. Diurnal lipid and mucus production in the staghorn coral *Acropora acuminata*. Mar. Biol. 60: 81–90.

Gore, M.A. 1984. Factors affecting the feeding behavior of a coral reef fish, *Chaetodon capistratus*. Bull. Mar. Sci. 35: 211–220.

Harmelin-Vivien, M.L. 1981. Trophic relationships of reef fishes in Tulear (Madagascar). Oceanol. Acta. 4: 365–374.

Harmelin-Vivien, M.L. & Y. Bouchon-Navaro. 1981. Trophic relationships among chaetodontid fishes in the Gulf of Aqaba (Red Sea). Proc. Fourth Internat. Coral Reef Symp. 2: 537–544.

Harmelin-Vivien, M.L. & Y. Bouchon-Navaro. 1983. Feeding diets and significance of coral feeding among chaetodontid fishes in Moorea (French Polynesia). Coral Reefs 2: 119–127.

Hiatt, R.W. & D. Strasburg. 1960. Ecological relationships of the fish fauna on coral reefs on the Marshall Islands. Ecol. Monogr. 30: 65–127.

Hixon, M.A. 1980. Food production and competitor density as the determinants of feeding territory size. Amer. Nat. 115: 510–530.

Hixon, M.A. 1982. Energy maximizers and time minimizers: theory and reality. Amer. Nat. 119: 596–599.

Hobson, E.S. 1974. Feeding relationships of teleostean fishes on coral reefs in Kona, Hawaii. U.S. Fish. Bull. 72: 915–1031.

Hourigan, T.F. 1987. The behavioral ecology of three species of butterflyfishes (Family Chaetodontidae). Ph.D. Dissertation, University of Hawaii, Honolulu. 496 pp.

Irons, D.K., A.J. Dee & J.D. Parrish. 1984. Johnston Atoll resource survey. Project report to the Department of the Army, U.S. Army Engineer District, Ft. Shafter, Honolulu, Hawaii. 723 pp.

Martin, P. & H.C. Kraemer. 1987. Individual differences in behavior and their statistical consequences. Anim. Behav. 35: 1366–1375.

Masuda, H., K. Amaoka, C. Araga, T. Uyeno & T. Yoshino (ed.). 1984. The fishes of the Japanese Archipelago. Tokai University Press, Tokyo. 437 pp.

Motta, P.J. 1985. Functional morphology of the head of Hawaiian and Mid-Pacific butterflyfishes (Perciformes, Chaetodontidae). Env. Biol. Fish. 13: 253–276.

Motta, P.J. 1988. Functional morphology of the feeding apparatus of ten species of Pacific butterflyfishes (Perciformes, Chaetodontidae): an ecomorphological approach. Env. Biol. Fish. 22: 39–67.

Neudecker, S. 1985. Foraging patterns of chaetodontid and pomacanthid fishes at St. Croix (U.S. Virgin Islands). Proc. Fifth Internat. Coral Reef Symp. 5: 415–420.

Ralston, S. 1981. Aspects of the reproductive biology and feeding ecology of *Chaetodon miliaris*, a Hawaiian endemic butterflyfish. Env. Biol. Fish. 6: 167–176.

Reese, E.S. 1973. Duration of residence by coral reef fishes on 'home' reefs. Copeia 1973: 145–149.

Reese, E.S. 1975. A comparative field study of the social behavior and related ecology of reef fishes of the family Chaetodontidae. Z. Tierpsychol. 37: 37–61.

Reese, E.S. 1977. Coevolution of corals and coral feeding fishes of the family Chaetodontidae. Proc. Third Internat. Coral Reef Symp. 1: 267–274.

Reese, E.S. 1981. Predation on corals by fishes of the family Chaetodontidae: implications for conservation and management of coral reef ecosystems. Bull. Mar. Sci. 31: 594–604.

SAS Institute Inc. 1985. SAS user's guide: statistics, version 5 edition. SAS Institute Inc., Cary. 956 pp.

Schoener, T.W. 1983. Simple models of optimal feeding-territory size: a reconciliation. Amer. Nat. 126: 633–641.

Schoener, T.W. 1987. Time budgets and territory size: some simultaneous optimization models for energy maximizers. Amer. Zool. 27: 259–291.

Sokal, R.R. & F.J. Rohlf. 1981. Biometry. W.H. Freeman and Company, New York. 859 pp.

Talbot, F.H. 1965. A description of the coral structure of Tutia Reef (Tanganyika Territory, East Africa), and its fish fauna. Proc. Zool. Soc. Lond. 145: 431–470.

Tricas, T.C. 1986. Life history, foraging ecology, and territorial behavior of the Hawaiian butterflyfish *Chaetodon multicinctus*. Ph.D. Dissertation, University of Hawaii, Honolulu. 248 pp.

Trivers, R.L. 1972. Parental investment and sexual selection. pp. 136–179. *In:* B.G. Campbell (ed.) Sexual Selection and Descent of Man, 1871–1971, Aldine Publishing Co., Chicago.

Forcipiger flavissimus from Hawaii. A male and female during courtship; the female is on the lower left side and her swollen abdomen is noticeable. Photo by P.S. Lobel.

Feeding habits of Japanese butterflyfishes (Chaetodontidae)

Mitsuhiko Sano
Department of Fisheries, Faculty of Agriculture, University of Tokyo, Yayoi 1-1-1, Bunkyo-ku, Tokyo 113, Japan

Received 17.11.1987   Accepted 16.5.1988

Key words: Stomach contents, Diets, Feeding groups, Scleractinian corals, Food resources, Reef, Southern Japan

Synopsis

Stomach content data from 32 species of Japanese butterflyfishes of the family Chaetodontidae were used to classify them into feeding groups and to determine their important food resources. Four major feeding groups were distinguished: (1) obligative coral feeders which prey exclusively or mostly on scleractinian corals, (2) facultative coral feeders that take both corals and other benthic organisms, (3) noncoralline invertebrate feeders which consume benthic invertebrates other than corals, and (4) zooplankton feeders. Ten species representing 31% of the butterflyfishes belong to the first category. The second and third categories include 13 (41%) and 8 (25%) species, respectively. The fourth category is represented by only one species which picks individual zooplankters, especially calanoid copepods, in midwater above the reefs. Facultative coral feeders consumed varying quantities of scleractinians (from 2 to 74% of food volume), along with a variety of benthic organisms including algae, alcyonarians, sea anemones, sedentary polychaetes, sponges, hydroids, etc. Noncoralline invertebrate feeders, on the other hand, tend to have low diversified diets, predominated by one prey item such as sea anemones, zoanthideans, polychaetes, or colonial ascidians. These dietary data suggest that scleractinian corals are the most important food resource for the Japanese butterflyfishes, and next important are sea anemones, sedentary polychaetes, alcyonarians, and algae.

Introduction

Butterflyfishes, Chaetodontidae, occur worldwide on tropical and subtropical reefs (Burgess 1978, Steene 1978, Allen 1980). Of the 49 species of 7 genera that occur in Japan (Masuda et al. 1984), many are found on coral reefs in the Ryukyu and Ogasawara (Bonin) islands (Yoshino et al. 1975, Zama & Fujita 1977, Yoshino & Nishijima 1981, Kuwamura et al. 1983, Masuda et al. 1984), where they constitute important components of the highly diverse reef systems (Sano et al. 1984a, 1987). Only few species, e.g. *Chaetodon nippon* and *C. auripes,* are abundant on the temperate reefs of the main islands of Honshu, Kyushu, and Shikoku. *C. daedalma* is endemic to southern Japan, where it is common around rocky reefs in certain areas of the Izu and Ogasawara islands (Masuda et al. 1984).

There is considerable information available on the feeding habits of butterflyfishes in the Indo-Pacific (Hiatt & Strasburg 1960, Talbot 1965, Hobson 1968, 1974, Reese 1975, 1977, Neudecker 1977, 1979, Ralston 1981, Harmelin-Vivien & Bouchon-Navaro 1981, 1983, Sano et al. 1984b, Kung & Ciereszko 1985, Tricas 1985, Cox 1986) and the Caribbean (Randall 1967, Birkeland & Neudecker

1981, Gore 1984, Lasker 1985, Neudecker 1985). Many species are known to feed on small benthic invertebrates and algae, with a preference to anthozoans, while a few show a plankton-picking habit. Motta (1982, 1985, 1988) investigated functional morphology of the head of Pacific butterflyfishes and related this to their feeding behavior.

Despite being conspicuous and numerous members of reef-fish communities in southern Japan, few workers have examined the feeding habits of Japanese butterflyfishes. The only major study is that of Sano et al. (1984b), who analyzed the stomach contents of 20 species from Okinawa-jima, one of the Ryukyu Islands. In the present study, I first analyze in detail the stomach contents of those Japanese butterflyfish species for which feeding habits remain unknown and of those for which the number of specimens examined in our previous work (Sano et al. 1984b) was scanty (<6 specimens). I then divide them into several feeding groups, and determine their important food resources, including, also, those species whose diets were already well examined by our previous work.

Material and methods

The stomach contents of 161 individuals from 22 species of chaetodontids were analyzed in detail. The species and numbers of specimens examined are listed in Table 1. All fishes except *Coradion altivelis* were collected using monofilament nets or spears while snorkeling or using SCUBA between 0900 and 1600 h. Most fishes were taken from fringing coral reefs consisting mostly of living coral *Acropora* and *Pocillopora* spp. and from rocky reefs partially with live coral growth at Sakiyama Bay of Iriomote-jima (Ryukyu Islands, 24°20′ N,

Table 1. The numbers, localities, and size of the 22 chaetodontid species collected from southern Japan in 1986 and 1987. All specimens had stomachs containing food. * = species for which the number of specimens examined by Sano et al. (1984b) was scanty (less than 6, mostly 2 or 3 specimens).

Species	Number of specimens				Range of standard length (mm)
	Iriomote-jima	Chichi-jima	Miyake-jima	Shiono-misaki	
Chaetodon argentatus	8				43–85
*Chaetodon baronessa**	9				72–91
*Chaetodon bennetti**		8			58–82
Chaetodon daedalma		14			84–154
*Chaetodon kleinii**	9				44–100
*Chaetodon melannotus**	9				63–94
Chaetodon meyeri	1				136
Chaetodon nippon			15		62–119
*Chaetodon ornatissimus**	9	1			74–134
Chaetodon punctatofasciatus	4				64–77
*Chaetodon rafflesi**	7				91–120
Chaetodon semeion	2				158 & 167
*Chaetodon speculum**	8				85–122
*Chaetodon ulietensis**	8				122–135
*Chaetodon unimaculatus**	5	5			52–143
Coradion altivelis				4	133–147
Forcipiger flavissimus	6	1			112–156
Hemitaurichthys polylepis	9				65–115
*Heniochus chrysostomus**	10				87–116
Heniochus monoceros	2				61 & 140
Heniochus singularius	2				173 & 183
Heniochus varius	5				83–116

123° 42' E) and Chichi-jima (Ogasawara Islands, 27° 05' N, 142° 12' E) during August and September in 1986 and 1987. Water depths range from 2 to 10 m. *Chaetodon nippon* was collected from volcanic sand areas containing numerous rocks and boulders of various sizes, many of which support coral outcroppings, at Igaya Bay of Miyake-jima (Izu Islands, 34° 05' N, 139° 30' E) in June 1986. Water depths at the Miyake-jima site range from 5 to 20 m. *Coradion altivelis* was obtained by bottom gill net fishing from rocky reefs ranging in depth from 20 to 40 m, where small patches of coral reef develop, at Shionomisaki of Kii Peninsula (central Honshu, 33° 28' N, 135° 48' E) during January and March in 1987. Detailed descriptions and maps of the collection sites appear in Sano et al. (1987) for Iriomote-jima, Kuwamura et al. (1983) for Chichi-jima, Tribble & Randall (1986) for Miyake-jima, and Tamura et al. (1966) for Shionomisaki.

Immediately after collection, specimens were placed in 10% formalin and the stomach contents preserved by injecting concentrated formalin directly into the body cavity. The standard length was measured for each specimen in the laboratory. Food items from the stomach contents were sorted and identified under a low-power binocular microscope. The percentage volume of each sorted item was visually estimated on a section paper to reduce error (see Sano et al. 1984b). The relative importance of each food item to the diet of each species was expressed by two methods: (1) percentage occurrence of the item and (2) mean percentage of the item in the diet volume. The latter was calculated by dividing the sum total of the individual volumetric percentage for the item by the number of specimens examined.

The feeding groups defined in this paper include the 22 species of the present study and also 10 species reported by Sano et al.(1984b) from Okinawa-jima: *Chaetodon auriga, C. auripes, C. citrinellus, C. ephippium, C. lineolatus, C. lunula, C. plebeius, C. trifascialis, C. trifasciatus,* and *C. vagabundus*. These 10 species were adequately sampled during the previous study and so no additional specimens were collected during the present study. Some species considered by Sano et al. (1984b), however, were inadequately sampled (<6 specimens) so only the present dietary data are used because a sufficient number was obtained in this study (Table 1).

To compare specialist or generalist tendencies among species, the Shannon diversity index H' (Shannon & Weaver 1949, Berg 1979) was calculated using mean volumetric percentage of food. The natural logarithm was used in this calculation, because it increases the range of values and thus shows differences more clearly (Kotrschal & Thomson 1986). Despite some species with inadequate sample sizes, the index was computed because it is difficult to collect a sufficient number of specimens of such species, which are rare in Japan.

Results

Diets

Only scleractinian corals with no skeletal material were comsumed as food by the following 6 species: *Chaetodon baronessa, C. bennetti, C. meyeri, C. ornatissimus, C. speculum,* and *Heniochus singularius. C. semeion* and *H. monoceros* fed on only colonial ascidians and errant polychaetes, respectively. *C. ulietensis* had stomachs full of food, including zoanthideans (% freq. = 100, % vol. = 83), alcyonarians (38, 17), and filamentous algae (50, <1). The major food item of *Coradion altivelis* was colonial ascidians (100, 71), and the remaining items were sponges (100, 18) and hydroids (50, 11). Diets of the other 12 species are shown in Table 2. Among 10 specimens of *Chaetodon unimaculatus* collected, 2 (134 and 143 mm in standard length) from Chichi-jima and 1 (111 mm) from Iriomote-jima contained small amounts of fish feces consisting mostly of copepods, along with large amounts of scleractinians and/or alcyonarians. Only this species ingested scleractinians with many skeletal fragments. All algae consumed by the Japanese butterflyfishes examined were noncalcareous.

Feeding groups

On the basis of the stomach content data, the 32

Table 2. Percentage frequency of occurrence and percentage volume of food items in the diets of the 12 chaetodontid species collected from southern Japan in 1986 and 1987. + = less than 1% volume.

Food items	Chaetodon argentatus		Chaetodon daedalma		Chaetodon kleinii		Chaetodon melannotus		Chaetodon nippon		Chaetodon punctatofasciatus		Chaetodon rafflesi		Chaetodon unimaculatus		Forcipiger flavissimus		Hemitaurichthys polylepis		Heniochus chrysostomus		Heniochus varius	
	%F	%V	%F	%V	%F	%V	%F	%V	%F	%V	%F	%V	%F	%V	%F	%V	%F	%V	%F	%V	%F	%V	%F	%V
Filamentous algae	100	56	93	9	89	38	33	+	60	2	50	+	57	1	10	+							40	+
Algal fronds					67	27																		
Sponges	25	+			11	3			27	3							14	3			10	1	60	6
Hydroids	63	5	86	31	22	9			13	1							14	1						
Alcyonarians	100	14			22	10	100	92	33	1	50	5	14	6	40	20					60	43		
Sea anemones							22	1	7	+														
Zoanthideans													100	91										
Scleractinians	63	17	100	44	22	11	33	2	33	5	100	94	14	+	100	74					60	32	80	43
Sipunculid introverts																					40	4	20	12
Gastropods			14	+					7	+									33	2				
Errant polychaetes			14	1					33	1											10	+		
Sedentary polychaete tentacles	63	8	64	7	11	1			100	84	25	+					100	90			80	15	20	1
Ostracods																			56	5				
Calanoid copepods					11	+			7	+	100	1							100	77				
Cyclopoid copepods																	29	+	67	6				
Harpacticoid copepods					11	+			7	+	25	+							22	1				
Cirripedians	25	+	29	+					13	+														
Gammaridean amphipods			21	+					47	3											20	+		
Caprellid amphipods																					20			1
Crab megalops																			22	3				
Unidentified crustaceans																					20		20	2
Appendicularians																			33	4				
Colonial ascidians			7	1																	60		60	34
Invertebrate egg masses			36	7					13	+			43	2	10	+	14	6	33	+				
Fish eggs					11	+							14	+					22	2	20		20	1
Fish feces															30	6								
Unidentified animal material					22	1	11	5													20	5		

Japanese butterflyfishes investigated can be segregated into 4 major feeding groups (Reese 1977, Anderson et al. 1981, Harmelin-Vivien & Bouchon-Navaro 1981, 1983):

(1) Obligative coral feeders. This group includes species that feed exclusively or mostly on scleractinian corals. Ten species representing 31% of the butterflyfishes belong to this group: *Chaetodon baronessa, C. bennetti, C. meyeri, C. ornatissimus, C. plebeius, C. punctatofasciatus, C. speculum, C. trifascialis, C. trifasciatus*, and *Heniochus singularius*.

(2) Facultative coral feeders which take both scleractinians and other benthic organisms. This category includes 13 species (41%): *Chaetodon argentatus, C. auriga, C. auripes, C. citrinellus, C. daedalma, C. ephippium, C. kleinii, C. melannotus, C. nippon, C. unimaculatus, C. vagabundus, Heniochus chrysostomus*, and *H. varius*. These species comsumed a wide variety of benthic organisms in various proportions in their diets. Scleractinians were ingested by all species, ranging in percentage of food volume from 2 to 74%. The other prey predominantly taken by several species were algae (1 to 65% of food volume), alcyonarians (1 to 92%), sea anemones (1 to 76%), and sedentary polychaetes such as terebellids, serpulids, and sabellids (1 to 84%). Sponges (2 to 61%) and hydroids (1 to 31%) were also important. All species which show an omnivorous habit are included in this group.

(3) Noncoralline invertebrate feeders that consume benthic invertebrates other than scleractinians. This group is made up of 8 species (25%): *Chaetodon lineolatus, C. lunula, C. rafflesi, C. semeion, C. ulietensis, Coradion altivelis, Forcipiger flavissimus*, and *Heniochus monoceros*. These species tend to have low diversified diets, predominated by one prey item. Sea anemones, zoanthideans, sedentary and errant polychaetes, and colonial ascidians were important dietary items for particular species.

(4) Zooplankton feeders. This category is represented by only one species, *Hemitaurichthys polylepis*, which picks individual zooplankters, especially calanoid copepods, in midwater above the reefs. This species ingested benthic invertebrates such as gastropods, but their amount in the diet was very small, less than 3% of food volume.

Feeding diversity

According to the species' feeding diversity (H'), three groups may be distinguished: (1) specialists, (2) low diversity feeders, and (3) high diversity

Table 3. Feeding diversity (H') of the 32 chaetodontid species collected from southern Japan, listed in order of decreasing values of H'. O = obligative coral feeders, F = facultative coral feeders, N = noncoralline invertebrate feeders, Z = zooplankton feeders, * = species showing an omnivorous habit, No. = number of specimens examined.

Species	No.	H'	Feeding group
Chaetodon citrinellus	9	1.879	F*
Chaetodon kleinii	9	1.608	F*
Chaetodon auriga	13	1.604	F
Chaetodon vagabundus	10	1.504	F*
Chaetodon daedalma	14	1.405	F*
Heniochus varius	5	1.369	F
Heniochus chrysostomus	10	1.337	F
Chaetodon argentatus	8	1.253	F*
Chaetodon ephippium	6	1.020	F
Hemitaurichthys polylepis	9	0.956	Z
Chaetodon auripes	10	0.867	F
Coradion altivelis	4	0.795	N
Chaetodon nippon	15	0.723	F
Chaetodon unimaculatus	10	0.714	F
Chaetodon lunula	9	0.656	N
Chaetodon ulietensis	8	0.456	N
Forcipiger flavissimus	7	0.415	N
Chaetodon rafflesi	7	0.379	N
Chaetodon melannotus	9	0.351	F
Chaetodon punctatofasciatus	4	0.254	O
Chaetodon lineolatus	6	0.154	N
Chaetodon semeion	2	0	N
Heniochus monoceros	2	0	N
Chaetodon baronessa	9	0	O
Chaetodon bennetti	8	0	O
Chaetodon meyeri	1	0	O
Chaetodon ornatissimus	10	0	O
Chaetodon plebeius	15	0	O
Chaetodon speculum	8	0	O
Chaetodon trifascialis	23	0	O
Chaetodon trifasciatus	28	0	O
Heniochus singularius	2	0	O

feeders or generalists (Table 3). The first group (Shannon index values near zero) includes all species of obligative coral feeders and 3 species of noncoralline invertebrate feeders. The second group (index values of 0.3 to 1.0) includes species whose diets tend to be predominated by one prey item. The 5 species of noncoralline invertebrate feeders not in the first group belong to this group, as do 4 facultative coral feeders and a zooplankton feeder. The third group (index values >1) consists of 9 species of facultative coral feeders. Species with an omnivorous habit tend to have relatively higher values of the feeding diversity (Table 3).

Important food

Scleractinian corals are the most important food resource for the Japanese butterflyfishes (Fig. 1). This prey was utilized as food in varying quantities (2 to 100% of food volume) by 23 species representing 72% of the butterflyfishes (i.e. obligative and facultative coral feeders). Sea anemones, sedentary polychaetes, alcyonarians, and algae are next in importance (Fig. 1). Other items of importance taken by 8 or more species included hydroids, sponges, and invertebrate and fish eggs. Colonial ascidians are a major part of the diets of *Chaetodon semeion*, *Coradion altivelis*, and *Heniochus varius*. Calanoid copepods and errant polychaetes dominated the diets of *Hemitaurichthys polylepis* and *Heniochus monoceros*, respectively. Zoanthideans are important to both *Chaetodon rafflesi* and *C. ulietensis*.

Discussion

Generally the butterflyfishes studied here had diets similar to reported diets of conspecifics elsewhere, but there were exceptions. For example, Hobson (1974) reported that *Chaetodon kleinii* (as *C. corallicola*) in Hawaii is a planktivore which feeds largely on copepods, but at Iriomote-jima this species preyed heavily on anthozoans, especially scleractinians and alcyonarians, along with considerable amounts of algae, suggesting that it is strictly a

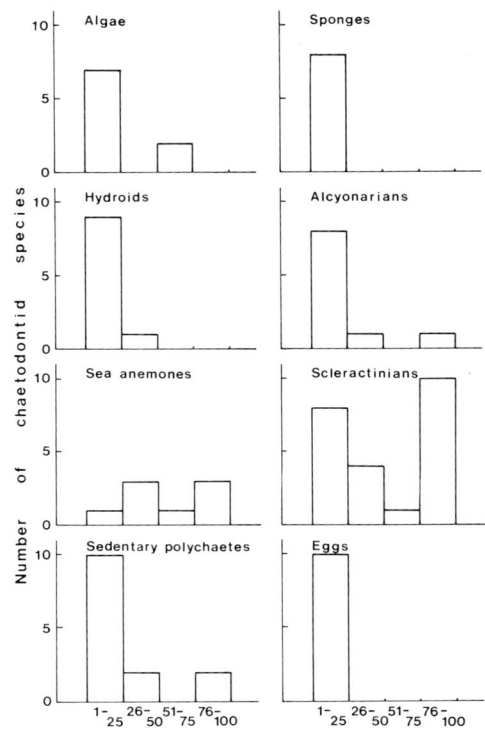

Fig. 1. Diagram to demonstrate comsumption of the 8 major prey items by the 32 chaetodontid species collected from southern Japan.

benthic omnivore. Motta's (1987) morphological and behavioral studies indicated that *C. kleinii* is equipped for planktivory to some degree, but is not very specialized in this habit. So, he concluded that planktivory may be a relatively recent event for this species. This may also be true of *Hemitaurichthys polylepis*, which captures primarily zooplankters but also feeds on benthic invertebrates in Hawaii (Hobson 1974: as *H. zoster*), French Polynesia (Harmelin-Vivien & Bouchon-Navaro 1983), and Japan (present work), although we lack a morphological study for this species.

Coprophagy may be important to some species. Small amounts of fish feces containing mostly copepods were consumed with scleractinians and/or alcyonarians by 3 of 10 specimens of *Chaetodon unimaculatus* in Japan. Robertson (1982) observed that at least 45 fishes associated with a coral reef at Palau fed on feces produced by a variety of reef fishes in addition to other food. However, most of these coprophagic fishes were herbivores and detritivores, and none of 7 corallivorous chaetodontid species including *C. unimaculatus* observed in his study ate feces. Stomach content data in this study provide evidence that *C. unimaculatus* in Japan sometimes shows a copraphagous habit.

The feeding relations of butterflyfishes studied here offer comparisons with results of workers in other Indo-Pacific areas (Table 4). My result fits with those elsewhere to some extent. In Hawaii, however, it appears that more plankton feeders are found than in Japan, whereas obligative and facultative coral feeders are much fewer. This may be due in part to the rareness of *Acropora* corals and underutilized zooplankters as food resources by other reef fishes in Hawaii (see Hourigan & Reese 1987). I observed that some Japanese obligative coral feeders such as *Chaetodon trifascialis*, *C. baronessa*, and *C. plebeius* are exclusively associated with reefs rich in *Acropora* corals on which they feed (see also Reese 1977, 1981), but these species do not occur in Hawaii where *Acropora* is rare (Gosline & Brock 1960, Grigg et al. 1981, Grigg 1983). Hourigan & Reese (1987) suggested that zooplankters may be abundant in Hawaii because of lacking many planktivore species which occur on other Indo-West Pacific reefs. Therefore, an abundant food resource may be partly responsible for more planktivorous butterflyfishes in Hawaii. Utilization of zooplankters by some Hawaiian butterflyfishes, especially *Chaetodon kleinii* and *C. miliaris*, may have been the result of escape from interspecific competition for benthic food such as corals (Ralston 1981), although the occurrence and effects of competition in reef fish assemblages are controversial (e.g. Sale 1980, Williams 1980, Doherty 1983, Gladfelter & Johnson 1983, Ebersole 1985). In French Polynesia, on the other hand, the apparent relative reduction in numbers is found for noncoralline invertebrate feeders, and the reverse direction is represented for facultative coral feeders. However, these trends are difficult to interpret. It may be considered, for example, that French Polynesian reefs carry a more extensive coral cover than other Indo-Pacific reefs and thus more facultative coral feeders can occur. This inference is questionable, however, because more obligative corallivores are not found in French Polynesia (Table 4).

Table 4. Comparisons of feeding groups among chaetodontid species in different locations. * = including omnivorous species that do not consume scleractinian corals.

Geographical locations	Total number of species	Obligative coral feeders	Facultative coral feeders	Noncoralline invertebrate feeders	Zooplankton feeders	Source
		No. (%)	No. (%)	No. (%)	No. (%)	
Pacific Ocean						
Japan	32	10 (31)	13 (41)	8 (25)	1 (3)	This study
Hawaii	16	3 (19)	3 (19)	6* (38)	4 (25)	Hobson (1974), Reese (1977)
Enewetak	17	4 (24)	6 (35)	5* (29)	2 (12)	Reese (1977)
Great Barrier Reef	21	8 (38)	9 (43)	4 (19)	0 (0)	Reese (1977), Anderson et al. (1981)
French Polynesia	20	5 (25)	13 (65)	1 (5)	1 (5)	Harmelin-Vivien & Bouchon-Navaro (1983), Bell et al. (1985)
Indian Ocean						
Madagascar	18	5 (28)	8 (44)	3 (17)	2 (11)	Harmelin-Vivien & Bouchon-Navaro (1983)
Red Sea						
Aqaba	8	2 (25)	4 (50)	1 (13)	1 (13)	Harmelin-Vivien & Bouchon-Navaro (1981)

There are noticeable trends in feeding diversity (Table 3). Noncoralline invertebrate feeders represent a tendency towards increasing dietary specialization than facultative coral feeders, as measured by a decrease in H'. In contrast, species which are facultative coral feeders and which show an omnivorous habit tend to have relatively higher values of H'. Similar trends can be recognized among chaetodontids in the Red Sea and in French Polynesia (Bouchon-Navaro 1986), but not in Hawaii (Hobson 1974).

The importance of sea anemones to Japanese butterflyfishes seems unmatched among butterflyfishes elsewhere, as these organisms have not been recognized as significant prey in the Marshall Islands (Hiatt & Strasburg 1960), Caribbean (Randall 1967), Hawaii (Hobson 1974), Red Sea (Harmelin-Vivien & Bouchon-Navaro 1981) or French Polynesia (Harmelin-Vivien & Bouchon-Navaro 1983). This may reflect the greater abundance of this prey in the Japanese areas, especially the Okinawa-jima study site (Sano et al. 1984b), although data that would verify this inference are lacking.

Acknowledgements

I am grateful to Makoto Shimizu for helpful suggestions and cooperation. The following persons and organizations kindly assisted in the collection of specimens: Hiroyoshi Kohno, Hiroyuki Yokochi, Akihiro Ishimaru, Hideo Sunagawa, Hirohiko Ohta, Jack T. Moyer, Hiroshi Misaki, Makoto Miki, Kenji Kato, Johnson Kimura, Oscar Sosa Nishizaki, Hiroyuki Tachikawa, Hitoshi Sasaki, Okinawa Regional Research Center of Tokai University, Ogasawara Fisheries Center of Tokyo, Ogasawara Marine Center, Fisheries Department of Ogasawara Branch of Tokyo Metropolitan Government, Ogasawara-jima Fishermen's Union, and Miyake-jima Fishermen's Union. Jack T. Moyer, Philip J. Motta, and two anonymous reviewers substantially improved earlier versions of this paper. My thanks are also extended to Tokiharu Abe, Kenji Mochizuki, Keiichi Matsuura, and Ken-ichi Hayashizaki for sharing valuable information with me. This study was made possible by a grant-in-aid from the Fujiwara Foundation for Natural History Education.

References cited

Allen, G.R. 1980. Butterfly and angelfishes of the world, Vol. 2. Wiley-Interscience, New York. 352 pp.

Anderson, G.R.V., A.H. Ehrlich, P.R. Ehrlich, J.D. Roughgarden, B.C. Russell & F.H. Talbot. 1981. The community structure of coral reef fishes. Amer. Nat. 117: 476–495.

Bell, J., M. Harmelin-Vivien & R. Galzin. 1985. Large scale spatial variation in abundance of butterflyfishes (Chaetodontidae) on Polynesian reefs. Proc. Fifth Internat. Coral Reef Congr. 5: 421–426.

Berg, J. 1979. Discussion of methods of investigating the food of fishes, with reference to a preliminary study of the prey of *Gobiusculus flavescens* (Gobiidae). Mar. Biol. 50: 263–273.

Birkeland, C. & S. Neudecker. 1981. Foraging behavior of two Caribbean chaetodontids: *Chaetodon capistratus* and *C. aculeatus*. Copeia 1981: 169–178.

Bouchon-Navaro, Y. 1986. Partitioning of food and space resources by chaetodontid fishes on coral reefs. J.Exp. Mar. Biol. Ecol. 103: 21–40.

Burgess, W.E. 1978. Butterflyfishes of the world. T.F.H. Publications, Neptune City. 832 pp.

Cox, E.F. 1986. The effects of a selective corallivore on growth rates and competition for space between two species of Hawaiian corals. J. Exp. Mar. Biol. Ecol. 101: 161–174.

Doherty, P.J. 1983. Tropical territorial damselfishes: is density limited by aggression or recruitment? Ecology 64: 176–190.

Ebersole, J.P. 1985. Niche separation of two damselfish species by aggression and differential microhabitat utilization. Ecology 66: 14–20.

Gladfelter, W.B. & W.S. Johnson. 1983. Feeding niche separation in a guild of tropical reef fishes (Holocentridae). Ecology 64: 552–563.

Gore, M.A. 1984. Factors affecting the feeding behavior of a coral reef fish, *Chaetodon capistratus*. Bull. Mar. Sci. 35: 211–220.

Gosline, W.A. & V.E. Brock. 1960. Handbook of Hawaiian fishes. University of Hawaii Press, Honolulu. 372 pp.

Grigg, R.W. 1983. Community structure, succession and development of coral reefs in Hawaii. Mar. Ecol. Prog. Ser. 11: 1–14.

Grigg, R.W., J.W. Wells & C. Wallace. 1981. *Acropora* in Hawaii, Part 1. History of the scientific record, systematics, and ecology. Pacif. Sci. 35: 1–13.

Harmelin-Vivien, M.L. & Y. Bouchon-Navaro. 1981. Trophic relationships among chaetodontid fishes in the Gulf of Aqaba (Red Sea). Proc. Fourth Internat. Coral Reef Symp. 2: 537–544.

Harmelin-Vivien, M.L. & Y. Bouchon-Navaro. 1983. Feeding diets and significance of coral feeding among chaetodontid fishes in Moorea (French Polynesia). Coral Reefs 2: 119–127.

Hiatt, R.W. & D.W. Strasburg. 1960. Ecological relationships of the fish fauna on coral reefs of the Marshall Islands. Ecol. Monogr. 30: 65–127.

Hobson, E.S. 1968. Predatory behavior of some shore fishes in the Gulf of California. U.S. Fish Wildl. Serv., Res. Rep. 73: 1–92.

Hobson, E.S. 1974. Feeding relationships of teleostean fishes on coral reefs in Kona, Hawaii. U.S. Fish. Bull. 72: 915–1031.

Hourigan, T.F. & E.S. Reese. 1987. Mid-ocean isolation and the evolution of Hawaiian reef fishes. Trend. Ecol. Evol. 2: 187–191.

Kotrschal, K. & D.A. Thomson. 1986. Feeding patterns in eastern tropical Pacific blennioid fishes (Teleostei: Tripterygiidae, Labrisomidae, Chaenopsidae, Blenniidae). Oecologia (Berl.) 70: 367–378.

Kung, S.-S. & L.S. Ciereszko. 1985. Occurrence of the wax cetyl palmitate in stomachs of the corallivorous butterflyfish *Chaetodon trifascialis*. Coral Reefs 4: 45–46.

Kuwamura, T., R. Fukao, T. Nakabo, M. Nishida, T. Yanagisawa & Y. Yanagisawa. 1983. Inshore fishes of the Ogasawara (Bonin) Islands, Japan. Galaxea 2: 83–94.

Lasker, H.R. 1985. Prey preferences and browsing pressure of the butterflyfish *Chaetodon capistratus* on Caribbean gorgonians. Mar. Ecol. Prog. Ser. 21: 213–220.

Masuda, H., K. Amaoka, C. Araga, T. Uyeno & T. Yoshino (ed.). 1984. The fishes of the Japanese Archipelago. Tokai University Press, Tokyo. 437 pp.

Motta, P.J. 1982. Functional morphology of the head of the inertial suction feeding butterflyfish, *Chaetodon miliaris* (Perciformes, Chaetodontidae). J. Morphol. 174: 283–312.

Motta, P.J. 1985. Functional morphology of the head of Hawaiian and Mid-Pacific butterflyfishes (Perciformes, Chaetodontidae). Env. Biol. Fish. 13: 253–276.

Motta, P.J. 1987. A quantitative analysis of ferric iron in butterflyfish teeth (Chaetodontidae, Perciformes) and the relationship to feeding ecology. Can. J. Zool. 65: 106–112.

Motta, P.J. 1988. Functional morphology of the feeding apparatus of ten species of Pacific butterflyfishes (Perciformes, Chaetodontidae): an ecomorphological approach. Env. Biol. Fish. 22: 39–67.

Neudecker, S. 1977. Transplant experiments to test the effect of fish grazing on coral distribution. Proc. Third Internat. Coral Reef Symp. 1: 317–323.

Neudecker, S. 1979. Effects of grazing and browsing fishes on the zonation of corals in Guam. Ecology 60: 666–672.

Neudecker, S. 1985. Foraging patterns of chaetodontid and pomacanthid fishes at St. Croix (U.S. Virgin Islands). Proc. Fifth Internat. Coral Reef Congr. 5: 415–420.

Ralston, S.V.D. 1981. Aspects of the reproductive biology and feeding ecology of *Chaetodon miliaris*, a Hawaiian endemic butterflyfish. Env. Biol. Fish. 6: 167–176.

Randall, J.E. 1967. Food habits of reef fishes of the West Indies. Stud. Trop. Ocean. 5: 665–847.

Reese, E.S. 1975. A comparative field study of the social behavior and related ecology of reef fishes of the family Chaetodontidae. Z. Tierpsychol. 37: 37–61.

Reese, E.S. 1977. Coevolution of corals and coral feeding fishes of the family Chaetodontidae. Proc. Third Internat. Coral Reef Symp. 1: 267–274.

Reese, E.S. 1981. Predation on corals by fishes of the family Chaetodontidae: implications for conservation and management of coral reef ecosystems. Bull. Mar. Sci. 31: 594–604.

Robertson, D.R. 1982. Fish feces as fish food on a Pacific coral reef. Mar. Ecol. Prog. Ser. 7: 253–265.

Sale. P.F. 1980. The ecology of fishes on coral reefs. Ocean. Mar. Biol. Ann. Rev. 18: 367–421.

Sano, M., M. Shimizu & Y. Nose. 1984a. Changes in structure of coral reef fish communities by destruction of hermatypic corals: observational and experimental views. Pacif. Sci. 38: 51–79.

Sano, M., M. Shimizu & Y. Nose. 1984b. Food habits of teleostean reef fishes in Okinawa Island, southern Japan. Univ. Mus., Univ. Tokyo Bull. 25: 1–128.

Sano, M., M. Shimizu & Y. Nose. 1987. Long-term effects of destruction of hermatypic corals by *Acanthaster planci* infestation on reef fish communities at Iriomote Island, Japan. Mar. Ecol. Prog. Ser. 37: 191–199.

Shannon, C.E. & W. Weaver. 1949. The mathematical theory of communication. University of Illinois Press, Urbana. 117 pp.

Steene, R.C. 1978. Butterfly and angelfishes of the world, Vol. 1. Wiley-Interscience, New York. 144 pp.

Talbot, F.H. 1965. A description of the coral structure of Tutia Reef (Tanganyika Territory, East Africa), and its fish fauna. Proc. Zool. Soc. Lond. 145: 431–470.

Tamura, T., T. Habe, F. Uchinomi, T. Tokioka, S. Fuse, C. Araga, S. Nishimura, H. Tanase, K. Tatsuki & T. Yamamoto. 1966. Marine parks in Wakayama Prefecture, Japan. Sci. Rep. Nat. Conser. Soc. Japan 27: 1–126. (In Japanese).

Tribble, G.W. & R.H. Randall. 1986. A description of the high-latitude shallow water coral communities of Miyake-jima, Japan. Coral Reefs 4: 151–159.

Tricas, T.C. 1985. The economics of foraging in coral-feeding butterflyfishes of Hawaii. Proc. Fifth Internat. Coral Reef Congr. 5: 409–414.

Williams, D.McB. 1980. Dynamics of the pomacentrid community on small patch reefs in One Tree Lagoon (Great Barrier Reef). Bull. Mar. Sci. 30: 159–170.

Yoshino, T. & S. Nishijima. 1981. A list of fishes found around Sesoko Island, Okinawa. Sesoko Mar. Sci. Lab. Tech. Rep. 8: 19–87.

Yoshino, T., S. Nishijima & S. Shinohara. 1975. Catalogue of fishes of the Ryukyu Islands. Bull. Sci. Engin. Div., Univ. Ryukyus, Mathem. Nat. Sci. 20: 61–118. (In Japanese).

Zama, A. & K. Fujita. 1977. An annotated list of fishes from the Ogasawara Islands. J. Tokyo Univ. Fish. 63: 87–138. (In Japanese).

Forcipiger longirostrus from Hawaii, the black phase. Photo by P.S. Lobel.

The brain organization of butterflyfishes

Roland Bauchot[1], Jean-Marc Ridet[1] & Marie-Louise Bauchot[2]
[1] *Laboratoire d'Anatomie Comparée, Université Paris 7, 2 Place Jussieu, 75251 Paris Cedex 05, France*
[2] *Laboratoire d'Ichthyologie générale et appliquée, Muséum National d'Histoire Naturelle, 23 rue Cuvier, 75231 Paris Cedex 05, France*

Received 8.3.1988   Accepted 23.9.1988

Key words: Encephalization, Brain morphology, Brain histology, Correlation with biological adaptations

Synopsis

The encephalization indices of angelfishes (Pomacanthidae) and butterflyfishes (Chaetodontidae) are typical of advanced perciform fishes: both families lie in the upper part of the polygon of teleost indices. The chaetodontids seem to be a little more encephalized than pomacanthids. The general morphology of the brains in both families is very similar: small olfactory bulbs, large optic tectum and a cerebellum which covers the brain structures in front of it like a cap. This morphology is shared by another family of the coral reef biotope, the Acanthuridae. The histological architecture is also typical of advanced teleosts, with a cortex-like pallium, a laminated nucleus geniculatus (= pretectalis superficialis), a complex valvula cerebelli and a corpus glomerulosum with a clear neuropile centre. The quantitative analysis of the main subdivisions of the brain, either from relative volumes or from indices, shows small olfactory bulbs (microsmy) but important telencephalic and diencephalic centres, large tectal centres (vision) and large cerebellum (precise locomotion). Many of these peculiarities are shared by other fishes inhabiting coral reefs. The differences between the two families seem to be primarily correlated with food habits: the angelfishes, which are sponge-feeders and may have an overweight due to the ballast of the sponge-skeleton in their digestive tract, and which do not need either such good vision or such precise locomotion to pick up their prey, could be a little less encephalized than the butterflyfishes.

Introduction

Butterflyfishes are inhabitants of coral reefs and are well known for their bright colors and for the precision of their swimming movements around coral heads. They belong to the family Chaetodontidae and have strong similarities with another family of tropical seas, the angelfishes or Pomacanthidae. The similarities of their morphology and way of life and the conservatism of the nervous structures has suggested that the two families are probably very similar also in the organization of the brain. Hitherto, except the study of Snow & Rylander (1982) on the optic system of chaetodontids, there has been no data concerning the brain of chaetodontids or pomacanthids in the literature.

Material and methods

Material

Table 1 lists 17 species of pomacanthid and 35 species of chaetodontid used in this study (named after the revision of Maugé & Bauchot 1985). The name, locality and number of specimens, standard length

Table 1. Brain-body weight relationships in Chaetodontidae and Pomacanthidae.

Family.
Genus: mean index +/− standard deviation (percentage of the mean)
Subgenus: mean index +/− standard deviation (percentage of the mean)
Species: locality (number of specimens), standard length (mm), bodyweight (g), brainweight (mg), encephalization index.

Pomacanthidae.

Centropyge: 110.4 +/− 15.1 (13.7%).

Species	(n)	SL	BW	BrW	EI
C. bicolor (Bloch, 1787). New Caledonia	(1)	95	35.6	167.2	117
C. bispinosus (Günther, 1860). Marshall, New Cal.	(2)	90	39.5	143.5	95
C. fisheri (Snyder, 1902). Marshall, Hawaii	(3)	72	15.4	89.7	100
C. flavissimus (Cuvier, 1831). Marshall	(1)	95	41.0	174.0	113
C. heraldi Schultz & Woods, 1953. Marshall, New Cal.	(2)	80	24.4	142.7	122
C. multispinis (Playfair, 1866). Red Sea	(3)	95	42.2	184.3	117
C. potteri Jordan & Metz, 1912. Hawaii	(2)	102	51.9	152.5	87
C. tibicen (Cuvier, 1831). New Caledonia	(1)	95	39.0	199.0	132

Genicanthus: 98.

G. caudovittatus (Günther, 1860). Réunion, Red Sea	(2)	180	211.2	331.8	98

Holacanthus: 98.0 +/− 11.3 (11.5%).

H. arcuatus (Linnaeus, 1758). Hawaii	(1)	145	138.6	298.0	106
H. xanthotis Fraser Brunner, 1951. Red Sea	(2)	180	264.8	334.6	90

Pomacanthus: 100.2 +/− 4.4 (4.4%).

P. imperator (Bloch, 1787). Marshall, Red Sea (2)	(2)	370	2055.7	747.1	96
P. maculosus (Forsskal, 1775). Red Sea	(2)	450	943.4	584.1	97
P. semicirculatus (Cuvier, 1831). New Caledonia	(1)	400	3234.4	879.9	99
P. sexstriatus (Cuvier, 1831). New Caledonia	(1)	600	11610.0	1237.3	102
P. striatus (Rüppell, 1835). Oman	(1)	420	3854.8	993.2	107

Pygoplites: 108.

P. diacanthus Boddaert, 1772. Marshall, Red Sea	(2)	220	553.1	538.4	108

Chaetodontidae.

Chaetodon: 130.

C. decussatus Cuvier, 1831. India	(1)	130	99.0	311.9	130

Chaetodontops: 163.7 +/− 22.9 (14.0%).

C. collare (Bloch, 1787). India	(1)	130	125.8	370.0	138
C. flavirostris (Günther, 1874). New Caledonia	(2)	135	113.5	464.8	182
C. lunula (Lacepède, 1803). Marshall, Hawaii	(2)	170	190.6	552.7	171

Citharoedus.
C. *Citharoedus:* 140.5 +/− 20.5 (14.6%).

C. C. ornatissimus (Cuvier, 1831). Marshall	(1)	170	297.2	492.7	126
C. C. reticulatus (Cuvier, 1831). Marshall	(2)	150	160.6	464.2	155

C. *Gonochaetodon:* 93.

C. G. baronessa (Cuvier, 1831). New Caledonia	(1)	120	102.6	225.7	93

Exornator.
E. *Exornator:* 136.0 +/− 4.6 (3.4%).

E. E. citrinellus (Cuvier, 1831). Marshall	(1)	120	72.6	290.5	141
E. E. kleinii (Bloch, 1790). Réunion, Hawaii	(2)	100	43.5	214.5	135
E. E. multicinctus (Garrett, 1863). Hawaii	(2)	100	38.0	196.3	132

vervolg tabel 1:

E. Rhombochaetodon: 133.5 +/− 9.2 (6.9%).					
E. R. mertensi (Cuvier, 1831). New Cal., Marshall	(2)	110	53.2	247.0	140
E. R. paucifasciatus (Ahl, 1923). Red Sea	(2)	110	62.0	241.6	127
Forcipiger: 148.5 +/− 9.2 (6.2%).					
F. flavissimus Jordan & McGregor, 1898. Hawaii, Mars.	(5)	160	80.8	338.1	155
F. longirostris (Broussonet, 1782). Hawaii	(1)	180	110.3	359.0	142
Hemitaurichthys: 132.					
H. polylepis (Bleeker, 1857). Marshall	(1)	130	96.8	313.1	132
Heniochus: 128.3 +/− 17.1 (13.3%).					
H. acuminatus (Linnaeus, 1758). Réunion	(2)	160	190.4	462.3	143
H. chrysostomus Cuvier, 1831. Marshall	(1)	150	174.2	456.2	147
H. intermedius Steindachner, 1843. Red Sea	(1)	160	291.4	381.8	99
H. monoceros Cuvier, 1831. Marshall	(1)	220	496.6	614.1	128
H. singularius Smith & Radcliffe, 1911. New Caledonia	(1)	230	614.1	635.8	123
H. varius (Cuvier, 1829). New Caledonia	(1)	180	283.7	497.7	130
Heterochaetodon.					
H. Burgessius: 126.					
H. B. miliaris (Quoy & Gaimard, 1824). Hawaii	(5)	130	89.8	288.4	126
H. Heterochaetodon: 152.					
H. H. dolosus (Ahl, 1923). Réunion	(2)	125	71.8	312.6	152
H. Lepidochaetodon: 133.0 +/− 12.7 (9.6%).					
H. L. quadrimaculatus (Gray, 1831). Marshall	(1)	110	57.0	259.4	142
H. L. unimaculatus (Bloch, 1787). Hawaii, Marshall	(2)	130	95.9	292.7	124
Megaprotodon: 143.0 +/− 18.4 (12.9%).					
M. plebeius (Cuvier, 1831). New Caledonia	(1)	110	49.4	221.4	130
M. trifascialis (Quoy & Gaimard). Marshall	(1)	130	86.6	351.8	156
Mesochaetodon.					
M. Corallochaetodon: 127.0 +/− 8.5 (6.7%).					
M. C. austriacus Rüppell, 1835. Red Sea	(2)	120	94.1	282.9	121
M. trifasciatus (Mungo Park, 1794). Hawaii, Marshall	(2)	120	62.5	255.3	133
M. Strongylochaetodon: 162.					
M. S. melannotus (Schneider, 1801). Red Sea	(1)	120	67.5	322.7	162
Nalbantius: 130.					
N. speculum (Cuvier, 1831). New Caledonia	(1)	120	73.7	271.3	130
Rabdophorus.					
R. Linophora: 167.0 +/− 26.9 (16.1%).					
R. L. auriga (Forsskal, 1775). Hawaii, Marshall	(4)	165	170.4	454.7	148
R. L. fasciatus (Forsskal, 1775). Red Sea	(2)	170	110.9	471.0	186
R. Rabdophorus: 145.					
R. R. ephippium (Cuvier, 1831). New Cal., Marshall	(3)	160	152.7	425.5	145
R. Oxychaetodon: 146.					
R. O. ulietensis (Cuvier, 1831). Marshall	(1)	140	90.2	335.3	146

(mm), body weight (g), brain weight (mg) and brain index is given for each species. The brain index was calculated by plotting the regression of brain and body weight values of specimens of these two families. The best adjustment, calculated for the family Pomacanthidae which has a larger brain weight amplitude, is given by the formula:

$$\log bw = -0.065 (\log Bw)^2 + 0.993 \log Bw - 0.55,$$

which is the formula of a quadratic curve in which bw is the brain weight and Bw the body weight. The index is the distance of each point, representing a species, from the curve to which is given the value 100. We used the Student test (t test) for comparing as well relative brain volumes as indices between the two families.

Histology

Sections of a brain from each species were cut in 10 μm, stained with cresylviolet and analyzed by photogrammy (Bauchot 1963). From each brain series 50 photos at equal intervals were taken between the anterior olfactory bulb and the posterior brainstem. After identification of the limits of the main brain structures, each structure part was cut out and weighed to determine area and then total volume. Comparison of the total volume of the brain after sectioning to its volume at the time of fixation enabled calculation of a retraction (shrinkage) coefficient and from it the true volumes of the parts of the fresh brain. Since the coefficient of retraction varied from 1.5 to 2.5 among brains, this method avoided errors before comparisons were made between species.

Results

Brain-body weight relationships in chaetodontids and pomacanthids

Figure 1 shows that almost all chaetodontids lie above the level of pomacanthids and seem to be somewhat more encephalized. The reasons for a higher encephalization in chaetodontids could be ethological. The fact that pomacanthids are somewhat bigger is not an explanation, since we use a quadratic curve which takes into account this diminution of brain index with increase in body weight. The explanation may lie in a better adaptation of chaetodontids to their environment, in greater locomotory ability or in some overweight of the digestive tract of pomacanthids due to their food habits (sponge-eating, for example). In reality, the t test shows that this difference in encephalization is not statistically significant. We need more pomacanthid material to be sure that there is greater encephalization in butterflyfishes.

There is rather little variation in the brain indices within each genus or subgenus. The greatest variation is that of *Linophora* with a range of 16.1% from the mean. Within the pomacanthids, the indices vary from 87 (*Centropyge potteri*) to 132 (*Centropyge tibicen*), and within the chaetodontids, they vary from 93 (*Citharoedus (Gonochaetodon) baronessa*) to 186 (*Rabdophorus (Linophora) fasciatus*). These values must be compared to those of other teleost fishes, which vary from 7 (*Moringua* sp) to 223 (*Coryphaena hippurus*). Figure 2, which locates the families Pomacanthidae and Chaetodontidae inside the polygon of brain indices for all teleosts, shows that they lie in the upper part of the polygon, as do almost all perciformes, the chaetodontids being a little higher situated than the pomacanthids.

Gross brain morphology of chaetodontids and pomacanthids

The morphology of the brain of these fishes is so uniform that we need only give the dorsal and lateral aspect of three species: *Rabdophorus (Linophora) fasciatus* and *Heniochus intermedius* for the chaetodontids, and *Centropyge fisheri* for the pomacanthids. Figure 3 gives the general aspect of their brains in dorsal and lateral views. They are very similar to each other and altogether very different from the brains of other teleosts, except those of the family Acanthuridae.

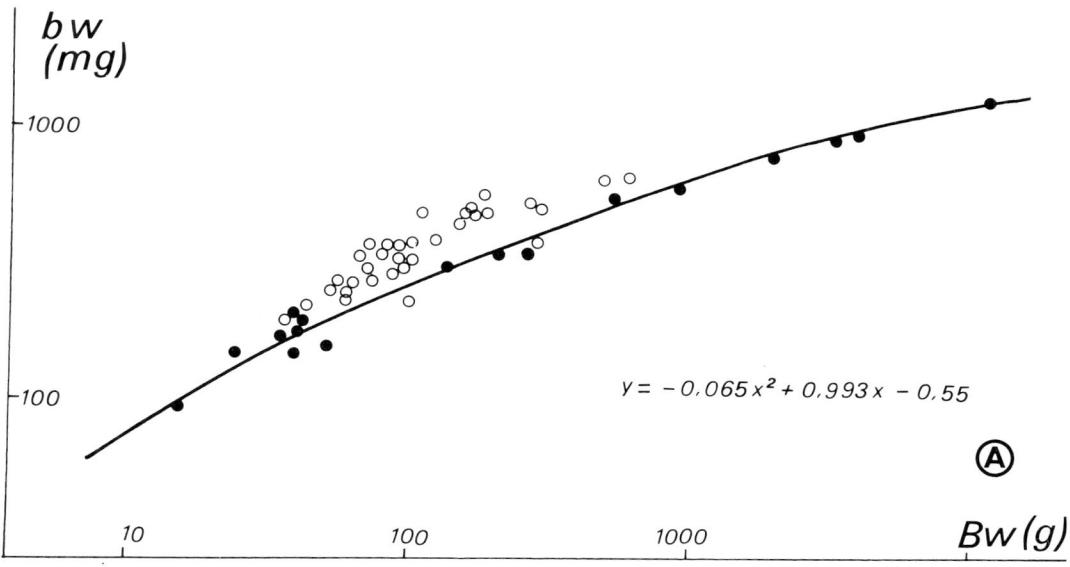
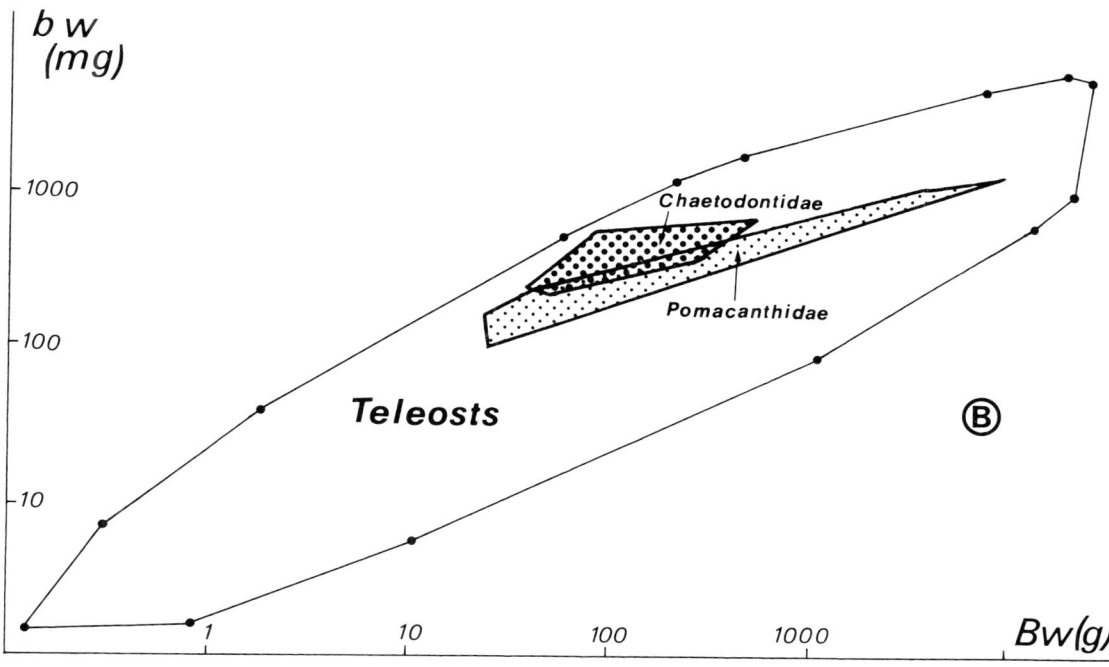

Fig. 1. A — Brain-body weight relationship using double logarithmic coordinates for Chaetodontidae (open circles) and Pomacanthidae (dark circles). The formula of the curve adjusted for pomacanthids is given; B — Brain-body weight relationship using double logarithmic coordinates and placing Chaetodontidae and Pomacanthidae within other teleosts [Bw = body weight (g), bw = brain weight (mg)].

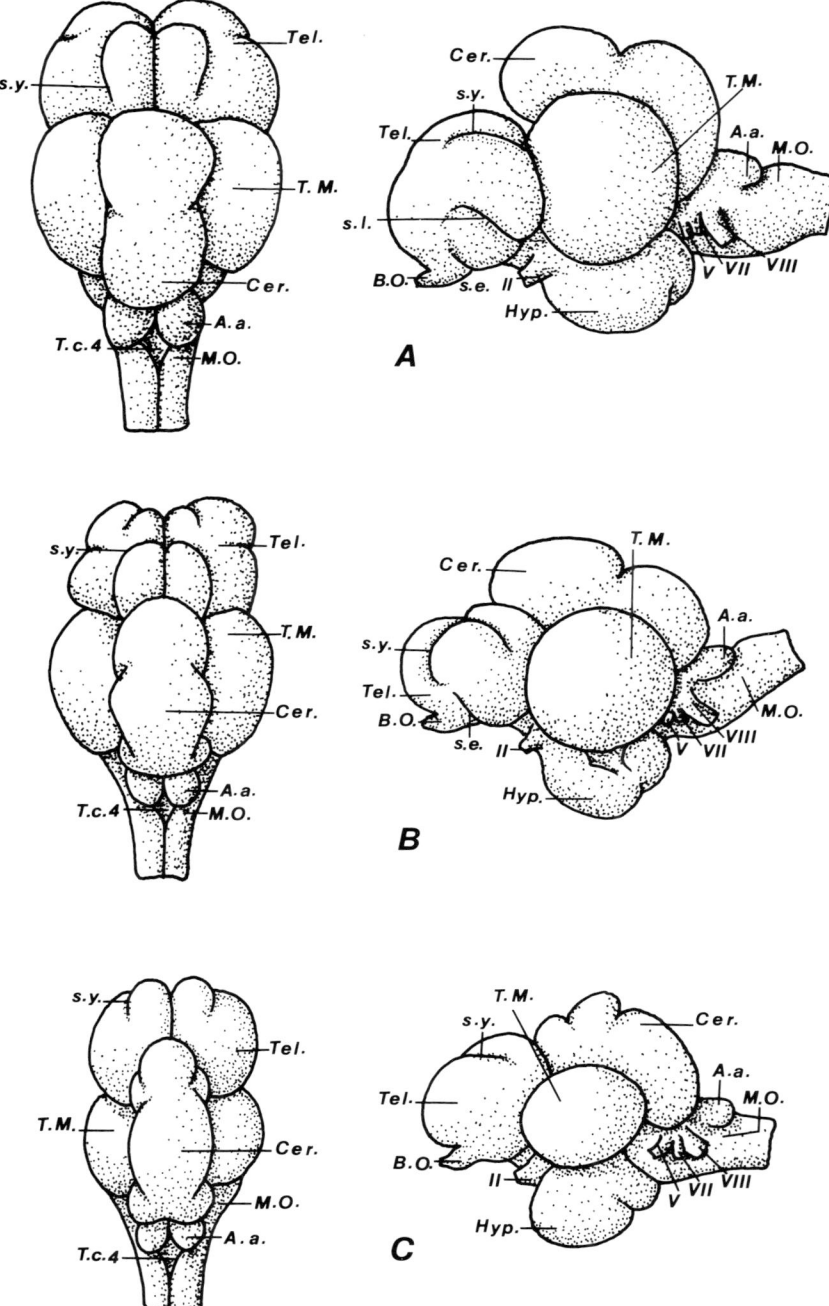

Fig. 2. The external morphology of the brain (left: dorsal view, right: lateral view) of: A – *Rabdophorus (Linophora) fasciatus;* B – *Heniochus intermedius;* C – *Centropyge fisheri.* (A.a. = acoustico-lateral area, B.o. = olfactory bulb, Cer. = cerebellum, Hyp. = hypothalamus, M.O. = medulla oblongata, s.e. = sulcus externus, s.l. = sulcus lateralis, s.y. = sulcus ypsiliformis, T.c.4 = choroid plexus of the 4th ventricle, Tel. = telencephalon, T.M. = optic tectum, II = optic nerve, V = trigeminal nerve, VII = facial nerve, VIII = acousticolateral nerve).

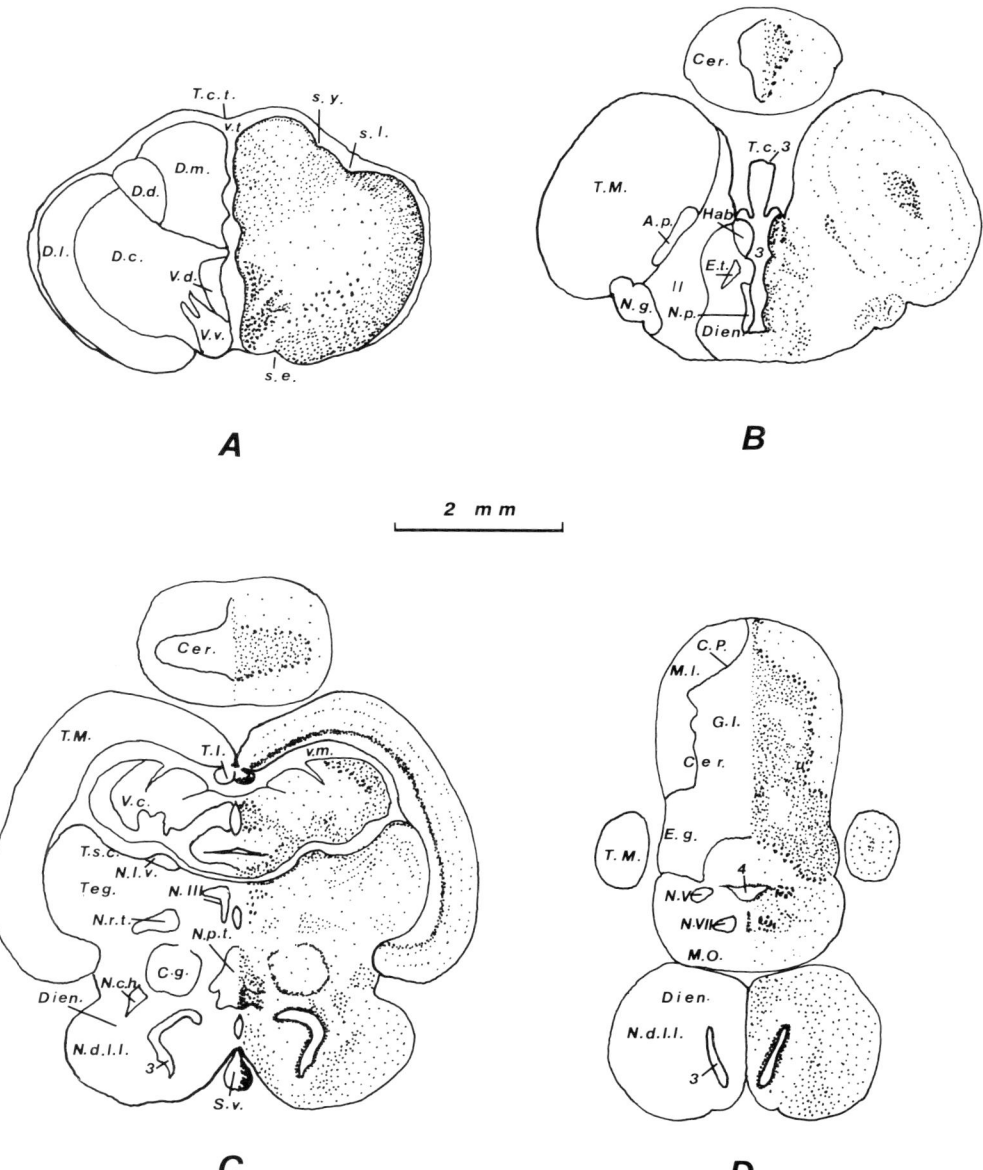

Fig. 3. Sections of 4 transverse levels of the brain of *Heterochaetodon (Burgessius) miliaris:* A – telencephalic level; B – diencephalic level; C – mesencephalic level; D – metencephalic (cerebellar) level (A.p. = area pretectalis, Cer. = cerebellum, C.g. = corpus glomerulosum, C.P. = Purkinje cells of the cerebellar cortex, D.c. = nucleus dorso-centralis, D.d. = nucleus dorso-dorsalis, D.l. = nucleus dorso-lateralis, D.m. = nucleus dorso-medialis, Dien. = diencephalon, E.g. = eminentia granularis, E.t. = eminentia thalami, G.l. = granular layer of the cerebellar cortex, Hab. = habenula, M.l. = molecular layer of the cerebellar cortex, M.O. = medulla oblongata, N.c.h. = nucleus cerebellaris hypothalami, N.d.l.l. = nucleus diffusus lobi lateralis, N.g. = nucleus geniculatus (N. pretectalis superficialis), N.l.v. = nucleus lateralis valvulae, N.p. = nucleus preopticus, N.p.t. = nucleus posterior tuberis, N.r.t. = nucleus ruber tegmenti, N.III = nucleus of the oculomotor common nerve, N.V. = motor nucleus of the trigeminal nerve, N.VII = motor nucleus of the facial nerve. s.e. = sulcus externus, s.l. = sulcus lateralis, S.v. = saccus vasculosus, s.y. = sulcus ypsiliformis, T.c.3 = choroid plexus of the 3rd ventricle, T.c.t. = choroid plexus of the telencephalic ventricle, Teg. = tegmentum, Tel. = telencephalon, T.l. = torus longitudinalis, T.M. = optic tectum, T.s.c. = torus semi-circularis, V.c. = valvula cerebelli, V.d. = nucleus ventrodorsalis, v.m. = mesencephalic ventricle, v.t. = telencephalic ventricle, V.v. = nucleus ventralis. 3 = third ventricle, 4 = fourth ventricle, II = optic nerve).

The general features are the shortness of the brain anteroposteriorly, its height being almost as great as its length, and the fact that the forebrain (Tel.) is very large, its width being almost as great as that of the midbrain (T.M., optic tectum). Another characteristic feature is the cerebellum (Cer.), which is relatively narrow and elongated and which covers the optic tectum (T.M.) and the posterior part of the telencephalon. The olfactory bulbs (B.O.) are small, indicating a probable microsmy. Behind the cerebellum the paired area acousticolateralis (A.a.) is visible. The forebrain is furrowed by the sulcus ypsiliformis (s.y.), the sulcus lateralis (s.l.) and the sulcus externus (s.e.). The ventral lobe of the hypothalamus (Hyp.) is also, as in almost all fishes, well developed. The general morphology of these three brains is very similar, with the exception of the smaller optic tectum in *Centropyge*. However, a better knowledge of the real size of the different parts of the brain needs a quantitative study from histological series.

Histological architecture of the brain of chaetodontids

Figure 4 shows four transverse sections of the brain of the chaetodontid, *Heterochaetodon (Burgessius) miliaris*. The general organization of the brain is so uniform in the chaetodontids, as is shown in the quantitative analysis, that this illustration can be accepted as characteristic of the family.

Figure 4A shows the telencephalon and namely the pallium as well as two sulci visible dorsally, the sulcus ypsiliformis (s.y.) and the sulcus lateralis (s.l.). It can be noted that the telencephalon of this species is specially everted since the attachment of the tela choroidea (T.c.t.) extends lateral and ventral. The great development of the laminated dorsolateral nucleus (D.l.) is characteristic of a high level of cortical organization.

Figure 4B shows the diencephalon, the front part of the optic tectum (T.M.) and the cerebellum (Cer.). The optic tract (II) is visible as well as the area pretectalis (A.p.) and the nucleus geniculatus (N.g.). This last nucleus, also named nucleus pretectalis superficialis (Braford & Northcutt 1983), is laminated, a feature characteristic of highly evolved fishes with good vision (Franz 1912, Ridet 1982). The anterior protrusion of the corpus cerebelli is characteristic of the brains of some families of teleosts (among reef fishes, Chaetodontidae, Pomacanthidae and Acanthuridae).

Figure 4C shows the mesencephalon, the anterior portion of the cerebellum (Cer.), the valvula cerebelli (V.c.) inside the mesencephalic ventricle (v.m.) and the rear part of the hypothalamus (notably the saccus vasculosus S.v.). In the valvula, the large lateral lobes are a part of the reflexe lamina. According to Bănărescu (1957) this type of valvula (his *Carangid-type*), with a big reflexe lamina, is characteristic of good swimmers (families such as Carangidae, Lutjanidae, Scombridae, but also as Acanthuridae and Scaridae). The main tegmental structures are visible, like the nucleus oculomotorius communis (N.III) and the corpus glomerulosum (C.g.). The histology of this last nucleus is typical of perciformes, with a central neuropile and a peripheral cortex-like cell zone.

Figure 4D shows the metencephalon and, dorsoventrally superimposed, the lobus lateralis hypothalami (N.d.l.l.), the medulla oblongata (M.O.) and motor nerve nuclei (N.V: trigeminal and N.VII: facial) and the corpus cerebelli with its three principal components: molecular layer (M.l.), Purkinje cells (C.P.) and granular layer (G.l.). This section includes the posterior part of the tectum (T.M.).

Quantitative analysis of the brain of chaetodontids and pomacanthids

The relative similarity of brain morphology in chaetodontids and pomacanthids makes it unnecessary to study the brain organization of more than a few representatives. These are listed in Table 2.

As the variability in brain parts in fishes is about 10%, any lesser differences are therefore not significant. It can be noted also that, since chaetodontids and pomacanthids are systematically close and have very similar brains, direct comparisons of relative volumes (in percentages of the volume of the whole brain) are probably valid. For comparison

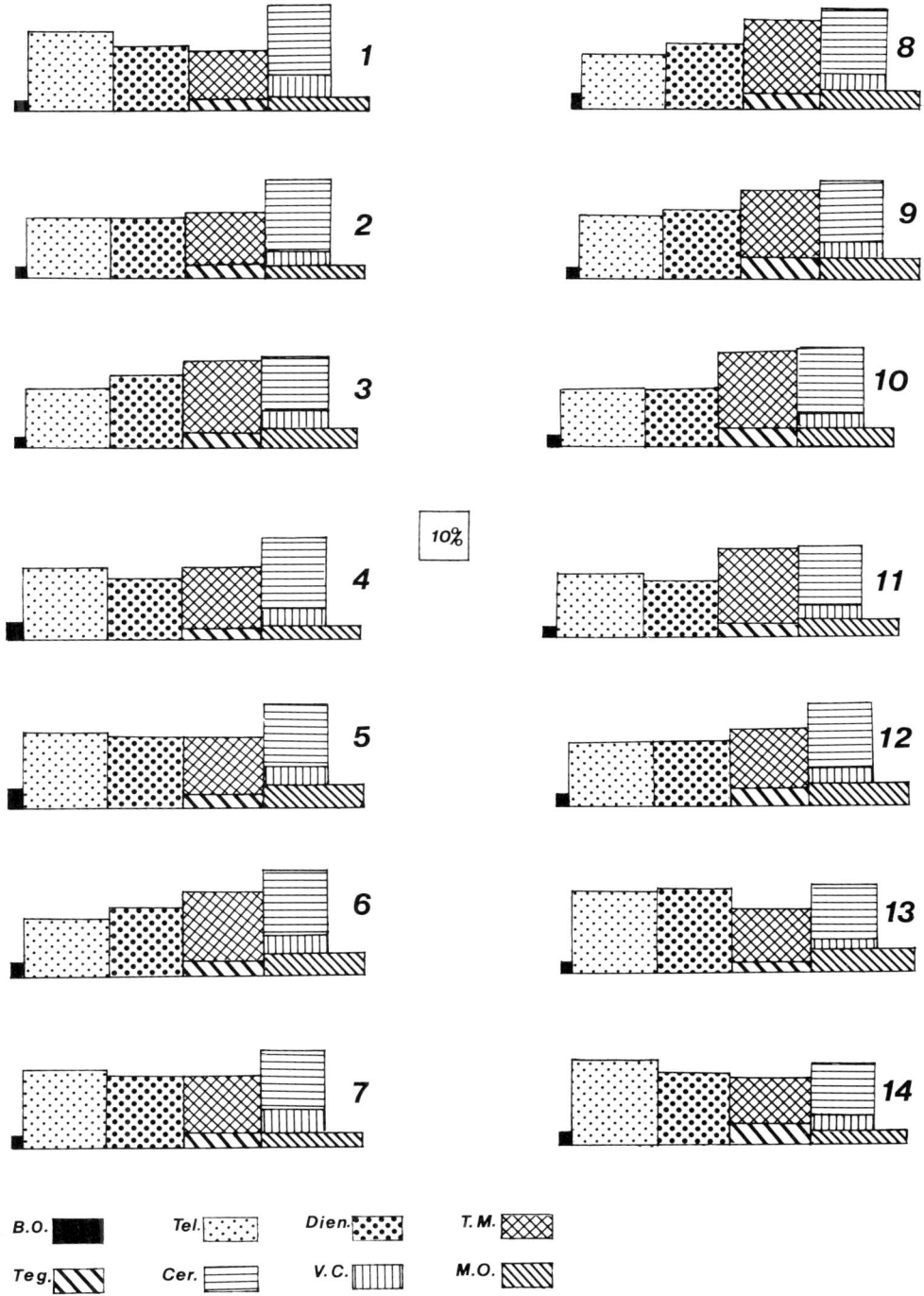

Fig. 4. Representation of the relative volumes of the main parts of the brains of some Chaetodontidae and Pomacanthidae. In this representation, the areas are proportionnal to the volumes. The numbers refer to the following species (see Table 3). Chaetodontidae: 01 – *C. flavirostris;* 02 – *C. reticulatus;* 03 – *E. kleinii;* 04 – *F. flavissimus;* 05 – *H. intermedius;* 06 – *H. B. miliaris;* 07 – *R. L. fasciatus;* 08 – *M. plebeius;* 09 – *M. C. trifasciatus;* 10 – *N. speculum;* 11 – *M. S. melannotus.* Pomacanthidae: 12 – *C. heraldi;* 13 – *G. caudovittatus;* 14 – *H. xanthotis* (B.O. = olfactory bulbs, Cer. = cerebellum, Dien. = diencephalon, M.O. = medulla oblongata, Teg. = tegmentum, Tel. = telencephalon, T.M. = tectum opticum, V.C. = valvula cerebelli).

with other teleost taxa, the calculation of indices was made from the same quadratic formula given above.

Table 3 gives the percentages by volumes of the main subdivisions of the brains studied and Table 4 the corresponding indices. Added to the values of chaetodontids and pomacanthids are some from other teleost species for comparison: *Labrus bergylta,* taken by one of us as a reference for a general study of the brain organization of teleosts (Ridet 1982), *Labroides dimidiatus* (Labridae), *Acanthurus triostegus* (Acanthuridae), *Scarus rubroviolaceus* (Scaridae), *Dascyllus aruanus* (Pomacentridae) and *Myripristis parvidens* (Holocentridae) which are all fishes associated with coral reefs.

(A) Relative volumes of brain structures (Table 3)
The olfactory bulb (OB) is always small in comparison with the mean value for teleosts (2.64%, Ridet 1982) but can be compared with the six other teleosts living in a coral reef environment. The mean value for chaetodontids (0.60%) is not statistically different from that for the pomacanthids (0.65%).

The telencephalon (TEL, without olfactory bulbs) is greater than its mean value in teleosts (15,2%, Ridet 1982) but not very different between

Table 2. Bodyweight Bw (g), brain weight bw (mg) and volume, in percentage of the whole volume of the brain, of the main subdivisions of the brain (see text for abbreviations).

Teleost mean	Bw	bw	OB	TEL	DI	TM	TL	TEG	CER	VC	MO	V
			2.64	15.2	15.9	22.0	0.70	6.35	16.8	3.57	16.8	6.36
Chaetodontidae.												
01 *C. flavirostris* (New Caledonia)	120.0	444	0.49	28.9	19.8	16.0	0.35	4.47	18.7	5.49	5.91	3.93
02 *C. reticulatus* (New Caledonia)	140.0	346	0.57	22.6	19.7	18.3	0.80	4.73	19.8	4.00	9.55	3.78
03 *E. kleinii* (Hawaii)	21.4	155	0.37	20.0	22.2	23.8	0.60	5.31	15.2	4.43	8.10	3.00
04 *F. flavissimus* (Hawaii)	52.6	227	0.81	24.4	19.9	20.3	0.54	4.26	18.7	4.94	6.12	2.79
05 *H. intermedius* (Red Sea)	240.0	352	0.86	26.3	20.8	18.6	0.64	2.94	16.7	5.49	7.72	5.00
06 *H. B. miliaris* (Hawaii)	53.3	213	0.71	19.8	20.8	22.5	0.56	5.42	17.0	5.15	8.17	4.93
07 *R. L. fasciatus* (Red Sea)	67.0	379	0.50	26.2	22.0	17.9	0.49	4.84	16.4	5.55	6.21	4.24
08 *M. plebeius* (New Caledonia)	17.6	131	0.72	18.6	19.8	23.8	0.70	5.62	17.8	4.28	8.72	3.61
09 *M. C. trifasciatus* (Tahiti)	45.5	219	0.34	21.1	20.8	21.6	0.60	6.83	16.5	4.25	8.03	7.33
10 *N. speculum* (New Caledonia)	18.0	133	0.59	20.9	17.7	25.0	0.63	5.74	17.7	4.56	7.17	4.18
11 *M.S. melannotus* (Red Sea)	20.0	176	0.59	22.9	18.4	25.1	0.80	5.42	15.2	3.66	8.00	4.34
Mean			0.60	22.9	20.2	21.2	0.61	5.05	17.2	4.71	7.61	4.28
St. deviation %			21.7	14.4	7.4	16.0	23.0	13.7	8.7	14.0	15.2	28.7
Pomacanthidae.												
12 *C. heraldi* (New Caledonia)	7.1	77	0.72	23.3	21.1	18.8	0.54	5.26	17.8	3.69	8.87	3.00
13 *G. caudovittatus* (Réunion)	73.0	255	0.61	28.3	25.0	16.6	0.48	3.54	14.4	3.34	7.81	4.32
14 *H. xanthotis* (Red Sea)	82.0	219	0.61	30.0	23.0	14.4	0.38	7.06	13.5	4.70	6.27	3.37
Mean			0.65	27.2	23.0	16.6	0.47	5.29	15.2	3.91	7.65	3.56
St. deviation %			9.2	12.9	8.7	13.3	17.0	33.3	15.1	18.2	17.1	19.1
General mean			0.61	23.8	20.8	20.2	0.58	5.10	16.8	4.54	7.62	4.13
St. deviation %			24.6	15.1	9.1	17.3	22.4	22.0	10.7	15.9	14.8	27.8
Labrus bergylta			0.60	22.8	22.7	21.9	0.20	4.00	16.3	3.00	8.40	3.60
Labroides dimidiatus			0.40	21.4	25.9	21.2	0.10	5.80	13.8	2.90	8.60	3.40
Dascyllus aruanus			0.50	16.9	21.1	27.7	0.80	7.20	14.3	2.60	9.00	3.60
Scarus rubroviolaeus			0.20	28.0	26.5	16.7	0.10	4.60	13.3	3.50	7.20	6.60
Acanthurus triostegus			0.70	32.1	16.6	14.8	0.50	3.50	22.3	2.00	7.50	3.80
Myripristis parvidens			0.30	23.6	15.3	21.5	1.10	5.10	20.4	5.40	7.30	4.30

chaetodontids (22.9%) and pomacanthids (27.2%), the variation in mean percentage being comparable (respectively 14.4% and 12.9%). As in teleosts in general, the volume of the diencephalon follows that of the telencephalon; it is not statistically different in pomacanthids (23.0%) and in chaetodontids (20.2%) but much bigger than in teleosts in general (15.9%).

The optic tectum or tectum mesencephali (TM), although not statistically different between the two families, is slightly larger in chaetodontids (21.2%) than in pomacanthids (16.6%), with a comparable variation (between 13 and 16%). We need more material to be sure that the chaetodontids are endowed with a better acuity than pomacanthids. If we compare these values with those of other teleost families (mean relative value: 22.0%, Ridet 1982), they indicate a normal level of visual capabilities. Nevertheless, a comparison of indices is needed, here as elsewhere, to get a better idea of the real importance of the brain functions between brains of different sizes and organizations.

The torus longitudinalis (TL), which has the same importance in chaetodontids (0.61%) as in pomacanthids (0.54%) and is a little smaller than in teleosts in general (0.70%), seems to follow the variations of size of the optic tectum. The tegmentum (TEG) is not very different between the two families: 5.29% in pomacanthids and 5.05% in chaetodontids, but smaller than the mean value in teleosts (6.35%).

Table 3. Indices of the main subdivisions of the brains (see text for abbreviations).

	OB	TEL	DI	MES	CER	MO
Chaetodontidae.						
01 *C. flavirostris*	124	184	100	115	146	102
02 *C. reticulatus*	123	123	108	113	156	142
03 *E. kleinii*	83	114	128	148	135	126
04 *F. flavissimus*	199	149	123	134	177	103
05 *H. intermedius*	114	88	70	65	90	70
06 *H. B. miliaris*	131	92	98	116	124	103
07 *R. L. fasciatus*	140	185	156	144	186	120
08 *M. plebeius*	169	110	117	154	155	139
09 *M. C. trifasciatus*	66	106	105	127	124	109
10 *N. speculum*	144	129	110	169	164	120
11 *M. S. melannotus*	168	165	133	197	164	156
Mean	133	131	113	135	147	117
St. deviation %	28.6	26.7	19.5	25.2	19.0	20.5
Pomacanthidae.						
12 *C. heraldi*	180	148	134	136	165	154
13 *G. caudovittatus*	96	112	99	70	84	84
14 *H. xanthotis*	93	116	89	74	84	66
Mean	123	125	107	93	111	101
St. deviation %	39.8	16.0	22.4	39.8	42.3	45.5
General mean	131	130	112	126	140	114
St. deviation %	29.8	23.8	19.6	30.2	24.3	24.6
Labrus bergylta	100	100	100	100	100	100
Labroides dimidiatus	98	125	143	138	116	81
Dascyllus aruanus	85	75	90	138	89	84
Scarus rubroviolaceus	38	117	114	78	83	82
Acanthurus triostegus	127	150	80	80	135	95
Myripristis parvidens	73	130	84	133	168	68

The cerebellum (CER), though a little bigger in chaetodontids (17.0%) than in pomacanthids (15.2%) is not statistically different and is very near the mean value of teleosts (16.8%). The relative volumes vary from 3.8% in eels to 32.7% in a catfish, but here also the use of indices is needed for comparisons with other teleost families. The valvula cerebelli (VC), which is not significantly different in chaetodontids (4.60%) and in pomacanthids (3.91%), is a little larger than in teleosts as a whole (3.57%). The theory of Bănărescu (1957) implies that the former have a better locomotory efficiency than the general teleostean level.

The medulla oblongata (MO, brainstem) is small in both families (7.61% in chaetodontids and 7.65% in pomacanthids) compared to the mean value in teleosts (16.8%). This can be considered as a general evolutionary feature, since primitive families have larger brainstem centres compared to the size of the telencephalon. It is even possible to define a general index of evolution inside teleosts by the ratio of the volume of the telencephalon (where are located some integrative parts of the brain) to that of the medulla oblongata, taken as a reference. This index (3.01 in chaetodontids and 3.56 in pomacanthids) must be compared to that of some other teleost families: Lophiidae (0.16), Cyprinidae (0.18), Clupeidae (0.30), Chanidae (0.30), Argentinidae (0.31) for one part and Serranidae (3.14), Balistidae (3.34), Scaridae (3.89), Diodontidae (4.23) and Acanthuridae (4.28) for the other part.

Finally, the volume of the ventricles (V, 4.28% in chaetodontids and 3.56% in pomacanthids) is relatively small in comparison with the mean value of teleosts (6.36%) and recalls the value found in mammals (less than 2%). The relative volumes of the main parts of the brains of chaetodontids and pomacanthids are given in Figure 5 (the areas of the different parts of the brain are proportional to their volumes).

(B) Comparisons using indices (Table 4)
These indices were compared to the value 100 given to the species *Labrus bergylta* chosen by Ridet (1982) as a reference for comparison in his study of teleosts, due to its generalized organization.

For olfactory bulbs, the indices are greater than 100 for all chaetodontids except *E. kleinii* (83) and *M. trifasciatus* (66). In pomacanthids, the value of *C. heraldi* (180) is almost double of that of the other two species and near that of *Forcipiger* (199), *M. plebeius* (169) or *M. melannotus* (168). The significance of these indices is easy to understand: while the relative volumes could suggest that these fishes are poorly olfactive in comparison with other teleosts, the indices show that they have actually a level of olfactory activity similar to most fish species, the low relative volume being due to the larger volumes of other parts of their brain. Nevertheless, these indices must be compared to those of other species or genera: *Labrus bergylta* (100), *Scarus* (38), *Dascyllus* (85), *Acanthurus* (127) and *Labroides* (98); all can be said microsmatic when compared to *Anguilla anguilla* (492), *Soleo lascaris* (794), *Gymnothorax marginatus* (1046), *Fistularia petimba* (1060) or *Caranx sexfasciatus* (1244). Concerning olfaction, the coral reef environment is very different from the benthic or the pelagic ones, in which fishes have larger olfactory centres (Ridet & Bauchot 1984).

The indices for the telencephalon are comparable in chaetodontids (mean: 131) and in pomacanthids (mean: 125) but higher than in *Labrus* (100). As a good index of the general evolutionary level of species, the telencephalon shows that pomacanthids and chaetodontids are typical perciformes, with well developed telencephalic centers. Ridet (1982) showed that high telencephalic indices are characteristic of every fish family living in the near vicinity of coral reefs.

The diencephalon indices are about the same as the telencephalon ones. The chaetodontid mean (113), though a little higher than that for pomacanthids (107) but not statistically different, is higher than that of *Labrus* (100). It is difficult to say if these results confirm the theory of Senn (1981) that a large diencephalon (and notably the hypothalamus) is linked to a high degree of aggressiveness since chaetodontids and pomacanthids are generally less aggressive (Reese 1975).

The indices of the mesencephalon are interesting since they can be taken as a means to measure, at least in part, the visual abilities of fishes (Snow &

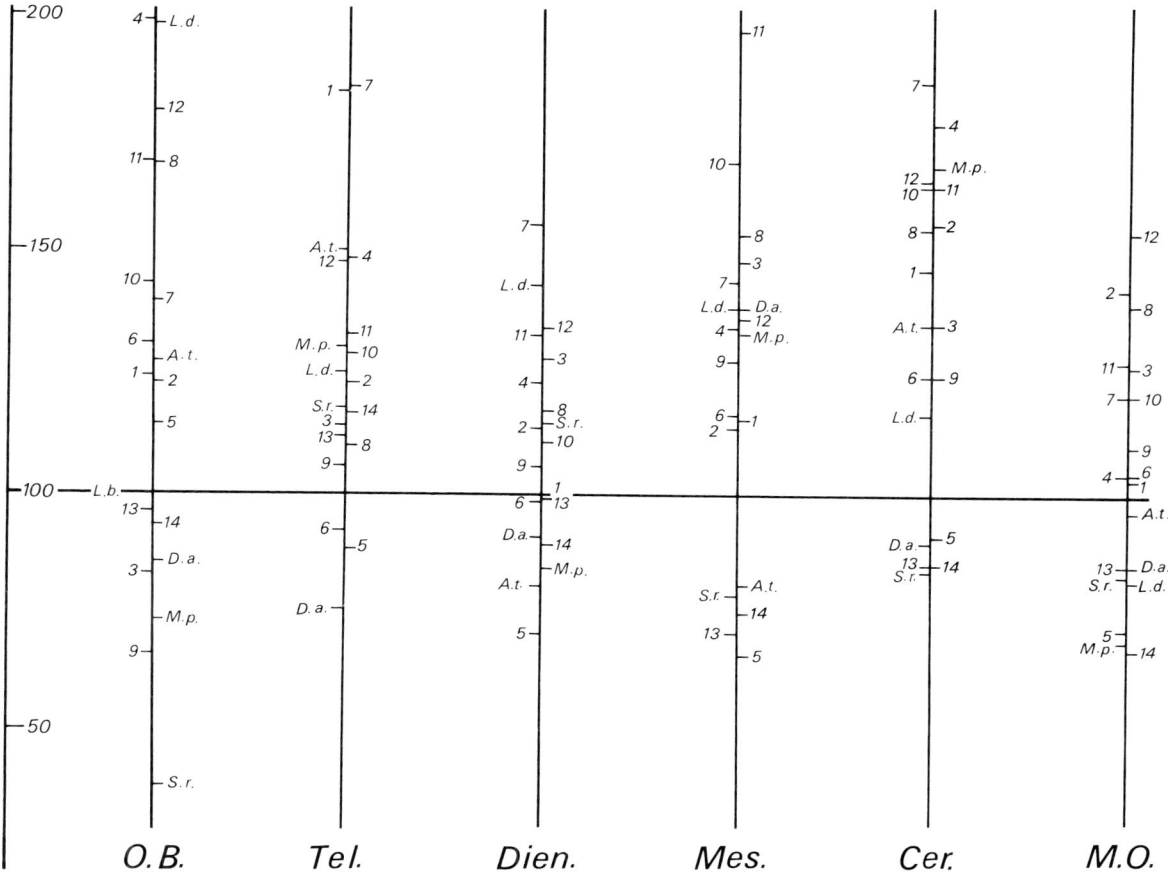

Fig. 5. Comparison of the indices of the main structures of the brains of Chaetodontidae and Pomacanthidae with those of other teleosts. The reference 100 corresponds to the generalized species *Labrus bergylta*. Numbers and letters refer to the following species. Chaetodontidae: 1 – *C. flavirostris;* 2 – *C. reticulatus;* 3 – *E. kleinii;* 4 – *F. flavissimus;* 5 – *H. intermedius;* 6 – *H. B. miliaris;* 7 – *R. L. fasciatus;* 8 – *M. plebeius;* 9 – *M. C. trifasciatus;* 10 – *N. speculum;* 11 – *M. S. melannotus.* Pomacanthidae: 12 – *C. heraldi;* 13 – *G. caudovittatus;* 14 – *H. xanthotis.* Other species: A.t. – *Acanthurus triostegus;* D.a. – *Dascyllus aruanus;* L.b. – *Labrus bergylta;* L.d. – *Labroides dimidiatus;* M.p. – *Myripristis parvidens;* S.r. – *Scarus rubroviolaceus* (Cer. = cerebellum, Dien. = diencephalon, Mes. = mesencephalon, M.O. = medulla oblongata, O.B. = olfactory bulbs, Tel. = telencephalon).

Rylander 1982). The differences between the two families, although not statistically significant, are larger (mean of chaetodontids 135, compared to that of pomacanthids 93). The variability inside each family is also larger (*Genicanthus:* 70, *Centropyge:* 136 for pomacanthids; *Heniochus:* 65, *M. S. melannotus:* 197 for chaetodontids). A lot of different adaptations (food habits, diurnal/nocturnal activity, social life, protection against predators) may explain these differences.

The cerebellum indices can give a good idea of the locomotory aptitudes of the fishes. The chaetodontids (mean: 147) are above the level of the pomacanthids (mean: 111), and this difference may be due to the better swimming capacities of the first. The variations are similar in the two families (from *Heniochus* 90 to *R. L. fasciatus* 186 in chaetodontids and from *Genicanthus* and *Holacanthus* 84 to *Centropyge* 165). Very low indices are found in benthic fishes like *Antennarius drombus* (8) or *Gymnothorax marginatus* (14) and much higher indices in pelagic fishes like *Sphyraena jello* (171) or *Caranx sexfasciatus* (313).

The medulla oblongata indices are near the reference value of teleosts (100 in *Labrus bergylta*): mean 117 for chaetodontids and 101 for pomacanthids. The variation is similar in the two families (from 70 in *Heniochus* to 156 in *M. S. melannotus* for chaetodontids and from 66 in *Holacanthus* to 154 in *Centropyge* for pomacanthids. As for the olfactory bulbs, it can be noted that the indices are normal whereas the mean relative volumes showed an apparent lower size of these centers. The indices of other reef fishes are comparable.

Discussion

Numerous papers have dealt with various biological aspects of the chaetodontids and pomacanthids, namely feeding behavior (Birkeland & Neudecker 1981, Earle 1972, Hobson 1975, Motta 1988, Randall 1967, 1974, 1975, Randall & Hartman 1968, Thresher 1979), social relationships (Reese 1975a, b, Smith & Tyler 1972), nycthemeral rhythm (Collett & Talbot 1972) or more generally with the biological adaptations of these fishes (Collette & Earle 1972, De Graaf 1977, Hiatt & Strasburg 1960, Keenleyside 1979, Klausewitz 1972). This list is clearly not exhaustive, but these papers provide sufficient data for discussion. In fact, there are too few data dealing with each species, and the data are often so general that it is not possible to understand behavioral interspecific differences. As the differences between the two families, as well for comparative volumes as for indices, are never statistically significant, we may better show the features common to both families than the differences between them.

Pomacanthids and chaetodontids are diurnal fishes which probably find their food more from visual than from olfactory cues: this behavior may explain their high index of tectal (visual) centers and their relatively poor index of olfactory centers. Their diets are quite disparate: pomacanthids are overall sponge-feeders while chaetodontids are generally polyp-feeders (or feed on branchial tentacles of worms or on small invertebrates living in the crevices of the coral reefs). Larger visual centers could allow chaetodontids to pick very delicately at polyps without taking the skeleton, which the gastric contents of pomacanthids show an important percentage of sponge-skeleton, often coated with mucus to avoid injuries to the digestive tract. The less precise sponge-feeding versus precise polypfeeding can be evoked to explain the general inferior encephalization level of pomacanthids (if it is verified). Conceivably, but we have no new material to verify it, the pomacanthids have an overweight due to the high percentage of non-edible material in their digestive tract, an overweight that the chaetodontids, whose stomachs very rarely contain coral skeletal particles, lack.

The locomotory habits of the two families are primarily the same (Webb 1982). The more precise way that chaetodontids pick up their food could explain a slightly higher cerebellar index, if verified from a larger study material. The social behavior of chaetodontids seems to be better known than that of pomacanthids, but we assume that the latter is not basically different. A few species are territorial; other species are home-rangers or more rarely erratic. All stay within their territory or their range for months if not years, a behavior which supposes a good knowledge of their environment and thus a good encephalization index (Reese 1975).

Chaetodontids are said to be less aggressive intra- or interspecifically; some of them live in groups or in sexual pairs which seem to last a very long time (Reese 1975). The species that have pair-bonds may pair strongly or weakly. There is no real differences between these two groups (compare the strongly paired *C. ornatissimus* (126), *E. multicinctus* (132), *H. unimaculatus* (124) or *M. trifasciatus* (133) to the weakly paired *C. reticulatus* (155), *E. citrinellus* (141) or *H. quadrimaculatus* (142). *Heterochaetodon miliaris*, which is often seen in schools, does not have a lower index (126) than the paired species. Social relationships seem then to have but a small influence on the level of encephalization in butterflyfishes. Yet, fishes living in schools or being strongly paired could be less cautious of their environment or of the presence of predators and thus have lower encephalization indices.

Angelfishes and butterflyfishes, midwater coral reef browsers, seem to have a similar brain organization, probably also shared with surgeonfishes.

The study of additional species of these families is needed to confirm the results obtained. On the other hand, comparison of brains of these families with those of coral reef fish families of very different biology (bottom dwellers, predators, etc.) could allow the identification of brain structures linked to modes of life.

Acknowledgements

We are thankful to Paul Guézé, who helped us in Réunion, to Jack Randall, who helped us during our trips collecting fishes and who identified most of the fishes of this study, to Roland Platel, André Maugé and Monique Diagne for their help in many places and to René Galzin who sent us butterflyfishes from Polynesia. Peter Whitehead (British Museum Natural History, London) and Tyson Roberts (ORSTOM, Paris) helped improve the English in the final version of the manuscript.

References cited

Bǎnǎrescu, P. 1957. Vergleichende Anatomie und Bedeutung der Valvula cerebelli der Knochenfische. Revue Roumaine de Biologie 2: 255–276.

Bauchot, R. 1963. L'architectonique comparée, qualitative et quantitative, du diencéphale des Insectivores. Mammalia 27 Suppl. 1: 1–400.

Birkeland, C. & S. Neudecker. 1981. Foraging behavior of two Caribbean chaetodontids: *Chaetodon capistratus* and *C. aculeatus*. Copeia 1981: 169–178.

Braford, M.R. & R.G. Northcutt. 1983. Organization of the diencephalon and pretectum of the ray-finned fishes. pp. 117–163. In: R.E. Davis & R.G. Northcutt (ed.) Fish Neurobiology, Volume 2: Higher Brain Areas and Functions, University of Michigan Press, Ann Arbor.

Collette, B.B. & S.A. Early. 1972. Results of the tektite program: ecology of coral reef fishes. Nat. Hist. Mus. Los Angeles County Sci. Bull. 14: 1–180.

Collette, B.B. & F.H. Talbot. 1972. Activity patterns of coral reef fishes with emphasis on nocturnal-diurnal changeover. pp. 98–124. In: B.B. Collette & S.A. Earle (ed.) Results of the Tektite Program: Ecology of Coral Reef Fishes, Nat. Hist. Mus. Los Angeles County Sci. Bull. 14.

De Graaf, F. 1977. Encyclopédie des poissons d'aquarium marin. Elsevier-Séquoia, Paris-Bruxelles. 344 pp.

Earle, S.A. 1972. The influence of herbivores on the marine plants of great Lameshur bay, with an annotated list of plants. pp. 17–44. In: B.B. Collette & S.A. Earle (ed.) Results of the Tektite Program: Ecology of Coral Reef Fishes, Nat. Hist. Mus. Los Angeles County Sci. Bull. 14.

Franz, V. 1912. Beiträge zur Kenntnis des Mittlehirns und Zwischenhirns der Knochenfische. Folia Neurobiol. 6: 402–441.

Hiatt, R.W. & D.W. Strasburg. 1960. Ecological relationship of the fish fauna on coral reefs of the Marshall Islands. Ecol. Monogr. 30: 65–127.

Hobson, E.S. 1975. Feeding patterns among tropical reef fishes. Amer. Scientist 63: 382–392.

Keenleyside, M.H.A. 1979. Diversity and adaptation in fish behavior. Springer-Verlag, New York. 208 pp.

Klausewitz, W. 1972. Litoralfische der Malediven. II. Kaiserfische der Familie Pomacanthidae. Senckenberg. Biol. 53: 361–372.

Maugé, A. & R. Bauchot. 1984. Les genres et sous-genres de Chaetodontidés étudiés par une méthode d'analyse numérique. Bull. Mus. Nat. Hist. Nat. Paris 4° Sér. 6A: 453–485.

Motta, P.J. 1988. Functional morphology of the feeding apparatus of ten species of Pacific butterflyfishes (Perciformes, Chaetodontidae): an ecomorphological approach. Env. Biol. Fish. 22: 39–67.

Randall, J.E. 1967. Food habits of reef fishes of the West Indies. Stud. Trop. Oceanogr. 5: 665–847.

Randall, J.E. 1974. The effect of fishes on coral reefs. Proc. Second Internat. Coral Reef Symp., Brisbane 1: 159–166.

Randall, J.E. 1975. A revision of the Indopacific angelfish genus *Genicanthus* with descriptions of three new species. Bull. Mar. Sci. 25: 393–421.

Randall, J.E. & W.D. Hartman. 1968. Sponge-feeding fishes of the West Indies. Mar. Biol. 1: 216–225.

Reese, E.S. 1975. A comparative field study of the social behavior and related ecology of reef fishes of the family Chaetodontidae. Z. Tierpsychol. 37: 37–61.

Reese, E.S. 1975. Duration of residence by coral reef fishes on 'home' reefs. Copeia 1975: 145–149.

Ridet, J.M. 1982. Analyse quantitative de l'encéphale des Téléostéens: caractères évolutifs et adaptatifs de l'encéphalisation. Ph.D. Thèse, Univ. Paris VII, Paris. 306 pp.

Ridet, J.M. & R. Bauchot. 1984. L'olfaction chez les Téléostéens. Cybium 8: 15–25.

Senn, D.G. 1981. Morphology of the hypothalamus in advanced teleost fishes. Zool. J. Linn. Soc. 73: 343–350.

Smith, C.L. & J.C. Tyler. 1972. Space resource sharing in a coral reef fish community. pp. 125–170. In: B.B. Collette & S.A. Earle (ed.) Results of the Tektite Program: Ecology of Coral Reef Fishes, Nat. Hist. Mus. Los Angeles County Sci. Bull. 14.

Snow, J.L. & M.K. Rylander. 1982. A quantitative study of the optic system of butterflyfishes (Family Chaetodontidae). J. Hirnf. 23: 121–125.

Thresher, R.E. 1979. Possible mucophagy by juvenile *Holacanthus tricolor*. Copeia 1979: 160–162.

Webb, P.W. 1982. Locomotor patterns in the evolution of actinopterygian fishes. Amer. Zool. 22: 329–342.

Forcipiger longirostrum from Hawaii, the yellow phase. Photo by P.S. Lobel.

The eye muscles and their innervation in *Chaetodon trifasciatus* (Pisces, Teleostei, Chaetodontidae)

Roland Bauchot[1], Athanase Thomot[1] & Marie-Louise Bauchot[2]
[1] *Laboratoire d'Anatomie comparée, Université Paris VII, 2 Place Jussieu, 75251 Paris Cedex 05, France*
[2] *Laboratoire d'Ichthyologie générale et appliquée, Muséum National d'Histoire Naturelle, 57 rue Cuvier, 75231 Paris Cedex 05, France*

Received 22.2.1988   Accepted 3.9.1988

Key words: Extraocular muscles, Oculomotor nerves, Eye movements, Butterflyfishes

Synopsis

In *Chaetodon trifasciatus*, the large eye has the form of a thick disk rather than that of a globe. A deep cutaneous groove surrounds the eyeball, probably allowing rapid eye movements. The form and innervation of the three pairs of extraocular muscles are described. Each muscle is made of two types of fascicles of fibres, thick and thin. There is neither an anterior nor posterior myodome. The skull attachment of the obliques and of the inferior rectus is made on the thin sagittal ethmoidal membranous septum while that of the other recti occurs on osseous pieces of the skull. The attachment on the eyeball is made on the cartilaginous sclera. The ratio of the lengths of the antagonist muscles, superior vs. inferior oblique, superior vs. inferior rectus and medial vs. lateral rectus, is about 1.43 : 1. The three oculomotor nerves (III: common oculomotor, IV: trochlear and VI: abducens) as well as the ciliary system are described. For the following reasons, an analogy between the lateral rectus of *Chaetodon trifasciatus* and the lateral rectus + retractor bulbi of other vertebrates is indicated: (1) the nucleus of nerve III (which innervates four muscles) has four sectors, while that of IV (which innervates only the superior oblique) is made of one sector; (2) nerve VI consists of two roots corresponding to two groups of nerve cells of its motor nucleus and (3) in other vertebrates, nerve VI innervates both the lateral rectus and the retractor bulbi.

Introduction

The chaetodontids are strongly paired, home ranging or territorial fishes that recognize each other as individuals, vision being the primary sensory modality (Reese 1975). Many species feed on or about corals, picking small prey items such as individual coral polyps with their delicate mouths (Motta 1985, 1988, Hobson 1974). Furthermore, the locomotory adaptations of these fishes are suited for precise maneuvering among the coral reefs (Webb 1984). All of these led us to study the extraocular muscles and the innervation of the eye in butterflyfishes. In particular, we wanted to know if butterflyfishes are capable of very precise eye movements that are coordinated with their precise swimming and feeding maneuvers. In spite of many works dealing with the anatomical organization of eye movements in fishes (Allis 1897, 1902, Van Gehuchten 1895, Oliva 1965, Freihofer 1978, Easter 1979, Graf & McGurk 1985, Thomot & Bauchot 1985, Kassem et al. 1988), there are no data concerning the relationship of eye movements and ocular anatomy in the Chaetodontidae.

Material and methods

We used adults and juveniles of *Chaetodon trifasciatus*, one of the species studied by Reese (1975) for this study. In order to obtain good histological series and to avoid too long and powerful decalcifications, we looked for the tinest juvenile with a typical adult skull organization (ossification present but light), a specimen of 16 mm total length from the Mataiva Atoll in Tuamotu Islands on which our description is based.

The study of histological series (after fixation in Bouin's mixture, 10 μm sections and staining with Azan trichromic method) and the use of graphical reconstruction (based on manual drawings with a camera lucida) allowed us to establish the disposition of each extraocular muscle from origin to insertion and to follow the pathway of cranial nerves III, IV and VI from their central nervous system nuclei to innervation of the muscles. All measures are given after corrections were made, taking into account the shrinkage due to fixation and dehydration (Bauchot 1967, Bauchot & Platel 1971).

Results

In *Chaetodon trifasciatus* as in other butterflyfishes, the large eyes are narrow along their optical axis. The eyes have the form of a thick disc rather than that of a globe. There is very little space between them in the sagittal plane for the extraocular muscles and the nerves which innervate them. A deep cutaneous groove is found all around the eye so that the epidermal layer covers not only the cornea but also the peripheral margin of the eyeball (Fig. 3a, b, c). The presence of this groove and the loose organization of the conjunctival tissue beneath it probably facilitate the movements of the eye in its orbit.

Extraocular muscles

(1) *Oblique muscles*
There is no anterior myodome. The oblique muscles have their skull attachment on the thin interorbital septum of the ethmoid, in the dorsorostral region of the orbit, with the superior oblique attachment being more anterior and dorsal than that of the inferior one (Fig. 3). The superior oblique (2 mm) is longer than the inferior (1.4 mm). Both are made of two types of muscle fascicles, thin and thick, with the superior oblique being more typical in this respect. The thin fascicles are generally situated near the periphery of the muscle. The distal region of the oblique muscles, in their zone of attachment to the eyeball, is always lateral to the recti muscles wherever they are superimposed.

Superior oblique muscle. – This muscle (Fig. 1, 2, 3a), in the zone of its attachment to the ethmoid septum, is rather bulky. It runs first posterodorsally, lying ventral to the olfactory nerve but in contact with it in the region where the nerve passes through its foramen. Posteriorly it is increasingly compressed and soon contacts the cartilaginous sclera on its mediodorsal margin. More posterolaterally, the muscle becomes further flattened where it closely follows the surface of the sclera, taking a horizontal direction. Distally, it covers and tightly contacts the horizontal and lateral region of the superior rectus, where it is flattened against the cartilaginous sclera. More posterolaterally, it lies between the superior cutaneous groove and the cartilaginous sclera, on which it attaches in the middorsal part of the eyeball. There are probably conjunctival bonds between the attachment regions of the superior oblique and rectus muscles, in the zone where they are superimposed and in tight contact.

Inferior oblique muscle. – In the region of its attachment on the interorbital ethmoid septum (Fig. 1, 2, 3a), the dorsal part of the inferior oblique is partially covered by the ventral margin of the superior oblique muscle. The inferior oblique muscle is flattened and ribbonlike, with a general caudoventral orientation. At first, it crosses the zone of attachment of the medial rectus medially on the cartilaginous sclera, without making tight contact. It then runs close to the medial inferior part of the cartilaginous sclera and, more ventrally, it covers

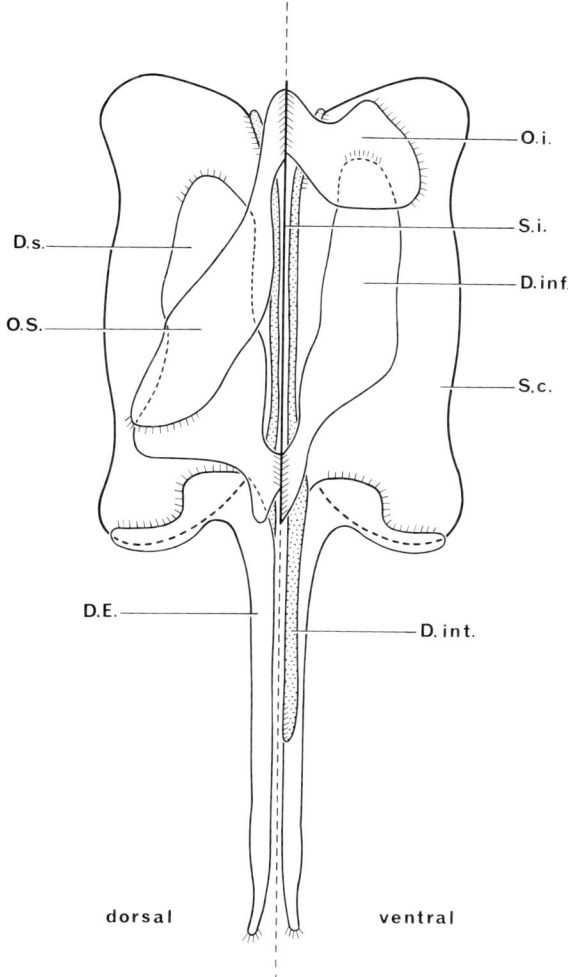

Fig. 1. Reconstruction, in dorsal and ventral views, made from transverse sections of the head of *Chaetodon trifasciatus*, showing the extraocular muscles. The medial rectus muscle is stippled. (D.E. = lateral rectus muscle, D.inf. = inferior rectus muscle, D.int. = medial rectus muscle, D.s. = superior rectus muscle, O.i. = inferior oblique muscle, O.S. = superior oblique muscle, S.c. = cartilaginous sclera, S.i. = interorbital septum).

the zone of attachment of the inferior rectus on the eyeball.

The two zones of attachment of the inferior oblique and the inferior rectus are flattened where they insert on the anteroventral part of the cartilaginous sclera, between it and the ventral cutaneous groove. The zone of insertion of the inferior oblique covers that of the inferior rectus posteriorly and, anteriorly, it attaches directly onto the cartilaginous sclera. There are probably conjunctival bonds between the zones of attachment of the two inferior muscles, oblique and rectus, in the zone where they are superimposed and in tight contact.

(2) Recti muscles

There is no proper posterior myodome. The inferior and superior recti muscles have their attachment on the skull at about the same transverse level but far from one another dorsoventrally. This region is far anterior to the level of the skull attachment of the medial and lateral recti. Although the latter run close to one another for a part of their pathway in a longitudinal cavity or tunnel at the base of the skull, they do not attach to it but continue caudally and have their zones of attachment separated and far from one another.

The recti muscles are composed, like the oblique, of two types of fibres, thin and thick. The thin ones are normally situated at the periphery of the muscle. The superior rectus (2.9 mm) is longer than the inferior (2.1 mm), while the medial rectus (3.2 mm) is longer than the lateral (2.2 mm). The ratio of the lengths of the muscles working in opposition is about the same (1 : 1.43) for the obliques as for the inferior superior or medial-lateral recti.

Superior rectus muscle. – The skull attachment of the superior rectus muscle (Fig. 1, 2, 3b, c) is situated on the parasphenoid, in the region where the bone begins to broaden laterally and dorsally, at about the transverse level of the posterior border of the eyeball (Fig. 3b). The proximal part of the superior rectus is laterally flattened in a narrow, vertical ribbon 0.05 mm broad. It lies parallel to the sagittal plane, between it and the eyeball, running from its zone of attachment on the parasphenoid toward the base of the skull. On this pathway, it first crosses the medial rectus laterally and the zone of attachment of the lateral rectus onto the eyeball medially. More dorsally, it crosses the proximal part of the inferior rectus laterally near its attachment on the skull and runs medially to the ciliary ganglion. At the base of the skull at the level of the

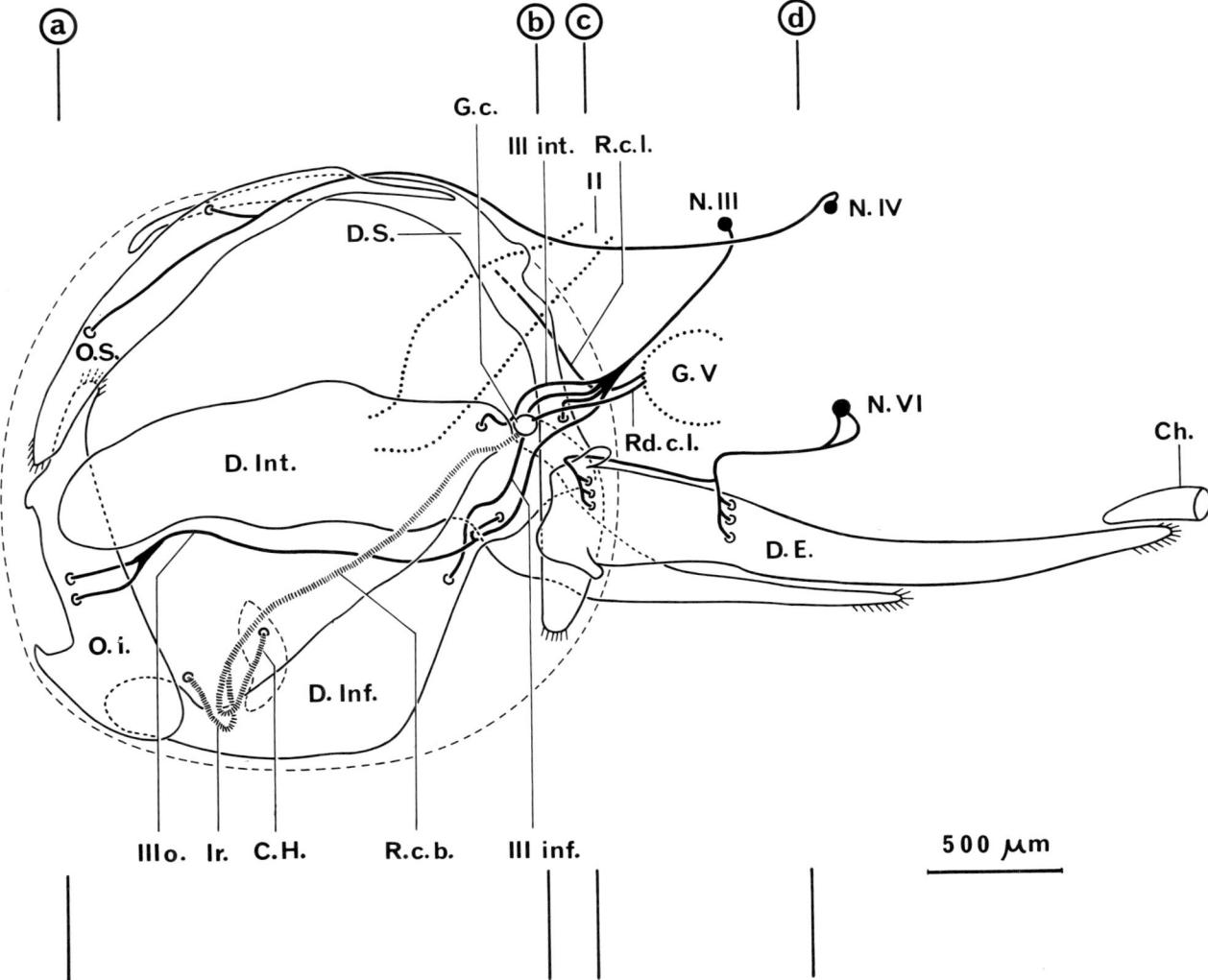

Fig. 2. Reconstruction in lateral view made from transverse sections of the head of *Chaetodon trifasciatus*, in the region of the extraocular muscles and oculomotor nerves. Parallel lines indicate the levels of the transverse sections of Figure 3. (Ch. = chorda, C.H. = branch of ramus ciliaris brevis innervating Haller's campanula, D.E. = lateral rectus muscle, D.inf. = inferior rectus muscle, D.int. = medial rectus muscle, D.s. = superior rectus muscle, G.c. = ganglion ciliaris, G.V. = ganglion nervi profundi, Ir. = branch of the ramus ciliaris brevis innervating the iris, N.III = motor nucleus of the common oculomotor nerve, N.IV = motor nucleus of the nervus trochlearis, N.VI = motor nucleus of the nervus abducens, O.i. = inferior oblique muscle, O.S. = superior oblique muscle, R.c.b. = ramus ciliaris brevis, R.c.l. = ramus ciliaris longus, Rd.c.l. = radix ciliaris longa. II = optic nerve, III inf. = branch of the common oculomotor nerve innervating the inferior rectus muscle, III int. = branch of the common oculomotor nerve innervating the medial rectus nerve, III o. = branch of the common oculomotor nerve innervating the inferior oblique muscle).

posterior border of the optic chiasm, it is more voluminous and makes conjunctival bonds, probably hanging devices (structure comparable to the trochlea of the superior oblique of many vertebrates) with the skull wall. It then turns progressively anterolaterally, becomes horizontal, running under the foramen of cranial nerve IV, to cross this nerve ventrally and finally pass under the zone of attachment of the superior oblique.

The distal parts of the two superior muscles,

oblique and rectus, are tightly superimposed and flattened on the cartilaginous sclera between the latter and the superior cutaneous groove. There are probably conjunctival bonds between the zones of attachment of the two muscles in the zone where they are superimposed and in tight contact. More anteriorly, the superior rectus is inserted between the superior cutaneous groove and the cartilaginous sclera onto which it attaches in the anterodorsal part of the eyeball.

Inferior rectus muscle. – The attachment of the inferior rectus muscle is on the posterior part of the thin interorbital ethmoidal septum (Fig. 1, 2, 3b, c), approximately at the same transverse level as the attachment of the superior rectus (slightly anterior to the level of the posterior border of the eyeball), under the base of the skull, where the septum is in close contact with the anteromedial margin of the prootic. In this region, the inferior rectus, which is medial to the superior rectus, is compressed laterally into a thick ribbon. It runs at first anteriorly in close contact with the sagittal septum, then partially covers, by its ventral margin, the posterior part of the medial rectus more ventral and contacts the sagittal septum in this region. More anteriorly, it runs medioventrally to the ciliary ganglion and then, becoming more distant from the sagittal plane, it crosses the medial rectus (which lies along the sagittal septum) laterally and comes in close contact with it. Then it flattens laterally, draws progressively nearer the eyeball and follows the fibrous sclera closely. More anteriorly, it follows the cartilaginous sclera, crosses the inferior oblique muscle, running between it and the cartilaginous sclera to which it attaches a little more anteriorly, in the anteroventral region of the eyeball. In this zone, the two distal portions of the inferior muscles, rectus and oblique, superimposed and in tight contact, are inserted between the inferior cutaneous groove and the cartilaginous sclera. There are probably conjunctival bonds between these two muscles in the region of their insertion.

Medial rectus muscle. – Three regions can be distinguished in this muscle (Fig. 1, 2, 3): a proximal, horizontally directed region, lying under the lateral rectus, corresponding to about 35% of the total length of the muscle; a distal region, also horizontal, lying in a plane slightly more dorsal than the preceding one and comprising about 45% of the total length, and a shorter middle region (about 20%) which is oblique and connects the first two. In this last region, the medial rectus crosses the superior and inferior recti muscles.

The attachment of the medial rectus muscle on the skull base is situated on the dorsal face of the parasphenoid, in the region where this bone encloses a tunnel-like cavity, formed dorsally and laterally by the prootics in front and by the basioccipital in the rear (Fig. 3d). The transverse level of this attachment is somewhat posterior, in the region of the vascular sac of the hypothalamus and near the anterior border of the sacculus of the membranous labyrinth. Here, the muscle is cylindrical, in contact dorsally with the lateral rectus muscle, being more voluminous at this point, and medially with its contralateral equivalent through a thin conjunctival septum. More anteriorly, the entire proximal region of this muscle is situated in the tunnel, ventral to the lateral rectus where it flattens progressively laterally and stays in contact with the interorbital septum while the lateral rectus becomes more distant from this septum.

In the intermediary portion, the medial rectus muscle crosses the superior rectus (which is vertical) medially and lies just under the zone of attachment of the inferior rectus to the skull (Fig. 3c). More anteriorly, at the level of the optic chiasm and always in contact with its contralateral, it proceeds more dorsally and crosses the inferior rectus muscle medially.

In its distal portion, the eyeballs are so close to the midsagittal plane in this region that the medial rectus muscle flattens continuously, being compressed between the interorbital septum and the sclera. The medial rectus attaches on the cartilaginous sclera in the anteromedial region halfway up the eyeball, crossing the inferior oblique laterally (Fig. 3a).

Lateral rectus muscle. – This is the most posterior and the thickest of all the extraocular muscles (Fig. 1, 2, 3c, d). Its general direction is horizontal and its

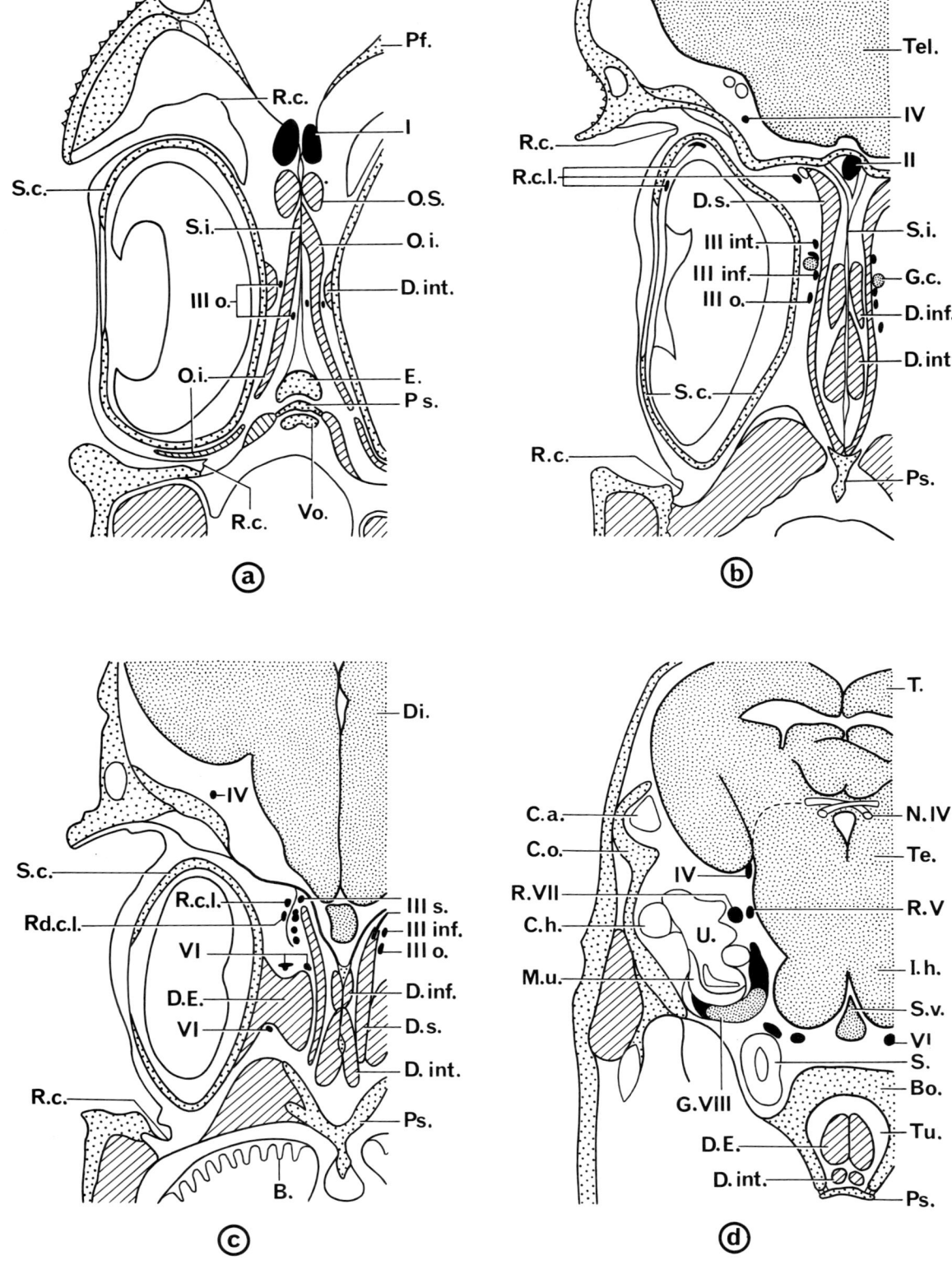

Fig. 3. Transverse sections of the head of *Chaetodon trifasciatus* at different levels, from rostral (a) to caudal (d), in the region of the extraocular muscles and associated nerves. The muscles are crosshatched, nerves are in solid black, cartilage and bone are lightly stippled and central nervous system and ganglia densely stippled. The levels of the sections are shown in Figure 2. (B = branchia, Bo = basioccipital, C.a. = anterior vertical semicircular canal of membranous labyrinth, C.h. = horizontal semicircular canal of the membranous labyrinth, C.o. = cartilaginous otic capsule, D.E. = lateral rectus muscle, Di. = diencephalon, D.inf. = inferior rectus muscle, D.int. = medial rectus muscle, D.s. = superior rectus muscle, E. = ethmoid, G.c. = ganglion ciliaris, G.VIII = ganglion of the acoustic nerve, I.h. = inferior lobe of the hypothalamus, M.u. = macula utriculi, N.IV = motor nucleus of the nervus trochlearis, O.i. = inferior oblique muscle, O.S. = superior oblique muscle, Pf. = prefrontal, Ps. = parasphenoid, R.c. = cutaneous groove, R.c.l. = ramus ciliaris longus, Rd.c.l. = radix ciliaris longa, R.V. = root of the trigeminal nerve, R.VII = root of the facial nerve, S. = sacculus, S.c. = cartilaginous sclera, S.i. = interorbital septum, S.v. = vascular sac, T. = optic tectum, Te. = tegmentum, Tel. = telencephalon, Tu. = tunnel at the base of the skull, U. = utriculus, Vo. = vomer. I = olfactory nerve, II = optic nerve, III = common oculomotor nerve, III.inf. = branch of nerve III innervating the inferior rectus muscle, III.int. = branch of nerve III innervating the medial rectus nerve, III.o. = branch of nerve III innervating the inferior oblique muscle, III.s. = branch of nerve III innervating the superior rectus muscle, IV = nervus trochlearis innervating the superior oblique muscle, VI = nervus abducens innervating the lateral rectus muscle).

transverse section is circular in its posterior part, elliptic in the anterior one. Its attachment on the skull is on the posterior part of the basi-occipital, at a transverse level far more posterior than the caudal border of the labyrinth. At first, it lies near the sagittal plane, in contact with its contralateral (Fig. 3d). From this level forward, two thirds of this muscle lies in the longitudinal tunnel at the base of the skull. Approximately at midlength, it lies dorsal to the medial rectus (in the region of its attachment to the skull). In the tunnel, its ventrolateral margin covers the dorsal part of the medial rectus. It progresses away from the sagittal plane, crossing the superior rectus (which is vertical) laterally and the inferior rectus in the region where it attaches on the sagittal septum. A little more anteriorly, the wings of the parasphenoid narrow and the tunnel opens laterally; the lateral rectus then bends progressively laterally to attach on the cartilaginous sclera at the mid-height of the posterolateral part of the eyeball (Fig. 3c).

Extraocular motor nerves

In *Chaetodon trifasciatus* as in *Tridentiger trigonocephalus* (Kassem et al. 1988), the extraocular motor nerves are different in diameter and in length. The common oculomotor nerve (III), which innervates four of the six muscles, is the thickest, while the trochlearis (IV) and the abducens (VI), which innervate each one muscle, are thinner and of about the same diameter. The VIth cranial nerve is the shortest, while the pathway of the other two is about the same length. The motor nuclei of nerves III and IV lie at about the same horizontal level, while the nucleus of the IV is more posterior, in the dorsoposterior part of the tegmentum (Fig. 2). The nucleus of the VIth nerve is located approximately at the same transverse level as that of the IVth, but it lies more ventrally (in the ventrolateral part of the anterior medulla oblongata).

(1) *The common oculomotor nerve (III)*

Nucleus and intracerebral pathway. – Near the ventricular wall of the mesencephalon in the tegmentum, at the transverse level of the anterior regions of the cerebellar valvula and of the inferior lobes of the hypothalamus, lies a set of voluminous nerve cells organized in four groups corresponding to the four sectors of the motor nucleus of nerve III. From these groups the fibres run ventro-postero-laterally and reach the lateroventral border of the brain, between the base of the tegmentum and the inferior lobes of the hypothalamus, at a transverse level situated between the anterior borders of the cerebellar valvula and of the sacculus of the membranous labyrinth (Fig. 2).

Extracerebral pathway. – Just after its exit from the brain, nerve III runs anteriorly and contacts the ganglion of nerve complex V + VII, at the transverse level of the exit of the mandibular branch of the facial nerve from this ganglion (Fig. 2, 3a, b, c).

More anteriorly, it runs anteroventro-medially near the surface of the brain and reaches its foramen in the prootic, at a transverse level slightly in front of the level of the posterior border of the eyeball and half-way between the medial border of the eyeball and the sagittal plane. There, it divides into four vertically superimposed fascicles. Of these four fascicles, the shortest one, just after its separation from the other three, penetrates the superior rectus muscle on its vertical part under the skull, prior to where this muscle bends anterolaterally.

The other three fascicles, that lie lateral to this muscle (Fig. 3b), run anteroventrally and medially cross the ramus ciliaris longus more dorsally and the radix ciliaris longa more ventrally (Fig. 3c). The most dorsal of these three fascicles runs medially to the ganglion ciliaris and a little more anteriorly it penetrates the medial rectus muscle in the dorsal part of its distal region. The middle fascicle runs medially to the ganglion ciliaris with which it comes in tight contact, then runs ventrally and penetrates the inferior rectus on its inferior lateral side.

The lowermost of the three fascicles runs ventrally, crosses the superior rectus muscle laterally in its vertical part and a little further the inferior rectus and, having reached the ventral border of the latter, sends a mediodorsal ramus which runs between these two contiguous muscles and probably innervates the inferior rectus (Fig. 2). This fascicle then runs anteriorly near the sagittal plane and under the inferior border of the medial rectus. Before reaching the posterior border of the inferior oblique, this fascicle pierces a nervous thickening from which exit two branches, dorsal and ventral, which run anteroventrally and penetrate the inferior oblique respectively on the lateral and the medial sides.

(2) *The trochlear nerve (IV)*
This nerve innervates one extraocular muscle, the superior oblique. Its length is about the same as that of the IIIrd cranial nerve, but it is thinner.

Nucleus and intracerebral pathway. – The nucleus of cranial nerve IV, which is more caudally situated than that of the IIIrd, is formed of a group of voluminous nerve cells situated in the contralateral part of the brain, ventral to the cerebellar valvula and near the ventricular wall of the mesencephalon and the midsagittal plane (Fig. 2, 3d). Its transverse level lies between the anterior border of the valvula and the posterior border of the inferior lobes of the hypothalamus.

From the nucleus, the fibres first run posteriorly and medially, crossing the contralateral ones in the sagittal plane. They then continue postero-latero-ventrally, where they run behind and under the contralateral nucleus and, making a curve, go antero-latero-ventrally. The fibres exit from the brain in the dorsolateral part of the medulla oblongata, in the corner made by the latter and the inferior border of the tectum, at the transverse level of the exit of the root of the trigeminal nerve (V), i.e. a little behind the rostral border of the sacculus of the membranous labyrinth.

Extracerebral pathway. – After its exit from the brain, cranial nerve IV runs anteriorly while staying near the brain. It then proceeds lateroventrally to the exit of nerve III and progressively more medioventral, reaching its foramen in the prootic at the transverse level of the optic chiasm and a little dorsal to the posterodorsal border of the eyeball (Fig. 2, 3b, c).

Emerging from the skull, nerve IV lies at the posteromedial border of the superior oblique muscle. More anteriorly, it adheres to the medial side of this muscle, runs medioventrally and sends a dorsal ramus which follows the medial side of the muscle and penetrates it in the region where the muscle becomes horizontal and partially covers the laterodorsal part of the superior rectus. This ramus probably innervates the peripheral region where the muscle comprises small fascicles. More anteriorly, the IVth nerve, continually adhering to the medial side of the superior oblique muscle, follows the muscle progressively more medioventral and penetrates this muscle in the thick region near its zone of attachment on the skull, at the transverse level of the foramen of the olfactory nerve. This last ramus innervates the voluminous part of the muscle, formed of the thick fascicles.

(3) Abducens nerve (VI)

The VIth cranial nerve, which innervates only one muscle, the lateral rectus, is the shortest of the three extraocular motor nerves.

Nucleus and intracerebral pathway. – The motor nucleus of the VIth nerve, the most posterior and ventral of all nerves innervating the extraocular muscles, is composed of an oval group of voluminous nerve cells, elongated anteroposteriorly and situated in the lateroventral region of the anterior medulla oblongata, equidistant from the sagittal plane and the lateral and ventral border of the brain. Its transverse level lies a little posterior to the caudal border of the inferior lobes of the hypothalamus and more posterior than the nucleus of the IVth nerve (Fig. 2).

From the anteroventral part of the nucleus of the VIth nerve a fascicle of fibres exits anteroventrally and reaches the ventrolateral surface of the medulla oblongata. This rostral fascicle constitutes the first root of the VIth nerve. From the posterior part of the motor nucleus another fascicle of nerve fibres exits and runs posteriorly, then antero-medio-ventrally, and finally reaches the ventrolateral wall of the medulla oblongata posterior to the exit of the first root (Fig. 2). This second root of the VIth nerve is a little anterior to the first root of the auditory nerve (VIII).

Extracerebral pathway. – After its exit from the brain, the second (posterior) root runs anteriorly under the ventrolateral wall of the medulla oblongata, between it and the anterodorsal wall of the sacculus of the membranous labyrinth (Fig. 2, 3c, d). It meets the first (anterior) root near its exit. Together, they constitute cranial nerve VI, which runs anteriorly under the ventrolateral wall of the medulla in the rear and then under the inferior lobes of the hypothalamus (Fig. 3d). Nerve VI then proceeds to the transverse level of the motor nucleus of the IIIrd nerve and to the exit of the mandibular branch of the facial nerve (VII) from nerve ganglia V + VII. There, the VIth nerve curves ventrally, passes through its foramen in the prootic and, in the tunnel where the medial and lateral recti muscles are superimposed, sends a branch medially, above the dorsal side of the lateral rectus. It then continues ventrally and penetrates the lateral face of the lateral rectus by three rami. The medial branch curves anteriorly, follows the superior border of the lateral rectus and then curves laterally with the muscle, in the region where the vertical region of the superior rectus lies between the inferior and medial recti muscles (which are near the sagittal plane) and the lateral rectus. It then curves once more ventrally and penetrates the muscle by three rami, near the zone of its attachment onto the eyeball (Fig. 3c).

(4) Ciliary nerves and their rami

The ciliary nerves innervate the intraocular muscles and have anatomical connections with extraocular muscles and nerves (Fig. 2, 3b, c).

The truncus ciliaris exits from the antero-medial part of the ganglion of the nervus profundus (VI), situated a little posterior to the posterodorsal border of the eyeball. It is made of two fascicles of fibres: a dorsal one, the ramus ciliaris longus, and a ventral one, the radix ciliaris longa. It runs anteriorly and after a short pathway its two components divide under the foramen of cranial nerve III. The ramus ciliaris longus runs anterodorsally and then laterally, above the posterodorsal border of the cartilaginous sclera, then passes it in the mid-posterodorsal region of the eyeball and innervates the dorsal intraocular muscles (Fig. 3b). The radix ciliaris longa runs anteroventrally, lateral to the branches of cranial nerve III (Fig. 3b), and pierces the ciliaris ganglion through its posterodorsal border.

The ciliaris ganglion is situated near the superior rectus muscle in its vertical region, between it and the more lateral cartilaginous sclera. From the antero-ventral pole of this ganglion exits a fascicle of fibres, the ramus ciliaris brevis, which runs anteroventrally; laterally crossing the inferior rectus muscle and approaching the fibrous part of the sclera, it passes ventrally to the region where the optic nerves exit from the eyeball. In the eye, it runs anteroventrally, near the internal wall of the sclera, to the medioventral part of the eye, where it divides into two branches, dorsal and ventral. The first one runs laterally, passes the various layers of the retina in its middle ventral region, bends posteriorly and upwards and pierces the Haller's campanula (Fig. 2). The ventral branch runs in the same direction as that of the ramus ciliaris brevis, near the internal wall of the cartilaginous sclera. After reaching the mid-ventral region of the eye, it runs first laterally and dorsally, and then penetrates the iris.

Discussion

A deep cutaneous groove surrounds the eyeball, providing increased mobility of the eyeball within the orbit and probably allowing for rapid eye movements. This groove increases mobility by giving more amplitudes to the eye movements, notably in the vertical direction. Such an organization has not been previously described in other fishes. Many butterflyfishes utilize intricate eye movements when feeding within the interstices of the reef. In coral feeders in particular, these movements are rapid and numerous as they scan individual coral polyps during foraging (Motta, personal communication). It would be interesting to observe and relate the eye movements during social and feeding behaviors, particularly intricate coral feeding, to such adaptations of the eyeball.

The disposition of the extraocular muscles and the organization of the nerves innervating them in *Chaetodon trifasciatus* is similar to that of other osteichthyan fishes (e.g. *Tridentiger*, Kassem et al. 1988), with minor variations. For the muscles, these variations include the position and mode of attachment onto the skull and eyeball, the orientation of the recti and the topographical relationships between obliques and inferior and superior recti; for nerves, the location of the motor nuclei, the number of roots, the pathways and the different branches of each.

The characteristic of two types of fibres in the extraocular rectus and oblique muscles of *Chaetodon trifasciatus* is shared with a number of other teleostean fishes and even with some species from other classes of vertebrates. Recent studies on lamprey extraocular muscle distinguish either two (Nakado & Aoki 1982) or three (Witalinski & Labuda 1982) types of fibres. In *Carassius* and *Rana*, Kilarsky & Bigaj (1969) have shown two functionally different categories of fibres in the extraocular muscles: red or slow (S fibres) and white or fast (FF fibres). Kordylewski (1974 in *Gobio gobio*) and Davey et al. (1975 in *Carassius*) studied the ultrastructure of these two types of fibres. Kassem et al. (1988) noted the presence of two categories of fibres in the extraocular recti muscles of *Tridentiger trigonocephalus*. Two types of fibres in extraocular muscles were also described by Housley & Montgomery (1984) in a selachian, by Nowogrodska-Zagorska (1974) in amphibians and by Witalinski & Loesch (1976) in a snake. But Kaczmarski (1970) and Alvarado-Mallart (1972) in birds, as did Kaczmarski (1969) in a lizard, found three types of fibres. Finally, Pachter (1982) described six types of fibres in the extraocular muscles of a mammal. For Kilarski & Bigaj (1969, p. 202): 'Slow fibers would be responsible for a tonic muscle action in pursuit movements as well as for the slow phase of labyrinthine reflexes. White fibers would be responsible for extremely rapid jerking movements of the eyeball, such as occasionally be observed in fishes'. More cautiously, Davey et al. (1975, p. 146) write that: 'The connexion, if any, between the two types of muscle fibre and fast and slow movements of the eye produced naturally, is not known for any species'. Nevertheless, this complex muscle organization could be needed for precise and repetitive eye movements in fishes as in other vertebrates.

In *Chaetodon trifasciatus* the skull attachment of the two oblique muscles is not far from one another and is made onto the thin sagittal ethmoidal septum. There is no anterior myodome. The skull attachments of the recti are separate and there is no posterior myodome, as in *Tridentiger trigonocephalus* (Kassem et al. 1988). This dispersed mode of skull attachment of the recti extraocular muscles is rare in fishes. On the other hand, the absence of myodomes, anterior as well as posterior, is commonly found in teleosts (Daget 1964). The skull attachment of the oblique muscles onto a thin sagittal septum has been described in young *Anguilla anguilla* (Norman 1926), and that of the superior obliques in *Nannocharax fasciatus* (Daget 1961). The skull attachment of the oblique muscles on the ethmoidal plate was described by Kindred (1919) in

Ictalurus nebulosus, by Kassem et al. (1988) in *Tridentiger trigonocephalus* and that of the inferior obliques by Daget (1961) in *Nannocharax fasciatus*. In *Tridentiger trigonocephalus*, the superior obliques attach on the ethmoidal plate by three distinct fascicles, the most medial on the contralateral region. The zone of attachment of the oblique muscles on the sagittal septum in *Chaetodon trifasciatus* is rather dorsal, at a horizontal level near that of the anterodorsal border of the eyeball, while in *Tridentiger trigonocephalus* the attachment on the ethmoidal plate is more ventral, at the level of the anteroventral border of the eyeball.

In *Chaetodon trifasciatus*, the skull fixation of the two obliques and the inferior rectus on a conjunctival membrane, the sagittal interocular septum, and not on a rigid skeletal element, raises questions concerning the function of these muscles. This organization has been previously described in other species for the oblique muscles (Daget 1961, in *Nannocharax fasciatus*) and recti muscles (Omarkhan 1948 and Daget & D'Aubenton 1960 in Mormyroidei, Kassem et al. 1988 in *Tridentiger trigonocephalus*).

The attachment of the oblique and recti muscles on the eyeball in *Chaetodon trifasciatus* is on the cartilaginous sclera, while in *Tridentiger trigonocephalus* the superior oblique and the medial rectus attach on the fibrous sclera. This may be attributed to either species differences or to the period of development studied (the *Tridentiger* specimen being a juvenile). In the zone of attachment on the eyeball in *Chaetodon trifasciatus*, the oblique muscles and the corresponding flattened superior and inferior recti are connected by conjunctival bonds between the sclera and the cutaneous groove surrounding the eyeball.

In *Chaetodon trifasciatus* the orientation of the superior rectus muscle is not posteroventral-anterodorsal as in other species (cf. *Tridentiger trigonocephalus*, Kassem et al. 1988). This muscle bends at nearly a right angle; in the region of the turning, the muscle is thicker and makes conjunctival bonds (ligaments) with the neighbouring skull wall. This anatomical organization, which has not been described previously in other species of fishes, seems to be comparable to the trochlea of the superior oblique of many vertebrates.

The innervation of the extraocular muscles in *Chaetodon trifasciatus* is comparable to that of other vertebrates studied: the nervus trochlearis (IV) innervates the superior oblique, the nervus abducens (VI) innervates the lateral rectus, and the common oculomotor nerve (III) innervates the four other muscles.

The motor nuclei lie in the same parts of the brain as in other teleostean fishes, but their relative positions are somewhat different. The motor nucleus of nerve III is at the same horizontal level as the nucleus of nerve IV but more anterior (in *Tridentiger trigonocephalus* these two nuclei lie at the same transverse level, that of III being more ventral; Kassem et al. 1988). Similarly, in *Chaetodon trifasciatus*, the motor nucleus of nerve VI is at the same transverse level as the nucleus of nerve IV, while in *Tridentiger trigonocephalus* the motor nucleus of nerve VI is far more posterior (Kassem et al. 1988).

The organization of the motor nucleus of nerve III into four groups of nerve cells in *Chaetodon trifasciatus* is consistent with the arrangement seen in other vertebrate species and is probably related to the fact that this nerve innervates four muscles. The contralateral location of the motor nucleus of nerve IV, seen in *Chaetodon trifasciatus* as in *Tridentiger trigonocephalus* (Kassem et al. 1988), seems to be an architectural organization common to fishes.

The motor nucleus of nerve VI, which is oval in form and elongated and which lies antero-posteriorly in *Chaetodon trifasciatus*, is made of anterior and posterior sectors; from each of them exits a fascicle or root. Once out of the medulla oblongata, these two roots meet and form the VIth nerve. Further down the nerve, they part and innervate two distant regions of the lateral rectus muscle. The presence of two sectors in the motor nucleus of VIth nerve has been described in *Carassius auratus* by Graf & McGurk (1985), and the existence of two roots of the VIth nerve, though less frequent, has been described in other species of fishes (Herrick 1899, Pankratz 1930, Atoda 1936, Harrison 1981). In *Tridentiger trigonocephalus* nerve VI has only one root, but further down the nerve divides into two distinct fascicles, which innervate two distant regions of the lateral rectus muscle. Similar observations have been described for frogs (Schwalbe 1879) and in the cat (Destombes et al. 1979), where nerve VI innervates the two muscles corresponding to the hyoid somite. This organization suggests an analogy between the lateral rectus of *Chaetodon trifasciatus* (and *Tridentiger trigonocephalus*, Kassem et al. 1988) and the lateral rectus plus retractor bulbi in other vertebrate groups.

The mobility of butterflyfish eyes in the orbits, occurring during swimming and feeding, is not easily observed by divers. Aquarium experiments could allow the measurement of the speed of these eye movements and the determination of the stimuli causing them. It also could be interesting to determine whether the phenomena of saccadic movements (ocular pursuit) and of nystagmus reflex (vestibulo-ocular coordination) are present in butterflyfishes.

Acknowledgements

We thank René Galzin (Ecole Pratique des Hautes Etudes, Museum Antenna in French Polynesia), who brought back the material on which our description is based, as well as Avi Baranes (Steinitz Marine Laboratory in Elat, Israel) and Jack Randall (Bishop Museum in Honolulu, Hawaii), for providing additional material. We thank also Monique Diagne for her help in the preparation of the histological series and Maxime Martinez-Loscos for the illustrations. Peter Whitehead (British Museum, Natural History, London) and Tyson Roberts (ORSTOM, Paris) helped to improve the English in the final version of the manuscript.

References cited

Allis, E.P. 1897. The cranial muscles and cranial and first spinal nerves in *Amia calva*. J. Morphol. 12: 498–809.
Allis, E.P. 1902. The lateral sensory canals, the eye muscles and the peripheral distribution of certain cranial nerves of *Mustelus laevis*. Quart. J. Micr. Sci. 45: 87–236.
Alvarado-Mallart, R.M. 1972. Ultrastructure of muscle fibres of an extraocular muscle of the pigeon. Tissue & Cell 4: 327–339.
Atoda, K. 1936. The innervation of the cranial nerves of the catfish *Parasilurus asotus*. Science Reports of Tohoku University 11: 91–115.
Bauchot, R. 1967. Les modifications du poids encéphalique au cours de la fixation. Journal für Hirnforschung 9: 253–283.
Bauchot, R. & R. Platel. 1971. Aspects quantitatifs de l'encéphale de *Scincus scincus* (L.) (Reptilia, Sauria, Scincidae). Etude de la variabilité intraspécifique. Zoologischer Anzeiger 187: 147–174.
Daget, J. 1961. Note sur les *Nannocharax* (Poissons Characiformes) de l'Ouest africain. Bulletin de l'Institut Français d'Afrique Noire, Série A 23: 165–181.
Daget, J. 1964. Le crâne des Téléostéens. Mémoires du Muséum National d'Histoire Naturelle, Nouvelle Série Zoologie 31: 163–342.
Daget, J. & F. D'Aubenton. 1960. Morphologie du chondrocrâne de *Mormyrus rume* C.V. Bulletin de l'Institut Français d'Afrique Noire, Série A 22: 1013–1052.
Davey, D.F., R.F. Mark, L.R. Marotte & V. Proske. 1975. Structure and innervation of extraocular muscles of *Carassius*. J. Anat. 120: 131–147.
Destombes, J., P. Gogan & A. Rouviere. 1979. The fine structure of neurones and cellular relationships in the abducens nucleus in the cat. Exper. Brain Res. 35: 249–267.
Easter, S.S. 1979. The growth and development of the superior oblique muscle and trochlear nerve in juvenile and adult goldfish. Anat. Rec. 195: 683–698.
Freihofer, W.C. 1978. Cranial nerves of a percoid fish, *Polycentrus schomburgkii* (Family Nandidae). A contribution to the morphology and classification of the Order Perciformes. Calif. Acad. Sci. 128: 1–78.
Graf, W. & J.F. McCurk. 1985. Peripheral and central oculomotor organization in goldfish, *Carassius auratus*. J. Comp. Neurol. 239: 391–401.
Harrison, G. 1981. The cranial nerves of the Teleost *Trichiurus lepturus*. J. Morphol. 167: 119–134.
Herrick, C.J. 1899. The cranial and first spinal nerves of *Menidia*. Arch. Neurol. Psychopathol. 2: 21–319.
Hobson, E.S. 1974. Feeding relationships of teleostean fishes on coral reefs in Kona, Hawaii. U.S. Fish. Bull. 72: 915–1031.
Housley, G.D. & J.C. Montgomery. 1984. The structure of the external rectus eye muscle of the carpet shark, *Cephaloscyllium isabella*. J. Anat. 138: 643–655.
Kaczmarski, F. 1969. The fine structure of extraocular muscles of the lizard *Lacerta agilis* L. Zeitschrift für mikroskopisch-anatomische Forschung 80: 517–531.
Kaczmarski, F. 1970. The fine structure of extraocular muscles of the tree sparrow *Passer montanus* L. Zeitschrift für mikroskopisch-anatomische Forschung 82: 523–536.
Kassem, M., A. Thomot & R. Bauchot. 1988. The eye muscles and their innervation in *Tridentiger trigonocephalus* (Gobiidae). Jap. J. Ichthyol. (in press).
Kilarski, W. & J. Bigaj. 1969. Organization and fine structure of extraocular muscles in *Carassius* and *Rana*. Z. Zellf. 94: 194–204.
Kindred, J.E. 1919. The skull of *Ameiurus*. Illinois Biological Monographs 5. 120 pp.

Kordylewski, L. 1974. The anatomy and the fine structure of extraocular muscles of the gudgeon, *Gobio gobio* (L.) Acta Anat. 87: 597–614.

Motta, P.J. 1985. Functional morphology of the head of Hawaiian and Mid-Pacific butterflyfishes (Perciformes, Chaetodontidae). Env. Biol. Fish. 13: 253–276.

Motta, P.J. 1988. Functional morphology of the feeding apparatus of ten species of Pacific butterflyfishes (Perciformes, Chaetodontidae): an ecomorphological approach. Env. Biol. Fish. 22: 39–67.

Nakado, T. & S. Aoki. 1982. An electron microscopic study on the extraocular muscles of a lamprey, *Lampetra japonica*. Anat. Rec. 202: 1–7.

Norman, J.R. 1926. The development of the chondrocranium of the eel (*Anguilla vulgaris*), with observations on the comparative morphology and development of the chondrocranium in bony fishes. Phil. Trans. Roy. Soc. London B 214: 369–464.

Nowogrodzka-Zagórska, M. 1974. The organization of neuromuscular junctions in the extraocular muscles of Anura. Zeitschrift für mikroskopisch-anatomische Forschung 88: 781–795.

Oliva, O. 1965. Contribution to the topographical anatomy of eye-muscle in Chondrostean fishes. Věst. Česk. Spol. Zool. 29: 118–126.

Omarkhan, M. 1948. The morphology of the chondrocranium of *Gymnarchus niloticus*. J. Linn. Soc. London 41: 452–481.

Pachter, B.R. 1982. Fibre composition of the superior rectus extraocular muscle of the rhesus macaque. J. Morphol. 174: 237–250.

Pankratz, D.S. 1930. The cranial nerve components in the toadfish (*Opsanus tau*). J. Comp. Neurol. 50: 247–285.

Reese, E.S. 1975. A comparative field study of the social behavior and related ecology of reef fishes of the family Chaetodontidae. Z. Tierpsychol. 37: 37–61.

Schwalbe, G. 1879. Das Ganglion oculomotorii. Ein Beitrag zür vergleichenden Anatomie der Kopfnerven. Jen. Z. Naturwiss. 13: 16–27.

Thomot, A. & R. Bauchot. 1985. The motor ocular nerves in larval *Polypterus senegalus*. Fortschritt Zoologie 3: 611–613.

Van Gehuchten, A. 1895. De l'origine du pathétique et de la racine supérieure du trijumeau. Bull. Acad. Roy. Sci. Belgique 29: 417–441.

Webb, P.W. 1984. Form and function in fish swimming. Sci. Amer. 251: 72–82.

Witalinski, W. & A. Loesch. 1975. Structure of muscle fibres and motor end plates in extraocular muscles of the grass snake *Natrix natrix* L. Z. mikr-anat. Forsch. 89: 1133–1146.

Witalinski, W. & H. Labuda. 1982. Extraocular muscles in the lamprey *Lampetra fluviatilis* L. I. Muscles fibres. Acta Anat. 114: 165–176.

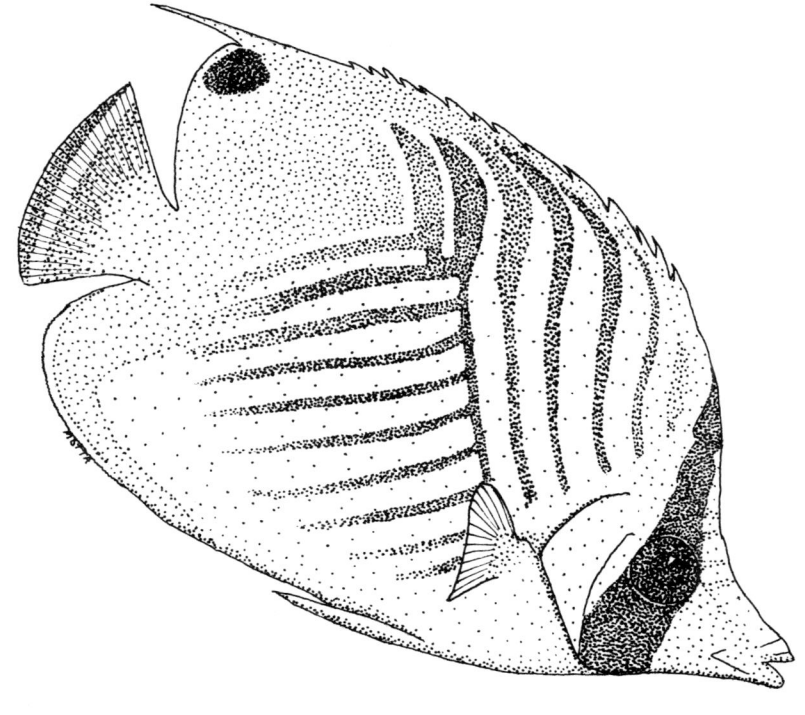

Chaetodon auriga

The membranous labyrinth and its innervation in *Chaetodon trifasciatus* (Pisces, Teleostei, Chaetodontidae)

Roland Bauchot[1], Athanase Thomot[1] & Marie-Louise Bauchot[2]
[1] *Laboratoire d'Anatomie comparée, Université Paris VII, 2 Place Jussieu, 75251 PARIS Cedex 05, France*
[2] *Laboratoire d'Ichthyologie générale et appliquée, Muséum National d'Histoire Naturelle, 57 rue Cuvier, 75231 PARIS Cedex 05, France*

Received 20.3.1988   Accepted 24.9.1988

Key words: Butterflyfish, Two maculae neglectae, Saccus endolymphaticus, Acoustic ganglion

Synopsis

In the butterflyfish *Chaetodon trifasciatus,* the labyrinth is characterized by its elevated form and especially the size of the vertical canals, the almost circular form of the horizontal canal and its posterior opening not directly in the utriculus but in the common pillar of the two vertical canals. There is an almost complete separation between utriculus and sacculus which are only linked by a virtual pore. The lagena, which is medially situated to the posterior part of the sacculus, is separated from it by an incomplete vertical wall. There are two maculae neglectae, the anterior macula being situated in the pore separating utriculus from sacculus and filling this pore, the posterior in a gutter of the floor of the utriculus. A long and narrow endolymphatic canal, originating from the sacculus close to the communication with the utriculus, follows the common pillar of the two vertical canals and widens into an endolymphatic sac at the top of the membranous labyrinth. The innervation of the labyrinth is made by the acoustic ganglion, which is connected to the brain by two roots and elongated into three parts: the anterior part innervates the anterior and horizontal cristae and the utricular and saccular maculae; the middle part innervates the macula sacculae and the macula neglecta I; the posterior part innervates the macula neglecta II, the macula lagenae and the posterior crista. The important size of the vertical canals and the almost circular form of the horizontal canal may reflect very precise locomotory aptitudes.

Introduction

The behavior of butterflyfish in coral reefs and notably the precision of their swimming movements around coral heads or that of their feeding habits (picking at coral polyps for example) led us to study their labyrinth and its innervation. In spite of many studies dealing with the organization and/or the embryology of the ear in fishes (Aly et al. 1988, Corwin 1978, Dale 1976, Fay & Popper 1980, Hama 1969, Iwasaki 1937, Jenkins 1977, Kassem et al. 1988, Krause 1906, Lowenstein 1971, Noorden 1883, Platt 1977, 1983a, 1983b & 1987, Platt & Popper 1981, Popper 1976, 1977, 1979 & 1981, Popper & Northcutt 1983, Popper et al. 1982, Ramprashad et al. 1986, Retzius 1881a, b, Saidel & McCormick 1985, Thomot & Bauchot 1987, Von Frisch 1936, Wegner 1979, Wenig 1911, 1913), no study, except Popper 1977, concerns the ear of butterflyfish (Chaetodontidae).

Material and methods

We used five juveniles of *Chaetodon trifasciatus*, and notably a specimen of 16 mm total length from the Mataiva Atoll in Tuamotu Islands which is the base of our description and illustrations.

The study of histological series (after fixation in Bouin's mixture, 10 μm transverse sectioning and staining with the Azan trichromic method) and the use of graphical reconstruction from camera lucida drawings, enabled us to establish the morphology of the different parts of the labyrinth, the position and form of the sensory organs (cristae and maculae) and the innervation of these sensory organs from the acoustic ganglion. The comparison between the size of the fish specimen, before fixation and after histological procedure and reconstruction, allowed us to calculate a shrinkage coefficient. The dimensions given in the descriptive part are those of the fresh material and correspond to the 16 mm total length specimen.

Results

General morphology

The membranous labyrinth of *Chaetodon trifasciatus* has its ventrodorsal dimension (1.76 mm) greater than the anteroposterior one (1.56 mm), this latter being itself greater than the mediolateral one (1.24 mm). In this feature, it is close to the form of the membranous labyrinth of *Solea vulgaris* (Aly et al. 1988) and differs from that of *Tridentiger trigonocephalus*, in which the greatest size is anteroposterior (Kassem et al. 1988).

The membranous labyrinth is made of two parts almost totally separated from each other: the superior part comprises the three semi-circular canals, the common pillar or crus of the two anterior and posterior vertical canals, and the utriculus; the inferior part comprises the sacculus, its extension as the endolymphatic canal and sac, and the lagena. The anterior and posterior vertical canals are tall as is also their common pillar. The horizontal canal almost forms a circle of small diameter. The ampullae of the three canals are well differentiated (Fig. 1).

The sacculus and lagena, two sacs of unequal size, constitute a whole structure elongated in the ventroposterior region and of relatively small volume by comparison with the total labyrinth.

The utricular region

The ampulla of the anterior vertical canal is the anteriormost cavity of the membranous labyrinth, that is widely communicating dorsally with the anterior canal and medio-ventro-posteriorly with the utriculus. The crista of this ampulla is situated on its floor (Fig. 1).

The utriculus is almost as voluminous in its anterior part as the anterior ampulla and contains its macula oriented horizontally and obliquely on its floor (Fig. 2A); in this region, it communicates widely with the ampulla of the horizontal canal dorsolaterally. More medioposteriorly, the utriculus is mediolaterally compressed and constitutes a relatively narrow cavity, its ventral wall close to the dorsal wall of the sacculus (Fig. 2B) and its posterior part communicating dorsoposteriorly with the common pillar of the two vertical canals (Fig. 2D). Slightly posteriorly, the utriculus has a wide communication ventrally with the ampulla of the posterior vertical canal. The crista of this last cavity lies on its floor (Fig. 1).

The voluminous ampulla of the horizontal canal is dorsolateral to the wide part of the utriculus and bears its crista on its floor. The common pillar of the two vertical canals opens ventrally in the narrow part of the utriculus (Fig. 2D); being narrow, it rises dorso-medioposteriorly and closely follows the lateral wall of the medulla oblongata. At the horizontal level of the ventral part of the cerebellum, the common pillar opens into the confluence of the two vertical canals.

The anterior canal is long (1.7 mm) and oriented at 45° to the horizontal plane of the head. The posterior canal, shorter (1.3 mm), is almost vertical (Fig. 1). The horizontal canal, from its very wide anteriorly situated ampulla, goes far laterally and constitutes an almost complete circle of short radius (0.4 mm). Posteriorly, it turns forward, following the lateral wall of the compressed part of the

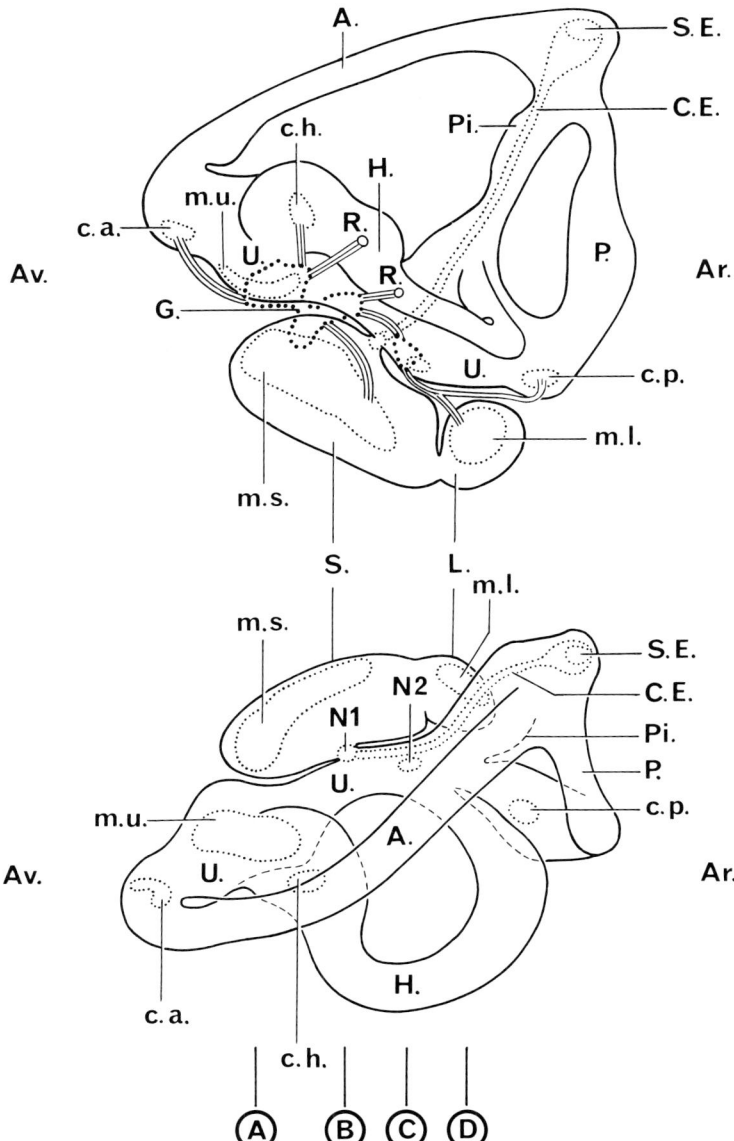

Fig. 1. Reconstruction, from histological sections, of the membranous labyrinth of *Chaetodon trifasciatus*: Above – lateral view, below – vertical view. Anterior region (Av) on the left, posterior (Ar) on the right. The position of the 4 sections of Figure 2 is illustrated (A = anterior vertical canal, c.a. – crista of the anterior canal, c.h. = crista of the horizontal canal, c.p. = crista of the posterior canal, C.E. = endolymphatic canal, G. = ganglion of the acoustic nerve (VIII), H. = horizontal canal, L. = lagena, m.l. = macula lagenae, m.s. = macula sacculi, m.u. = macula utriculi, N1 = macula neglecta I (sarazini), N2 = macula neglecta II (retzii), P. = posterior canal, Pi. = pillar common to the two vertical canals, R. = cranial nerve root of the acoustic nerve, S. = sacculus, S.E. = endolymphatic sack, U. = utriculus).

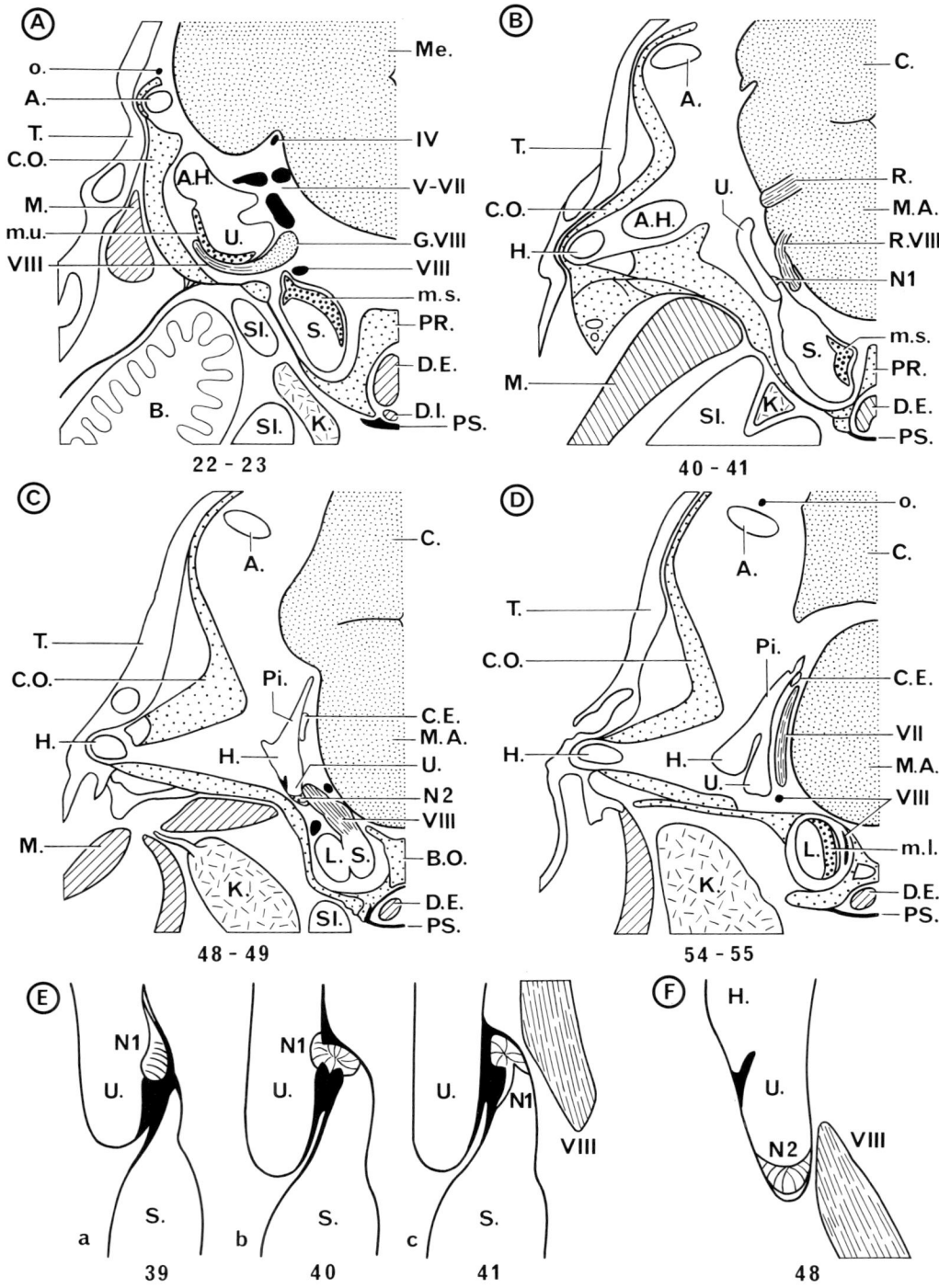

Fig. 2. Histological sections of the membranous labyrinth of *Chaetodon trifasciatus:* A – section at the level of the ampulla of the horizontal canal and macula utriculi, B – section at the level of the macula neglecta I (sarazini), C – section at the level of the macula neglecta II (retzii) and the connection between sacculus and lagena, D – section at the level of the opening of the horizontal canal into the common pillar of the two vertical canals and of the macula lagenae, E – structure of the macula neglecta I (sarazini) on three successive transverse sections (a, b, c) filling the pore of communication between sacculus and utriculus, F – structure of the macula neglecta II (retzii) at the floor of utriculus (A. = anterior vertical canal, A.H. = ampulla of the horizontal canal, B. = branchia, B.O. = basi-occipital cartilage, C. = cerebellum, C.E. = endolymphatic canal, C.O. = otic capsule (cartilaginous), D.E. = lateral rectus eye-muscle, D.I. = medial rectus eye-muscle, G.VIII = ganglion of the acoustic nerve, H. = horizontal canal, K. = kidney, L. = lagena, M. = muscles, M.A. = medulla oblongata (brain stem), Me. = mesencephalon, m.l. = macula lagenae, m.s. = macula sacculi, m.u. = macula utriculi, N1 = macula neglecta I (sarazini), N2 = macula neglecta II (retzii), o. = otic branch of the facial nerve (VII), Pi. = pillar common to the two vertical canals, PR. = prootic cartilage, PS. = parasphenoid bone, R. = cranial nerve root, R.VIII = root of the acoustic nerve, S. = sacculus, SI. = vascular sinus, T. = tegument, U. = utriculus, IV = trochlear nerve, V–VII = complex trigeminal-facial nerves, VII = facial nerve, VIII = acoustic nerve).

utriculus laterally and opening into the common pillar, close to the region where this pillar leads into the utriculus (Fig. 2C).

The saccular region

The anterior part of the sacculus is situated at the transverse level of the macula utriculi, in a ventromedial position to the utriculus (Fig. 2A). The macula sacculi is almost vertical (Fig. 2A, B) and very large (0.6 mm as compared to the macula utriculi 0.36 mm and to the macula lagenae 0.2 mm) and covers a great part of the medial wall of the sacculus). Globulous in its anterior part, the sacculus lengthens a little more posteriorly and, at about half the length of its anteroposterior size, it dorsally forms a tube-like appendix which is a little compressed and adheres to the medial wall of the utriculus (Fig. 2B). More dorsally, this tube constitutes the endolymphatic canal (Fig. 2C, D).

In the region where the saccular tube-like appendix is in contact with the wall of the utriculus, a very small orifice represents the communicating pore between sacculus and utriculus (Fig. 2B). This very small pore, visible in one section only, is situated between the transverse levels of the two roots of nerve VIII.

About 0.01 mm in front of the pore of communication between sacculus and utriculus, in a little cavity of the vertical wall of the narrow part of the utriculus, can be found a little mass of cells. This mass lengthens posteriorly and passes through the pore, filling it totally. More posteriorly, it spreads into the tube-like appendix of the sacculus. This mass of cells constitutes the macula neglecta I, which makes the pore between sacculus and utriculus a virtual orifice, and fills a great part of the tube-like dorsal appendix of the sacculus (Fig. 2E). The total anteroposterior length of the macula neglecta I is approximately 0.1 mm.

At a transverse level slightly posterior to the rear part of the macula neglecta I, at the level of the second root of the nerve VIII, the macula neglecta II is found (Fig. 2C, F). It lies in a longitudinal gutter of the floor of the narrow part of the utriculus. Its length is about 0.1 mm; it ends at the transverse level of the anterior part of the lagena, being more ventrally situated.

The lagena has an anteroposterior size (0.26 mm) about one third of that of the sacculus (0.76 mm) and is situated in the posterior prolongation of the sacculus (Fig. 2D). Its anterior region is lateral to the posterior part of the sacculus and separated from this latter by an incomplete vertical wall (Fig. 2C). The macula lagenae is vertically oriented and covers the medial wall of the lagena almost completely (Fig. 2D).

In many teleost species the lagena is a diverticulum of the sacculus and its limits from this latter are difficult to set (such as in *Polypterus senegalus*, Thomot et al. 1987 or in *Solea vulgaris*, Aly et al. 1988). Sometimes, the lagena constitutes (like in *Tridentiger trigonocephalus*, Kassem et al. 1988) a sac communicating more or less largely with the sacculus but, contrary to the situation seen in *Chaetodon trifasciatus*, the rostral part of the lagena generally lies medial to the posterior part of the sacculus.

The dorsal tube-like appendage of the sacculus stretches out dorsally to the communicating pore between sacculus and utriculus and constitutes the endolymphatic canal. It follows the medial wall of the common pillar of the two vertical canals and adheres to it (Fig. 2C, D). At the confluence of the two vertical canals, it opens into the endolymphatic sac, which represents the most dorsal part of the membranous labyrinth (Fig. 1). At the top of this sac, on its lateral wall, a mass of voluminous cells may be the secretory organ of the endolymph. An organ of the same aspect has been described in *Polypterus senegalus* (Thomot & Bauchot 1987).

The innervation of the labyrinth

Nerve VIII, which innervates the labyrinth, has two roots, both originating from the octavolateral region of the brain in the medulla oblongata. These two roots end in a single ganglionary mass of irregular form. A comparable organization of the ganglion, spread out rather than compact, was described in the goldfish (Rosenbluth & Palay 1961, Sento & Furukawa 1987). This ganglionary mass is anteroposteriorly elongated in *Chaetodon* and always close to the inferior region of the labyrinth (Fig. 1, 2A).

The total length of the ganglion is about 0.31 mm. It can be subdivided into three parts: a section in front of the first root of nerve VIII (0.17 mm); one between the two roots of nerve VIII (0.1 mm) and a third behind the posterior root of nerve VIII (0.04 mm). The anterior part which is most voluminous is situated under the utriculus and gives off a first ramus of nerve fibers that runs anteriorly towards the crista of the anterior vertical canal. Another ramus runs dorsally towards the macula utriculi and another one runs ventrally towards the macula sacculi. A fourth ramus of fibers runs towards the crista of the horizontal canal. The middle part is very thin and situated at the transverse level of the pore of communication between sacculus and utriculus. It sends a ventral ramus to the caudal part of the macula sacculi and two short rami to the macula neglecta I. The posterior part is a little more globulous than the middle part and sends a short dorsal ramus to the macula neglecta II, then a ventral ramus to the macula lagenae and last a third ramus running to the crista of the posterior vertical canal.

Discussion

The large vertical dimension of the upper part of the labyrinth, due essentially to the large size of the vertical canals and of the common pillar, may be due to the elevated body form of the butterflyfishes and could have no relation to a special development of these canals for a better adaptation of the swimming movements of the fish around coral heads and inside coral crevices. Nevertheless the large size of the ampullae of these canals, and the dimensions of their cristae, which are the sensory organs, linked through fibers of the nerve VIII to the cerebellum, lead us to believe that these canals play an important role in the precise swimming adjustment to the vertical plane. The form of the horizontal canal, linked to an ampulla of a greater volume than those of the vertical canals and forming an almost perfect circle of small radius, seems also to be correlated to the precise movements of this fish while picking at its food (coral polyps or worm tentacles, for example). The very large arcs of the semi-circular canals, and notably of the horizontal one, present in *Chaetodon* as in the pike or the goosefish (Retzius 1881b), could be linked to the rapid turning movements of these fish while picking their food, the frequency sensitivity of the cristae being determined by the fluid dynamic responses of the canal (Ramprashad et al. 1986). All swimming movements linked to rotations in the horizontal plane or to precise adjustment of the body in relation to the vertical plane have thus probably good detectors and explain the locomotory aptitudes of the butterflyfish.

The maculae, which are the sensory organs that detect gravity and sound, can be analyzed in relation to the biological adaptations of these fishes. The macula utriculi, situated at the floor of the utriculus, is relatively small compared to the macula sacculi, and not much larger than the macula lagenae. The great size of the macula sacculi could be linked to the fact that it probably plays a role, in butterflyfishes as in other teleosts, not only in the detection of linear acceleration, but also in the reception of sound (Lowenstein 1971, Popper 1976, 1983).

The presence of two maculae neglectae was rarely described in fishes. Retzius (1881) described the first one in *Polypterus bichir*. Thomot & Bauchot (1987) could find it in *Polypterus senegalus* only during larval stages. Iwazaki (1937) found two maculae neglectae in the carp, described by Retzius for what is present on the floor of the utriculus and by Sarazin for what lies at the roof of the sacculus. According to Iwazaki, it is clear that the macula neglecta I of *Chaetodon trifasciatus* corresponds to the macula sarazini while the macula neglecta II is the macula retzii. Two maculae neglectae were also described in the goldfish (Platt 1977), in a flatfish (Platt 1983b) and reported for *Amia* (Saidel et al. 1985). The position of the macula neglecta I in the pore of communication between sacculus and utriculus is difficult to interpret. The presence and significance of these maculae neglectae in adult fishes is not well documented, and it remains to be shown whether they may have gravistatic or acoustic functions. In sharks, the macula neglecta can grow to a large size and was shown to detect vibration (Corwin 1977, 1978, 1981).

There is no real communication between sacculus and utriculus in *Chaetodon trifasciatus*, since the macula neglecta completely fills the pore between them. In many teleosts, there is no communication between the upper part of the labyrinth (utriculus + semicircular canals) and the lower part (sacculus + lagena), while this connection is marked in *Polypterus* (Thomot & Bauchot 1987). The closing of this pore seems to occur very early during the development in butterflyfish.

The presence of the endolymphatic canal leading to the sac was difficult to detect in our material. This is probably due to an early regression of this organ in butterflyfish. On the contrary, this organ remains open and large in *Polypterus*, even in the adult (Retzius 1881). It is probable that this organ, in its upper part (sac), has a role of secretion of the endolymph; the remnants of a secretory epithelium in the sac seems to support this hypothesis.

The small size of butterflyfishes and the thickness of their skull will probably for some time to come impede measurement of nervous activity of their acoustic cristae and maculae during head movements. Such measurements, when they can be made, may confirm the auditory as well as vestibular functions of the macula sacculi of butterflyfishes. Another possibility is to study their behavior after surgical removal of parts of the labyrinth but the results may be difficult to interpret.

Acknowledgements

We thank René Galzin (Ecole Pratique des Hautes Etudes, Museum Antenna in French Polynesia) who supplied the material on which our description is based and also Avi Baranes (Steinitz Marine Laboratory in Elat, Israel) and Jack E. Randall (Bishop Museum in Honolulu, Hawaii), for providing additional material. We also thank Monique Diagne for her help in the preparation of the histological series and Maxime Martinez-Loscos for the final drawings of the figures. Philip J. Motta and Tyson R. Roberts helped improve the English in the final version of the manuscript.

References cited

Aly, A., A. Thomot, R. Bauchot & M.L. Bauchot. 1988. Evolution of the membranous labyrinth in soleid fish larvae. Progr. Zool. 35 (in press).

Corwin, J.C. 1977. Morphology of the macula neglecta in sharks of the genus *Carcharhinus*. J. Morphol. 152: 341–362.

Corwin, J.C. 1978. The relation of inner ear structure to the feeding behavior in sharks and rays. Scanning Electron Microscopy 1978: 1105–1112.

Corwin, J.C. 1981. Peripheral auditory physiology in lemon shark: evidence of parallel otolithic and non-otolithic sound detection. J. Comp. Physiol. 142: 379–390.

Dale, T. 1976. The labyrinthine mechanoreceptor organs of the cod *Gadus morhua* L. (Teleostei, Gadidae). A scanning electron microscopical study, with special reference to the morphological polarization of the macular sensory cells. Norw. J. Zool. 24: 85–128.

Fay, R.R. & A.N. Popper. 1980. Structure and function in teleost auditory systems. pp. 3–42. *In:* A.N. Popper & R.R. Fay (ed.) Comparative Studies of Hearing in Vertebrates, Springer Verlag, New York.

Hama, K. 1969. A study on the fine structure of the saccular macula of the goldfish. Z. Zellf. 94: 155–171.
Iwasaki, I. 1937. Entwicklungsgeschichtliche Untersuchungen über das häutige Labyrinth der Knochenfische. Jap. J. Med. Sci. I. Anat. 6: 301–419.
Jenkins, D.B. 1977. A light microscopic study of the saccula and lagena in certain catfishes. Amer. J Anat. 150: 605–629.
Kassem, M., A. Thomot & R. Bauchot. 1988. The membranous labyrinth of adult gobiid fish. Progr. Zool 35 (in press).
Krause, R. 1906. Entwicklungsgeschichte des Gehörorgans. pp. 83–138. In: O. Hertwig (ed.) Handbuch der vergleichenden und experimentelle Entwicklung der Wirbeltiere, Vol. 2, Jena.
Lowenstein, O. 1971. The labyrinth. pp. 207–240. In: W.S. Hoar & D.J. Randall (ed.) Fish Physiology, Vol. 5, Academic Press, New York.
Noorden, C. von 1883. Die Entwicklung des Labyrinthes bei Knochenfischen. Arch. Anat. Physiol. Anat. Abt. 7: 235–264.
Platt, C. 1977. Hair cell distribution and orientation in goldfish otolith organs. J. Comp. Neurol. 172: 283–298.
Platt, C. 1983a. The peripheral vestibular system of fishes. pp. 89–123. In: R.G. Northcutt & R.E. Davis (ed.) Fish Neurobiology, Vol. I, Brainstem and Sensory Organs, University of Michigan Press, Ann Arbor.
Platt, C. 1983b. Retention of generalized hair cell patterns in the inner ear of the primitive flatfish Psettodes. Anat. Rec. 207: 503–508.
Platt, C. 1987. Equilibrium underwater: signals, senses and steering in the vertebrates. pp. 413–429. In: J. Atema, R.R. Fay, A.N. Popper & W.N. Tavolga (ed.) Sensory Biology of Aquatic Animals, Springer-Verlag, New York.
Platt, C. & A.N. Popper. 1981. Fine structure and function of the ear. pp. 3–38. In: W.N. Tavolga, A.N. Popper & R.R. Fay (ed.) Hearing and Sound Communication in Fishes, Springer-Verlag, New York.
Popper, A.N. 1976. Ultrastructure of the auditory regions in the inner ear of the lake whitefish. Science 192: 1020–1023.
Popper, A.N. 1977. A scanning electron microscopic study of the sacculus and lagena in the ears of fifteen species of teleost fishes. J. Morphol. 153: 397–417.
Popper, A.N. 1979. Ultrastructure of the sacculus and lagena in a moray eel (Gymnothorax sp.). J. Morphol. 161: 241–256.
Popper, A.N. 1981. Comparative scanning electron microscopic investigations of the sensory epithelia in the teleost sacculus and lagena. J. Comp. Neurol. 200: 357–374.
Popper, A.N. 1983. Organization of the inner ear and auditory processing. pp. 125–178. In: R.G. Northcutt & R.E. Davis (ed.) Fish Neurobiology, Vol. I, Brainstem and Sensory Organs, University of Michigan Press, Ann Arbor.
Popper, A.N. & R.G. Northcutt. 1983. Structure and innervation of the inner ear of the bowfin, Amia calva. J. Comp. Neurol. 213: 279–286.
Popper, A.N., C. Platt & W.M. Saidel. 1982. Acoustic functions in the fish ear. Trends in Neuroscience 5: 276–280.
Ramprashad, F., J.P. Landolt, K.E. Money & J. Laufer. 1986. Comparative morphometric study of the vestibular system of the Vertebrata: Reptilia, Aves, Amphibia and Pisces. Acta Otol. Suppl. 427: 1–42.
Retzius, G. 1881a. Das membranöse Gehörorgan von Polypterus bichir Geof. und Calamoichthys calabaricus J.A. Smith. Biol. Unters. 4: 61–66.
Retzius, G. 1881b. Das Gehörorgan der Wirbelthiere. Vol. I. Das Gehörorgan der Fische und Amphibien. Samson & Wallin, Stockholm. 150 pp.
Rosenbluth, J. & S.L. Palay. 1961. The fine structure of nerve cell bodies and their myelin sheaths in the eighth nerve ganglion of the goldfish. J. Biophys. Biochem. Cytol. 9: 853–877.
Saidel, W.M. & C.A. McCormick. 1985. Morphology of the macula neglecta in the bowfin, Amia calva. Soc. Neuroscience Abstr. 383: 8.
Sento, S. & T. Furukawa. 1987. Intra-axonal labeling of saccular afferents in the goldfish Carassius auratus: correlations between morphological and physiological characteristics. J. Comp. Neurol. 258: 352–367.
Thomot, A. & R. Bauchot. 1987. The organogenesis of the membranous labyrinth of Polypterus senegalus Cuvier, 1829 (Pisces, Holostei, Polypteridae). Anat. Anz. 164: 189–211.
Von Frisch, K. 1936. Ueber den Gehörsinn der Fische. Biol. Rev. 11: 210–246.
Wegner, N.T. 1979. The orientation of hair cells in the otolithic organs and the papilla neglecta in the inner ear of the anabantid fish Colisa labiosa (Day). Acta Zool. 60: 205–216.
Wenig, J. 1911. Die Entwicklung des Ductus endolymphaticus bei den Knochenfischen. Anat. Anz. 38: 112–115.
Wenig, J. 1913. Untersuchungen über die Entwicklung der Gehörorgan der Anamnia. Gegenbaurs morphol. Jb. 45: 295–333.

Strengths and weaknesses in butterflyfish research: concluding remarks

Philip J. Motta
Department of Biology, University of South Florida, Tampa, FL 33620–5150, U.S.A.

In the last decade we have witnessed a vast increase in the number of scientific publications dealing with the chaetodontid butterflyfishes. Initially this research was primarily centered on their ecology and behavior but more recently the interests have broadened, as exemplified by this symposium. The interest in the group may be due to their conspicuousness, intriguing social behavior, cicumtropical distribution, and the fact that they are one if not the most important fish predator on corals. With isolated groups of researchers scattered throughout the world interest was expressed in trying to get many of those involved together. With this intention I convened this symposium and charged each participant with a variety of responsibilities: they were to present their latest data on butterflyfish; they were encouraged to be speculative in though to provoke new areas for study in butterflyfish biology and perhaps fish biology; they were asked to relate their findings to other fish groups; and lastly they were asked to suggest avenues for future research in their areas. The symposium as a whole was also intended to point out areas of strength and weakness in butterflyfish research; to bring the researchers together in one meeting to discuss findings, coordinate efforts, and increase communication; and the scientific exchange was to be brought to the scientific community as a whole.

Our symposium participants, representing a significant portion of the butterflyfish research community, have had many fruitfull hours of discussion. Resulting from these talks and presentations themselves I have taken the liberty of trying to summarize the findings.

First of all one notes that perhaps the greatest strength in chaetodontid research lies in the areas of feeding behavior and ecology, and secondarily social behavior. Timothy Tricas has addressed the question of optimal foraging by quantifying the feeding habits of *Chaetodon multicinctus*. The energy content, handling time, and defense mechanisms of various stony corals all contribute to prey choice in this species. Darby Irons has approached feeding by studying time budgets for another species, the solitary *Chaetodon trifascialis*. Similar to the results of Thomas Hourigan, she has found that males and females feed at different rates. Among other things this calls into question the habit of lumping feeding rates for the separate sexes. Mireille Harmelin-Vivien has demonstrated the importance of food resources on the community structure and recruitment of butterflyfishes on coral reefs. The specialized obligate coral feeders begin benthic feeding at settlement and scleractinian polyps are the necessary exclusive food resource for recruits among these species.

To draw patterns in feeding behavior and ecology one needs baseline feeding studies, and Mitsuhiko Sano has classified the feeding habits of 32 species of Japanese butterflyfishes, making them probably one of the best understood butterflyfish assemblages in terms of their feeding. Scleractinian corals are the most important food resource for the Japanese butterflyfishes, and this and the many other feeding studies point out the importance of healthy coral reefs. Both Ernst Reese, Thomas Hourigan, and now Darby Irons provide evidence that butterflyfishes may be used as indicator organisms for assessing reef health. This strong link between corals and chaetodontid fish assemblages was also demonstrated by the work of Yolande Bouchon-Navaro and Claude Bouchon with their work in the Red Sea. They found significant correlations between the density of chaetodontid fishes and the diversity of the coral community.

The determinates of social systems in vertebrates

is an area of keen interest. Thomas Hourigan has used the butterflyfishes as model organisms and determined that the distribution and quality of food resources are the major determinants of group size and mobility, and therefore specific social systems. James and Muriel Findley on the other hand, investigate the relationships between butterflyfish species richness and circumtropical patterns in distribution. They found no evidence of density compensation in richer communities despite what community theory predicts, but at the level of islands and regions, habitat breadth diminishes as species richness increases. Ernst Reese has tackled a much neglected area of fish biology, orientation behavior of reef fishes. The coral feeding chaetodontids follow stereotyped routes in their foraging, foraging paths that are based on learned locations of route specific landmarks.

The study of butterflyfish larval distribution, recruitment, and reproduction has attracted much attention because of the intense interest in these areas for reef fishes in general. Like many reef fishes they are territorial or home ranging, and broadcast spawners. What therefore is the fate of their larvae, how are they recruited onto the reef, and what factors influence the abundance and distribution of the adults? Jeffrey Leis has reviewed all the available literature on their larval stages and found among other things that chaetodontid larvae are uncommon in the plankton. Unexplainably, waters near reefs have the fewest larvae. Both size and age at settlement vary widely within the family and within the genus *Chaetodon*. Both Phillip Lobel and Patrick Colin describe spawning behavior. Lobel describes how intruders sneak spawns in one species *Chaetodon multicinctus,* and Colin reviews the literature on spawning among five western Atlantic species. He finds that smaller species may adopt the strategy of producing moderate numbers of eggs per day over a spawning season of at least a few months, while larger species may produce more eggs per day for a shorter period.

Remarkably, the systematics and biogeography of the butterflyfishes are sorely understudied, perhaps because it represents such monumental effort to analyse worldwide patterns. With the monograph of Burgess the biogeography and systematics of butterflyfishes have begun to be understood. Stanley Blum uses the most extensive cladistic analysis of the butterflyfishes to identify a set of vicariant events that have generated nearly a third of the present species diversity and he presents the geography of the inferred barriers.

Perhaps because the adaptive significance of coloration is so difficult to test there have been so few studies on this subject, for fish and other vertebrates. Butterflyfishes are often cited as examples of animals with bright or 'poster' coloration. The hypotheses have been numerous, as reviewed by Stephen Neudecker, but until now few conclusions have been drawn. Neudecker deduces that the primary selective force behind this bright coloration, particularly eyespots, has been predation, and color patterns in these fishes minimizes this threat and communicates social information.

The anatomy and physiology of butterflyfishes is another area of neglect. I have tried to pull together my data on the dentition of fifteen Pacific and western Atlantic species and relate this to the phylogeny of the group as presented by Stanley Blum. Despite what appears to be adequate time for evolution between the two faunas, many of the species retain the generalized tooth arrangement. However, as a whole the Pacific species show more specialized morphologies for hard coral feeding than do the western Atlantic species. Roland Bauchot, Thomot Athanase, Jean-Marc Ridet, and Marie-Louise Bauchot have presented the most comprehensive studies on butterflyfish central nervous systems and sensory systems. The eye muscles and their innervation are described in *Chaetodon trifasciatus* and related to their ecology, as is the membranous labyrinth and their precise locomotory aptitudes. Lastly, they present a comprehensive comparison of the brain of pomacanthids and chaetodontids, and speculate on the relationship to feeding in both families.

Now that we have presented some of our strengths what then are the weaknesses in our study of the butterflyfishes. We obviously know very little about the fate of their larvae and the life span prior to adulthood. How do juvenile butterflyfishes compete for space and food on the reef? There are almost no ontogenetic studies on butterflyfishes.

Are the social systems of the juveniles different from the adults, and how, and what factors influence their social systems? Similar to many other reef fishes we know very little about ontogenetic changes in morphology, in diet and feeding. We also know little about the recruitment processes of reef fishes and butterflyfishes in specific, as witnessed by the continuing debate about these processes.

The biogeography of this family is still not completely understood even though the distribution is quite well documented. We have also just begun to explore the peculiarities of the nervous and sensory systems as well as the functional morphology of the group as a whole. Pulling an example from my own work one notes that up until quite recently ichthyologists described butterflyfishes by their simple brush-like teeth, implying the dentitions were escentially alike. The groundwork of Warren Burgess and later myself demonstrated that simply because the teeth are small does not imply they are identical. Examining only ten species I have described a range of specializations in the dentition and jaw morphology. Where does it end? Each researcher can provide a plethora of avenues for continued studies, as they did at the end of every paper. In that respect we have accomplished another one of our intended goals. We have strengthened the base of our understanding of the group, contributed to the understanding of the coral reef fish community, and provided direction for future studies on these fishes.

Finally, this symposium has brought chaetodontid researchers together in person and in publication. There can only be greater communication and exchange from here on, and at the very least the published volume should serve as a communal source of scientific information on the group. Oftentimes such specialized symposia are unfortunately viewed with skepticism. But apart from all the aformentioned benefits, this symposium was the catalyst for seventeen publications involving twenty scientists. Similar symposia focusing on specific fish taxa should be encouraged and supported.

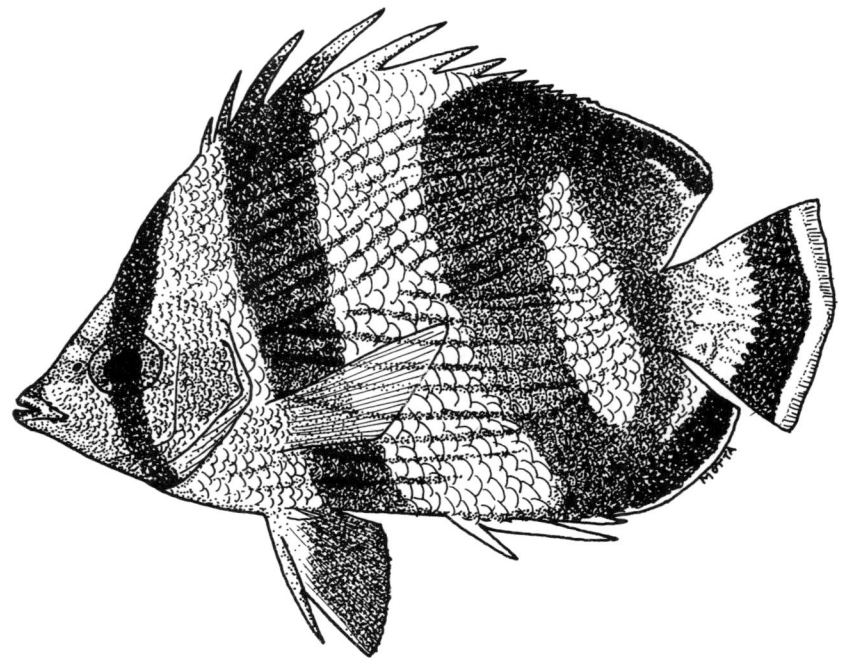

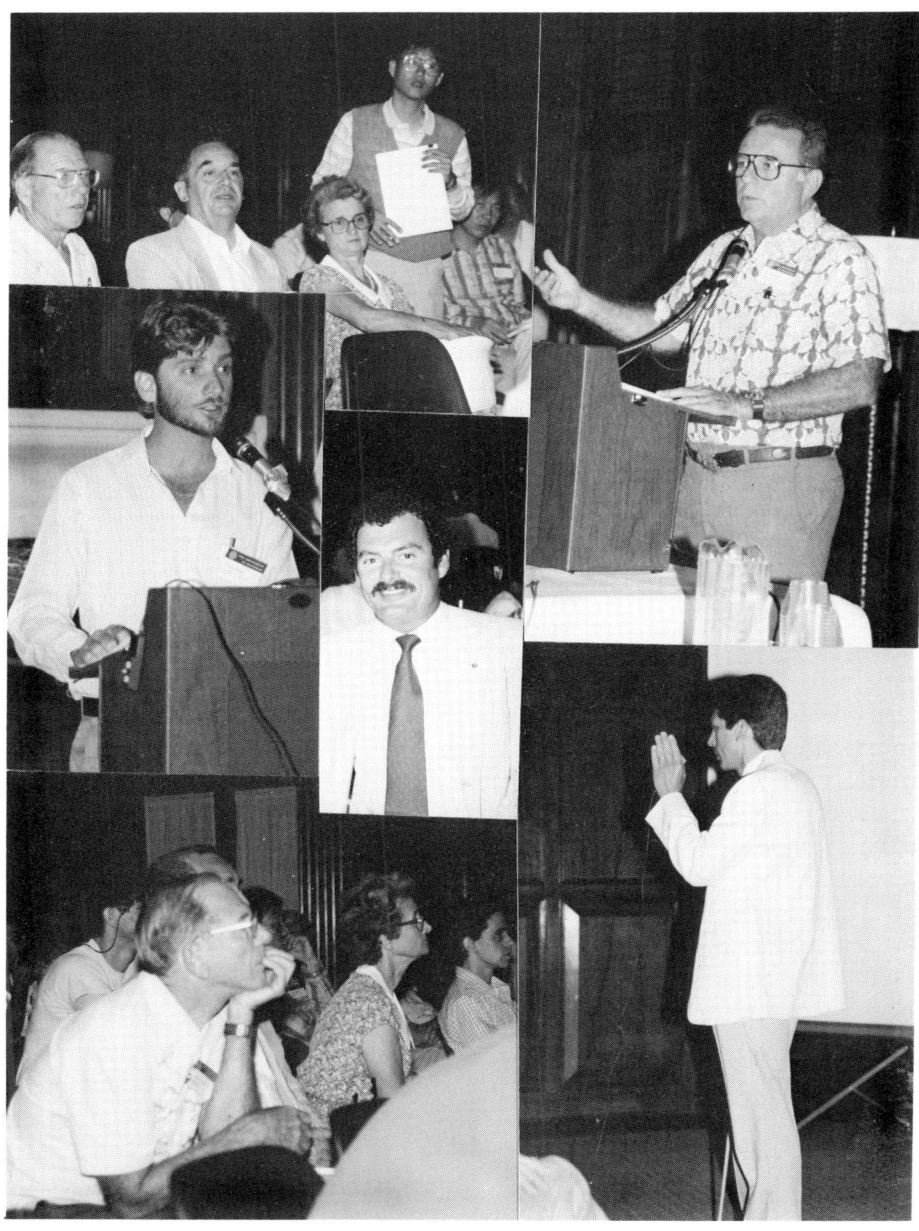

Clockwise around Stephen Neudecker (center), starting with the right top corner, some of the Symposium speakers and audience: Ernst Reese; Timothy Tricas; Roland Bauchot hidden behind Jack Randall and Marie-Louise Bauchot; Philip Motta delivering the closing remarks; and the attentive trio Randall – Bauchot – Bauchot and Hin Kiu Mok listen to questions from Mitsihiko Sano.

Some of the Butterflyfish Symposium banquet attendees in the Old German Restaurant of Ann Arbor (clockwise, starting upper right): while Ernst Reese concentrates on ordering meal his neighbor still admires a butterflyfish; Philip Motta, Darby Irons, Bruce Carlson, Jeffrey Leis and drinking Patrick Colins; Warren Burgess, Ralph Nursall, Colin Patterson, Stan Blum, Darby Irons looking on, and guests debate while Eugene Balon prepares to take a picture; framed between Mike Bruton and Lourdes Burgess out of focus, Timothy Tricas prepares slides while Philip Lobel and Michael Hin Kui Mok stare; soon Philip Lobel and Michael Mok are left to guard an empty chair; James Findley and Lourdes Burgess seem satisfied; and Robert Ross, Michael Mok and unidentified guests pose for the last picture.

Yolande Bouchon-Navaro and Claude Bouchon above Mireille Harmelin-Vivien raising a toast to the official symposium cap.

Chaetodon multicinctus from Hawaii. Probably the most thoroughly studied butterflyfish. Photo by P.S. Lobel.

Species and subject index

Abducens nerve, see cranial nerve VI
Acanthaster sp. 38
Acanthastrea echinata 52
Acanthurus leucosternon 71
   lineatus 70
   sp. 216
   triostegus 107, 108, 214, 215, 217
Acropora acuminata 191
   cytherea 58, 188, 189, 190, 191, 192
   digitifera 50
   erythraea 50
   granulosa 50
   hemprichii 50
   humilis 50
   hyacinthus 50, 82
   nasuta 50
   sp. 41, 47, 57, 69, 81, 160, 187, 188, 190, 191, 192, 196, 201
   squarrosa 50
   valida 50
Aggression 71, 72, 149, 216
   release hypothesis 143, 145, 155
Aggressive behavior 134
Alcyonarians 200
Algae 72, 176, 182, 197, 200
Allometry 167
Allopatry 11, 12, 15, 27, 28
Alpha diversity 33, 35, 36, 45
Alveopora allingi 51
   fenestrata 51
   ocellata 51
   verrilliana 51
Amia sp. 241
Amphichaetodon howensis 12
   melbae 12
   sp. 11, 12, 22, 88
Amphiprion sp. 26
Ampulla 236
Anguilla anguilla 216, 230
Antennarius drombus 217
Anterior canal 236
   myodome 230
Apomorphy 18, 167
Aposematic coloration 146, 153
   hypothesis 143, 146, 155
Area acousticolateralis 212
Area pretectalis 212
Ascidians 197, 199, 200
Astreopora myriophthalma 50

Atlantic 9, 33, 90, 166
Auditory nerve, see cranial nerve VIII
Australia 26, 27
autapomorphy 167

Barabattoia amicorum 51
Barriers 24, 25, 26, 27, 28
Basioccipital 225
Batesian mimicry 151, 152
Bathygobius soporator 80, 84
Beta diversity 33, 35, 36, 43, 45
Betta splendens 84
Biomass 7, 36, 92
Blastomussa merleti 52
Brain morphology 208
Broadcast spawning 61, 63, 66, 68
Burgessius miliarius 207
   sp. 207

Calanoides philippinensis 28
Callionymus enneactis 122
Calloplesiops altivelis 151, 199
Caloric content 173, 177, 180
Camouflage 144, 146, 147, 150
Caranx sexfasciatus 216, 217
Carassius auratus 119
Caribbean 40, 167
Cartilaginous sclera 222
Centropyge bicolor 206
   bispinosus 206
   fisheri 206, 208, 210
   flavissimus 206
   heraldi 206, 213, 214, 215, 216, 217
   multispinis 206
   potteri 206, 208
   sp. 140, 206, 217
   tibicen 206, 208
Cephalopholis argus 82
Cerebellar valvula 227, 228
Cerebellum 205, 208, 212, 215, 217, 236, 240
Chaetodon aculeatus 77, 88, 131, 132, 133, 134, 135, 137, 138, 139, 161, 162, 163, 165, 166, 167, 168
   adiergastos 13
   argentatus 14, 21, 22, 26, 78, 198, 199
   assarius 14, 21, 22, 23, 26, 27, 77, 149, 196
   aureofasciatus 14, 42, 77

   auriga 13, 39, 41, 42, 49, 53, 54, 56, 58, 73, 78, 89, 95, 103, 104, 105, 106, 107, 149, 160, 161, 162, 163, 165, 166, 167, 168, 197, 199, 234
   auripes 13, 78, 89, 149, 195, 197, 199
   austriacus 14, 17, 24, 26, 29, 43, 49, 53, 54, 56, 57, 58, 77, 107
   aya 89, 148
   baronessa 14, 42, 77, 95, 145, 196, 197, 199, 201
   bennetti 14, 39, 42, 147, 149, 151, 158, 196, 197, 199
   blackburnii 14, 89, 107
   burgessi 13, 148
   capistratus 13, 40, 41, 58, 73, 77, 89, 103, 104, 106, 107, 109, 131, 132, 134, 135, 137, 138, 139, 142, 144, 149, 151, 160, 162, 165, 166, 167, 168
   chrysurus 66, 71, 72, 73, 138
   citrinellus 13, 19, 22, 39, 41, 42, 77, 89, 90, 103, 104, 105, 106, 107, 149, 153, 154, 155, 197, 199
   collare 13, 77, 89, 95
   daedalma 14, 195, 196, 198, 199
   declivis 13, 27
   decussatus 13, 206
   dichrous 77, 147
   dolosus 14, 21, 22, 23, 26, 149
   ephippium 13, 39, 41, 42, 78, 89, 107, 151, 197, 199
   excelsa 89, 95
   falcifer 148
   falcula 13, 18, 19, 20, 22, 26, 27, 78
   fasciatus 13, 29, 43, 49, 53, 54, 56, 57, 58, 78, 145
   flavirostris 13, 42, 78, 213, 214, 215, 217
   flavocoronatus 13, 147
   fremblii 14, 65, 66, 68, 69, 70, 72, 73, 77, 138, 147, 149
   gardineri 13
   guezei 148
   guntheri 14, 21, 22, 23, 26, 27
   guttatissimus 14, 19, 20, 21, 26, 107, 112
   guyanensis 148
   hemichrysurus 89
   hoefleri 13

humeralis 13, 40, 77
intermedius 58
kleinii 14, 17, 18, 19, 26, 27, 29, 42, 73, 77, 95, 103, 104, 105, 106, 107, 138, 196, 198, 199, 200, 201
larvatus 14, 77, 145
lavocoronatus 148
leucopleura 13
lineolatus 13, 22, 23, 24, 26, 28, 29, 42, 78, 89, 95, 197, 199
litus 14, 27, 89, 147
lunula 13, 23, 29, 39, 41, 42, 43, 78, 89, 95, 103, 104, 105, 106, 107, 149, 197, 199
madagascariensis 14, 21, 22, 26, 28, 103, 104, 105, 106, 107
marcellae 148
marleyi 10, 13
melannotus 13, 14, 23, 25, 26, 28, 29, 42, 49, 53, 54, 56, 57, 58, 77, 89, 95, 107, 145, 149, 155, 196, 198, 199
melapterus 17, 24, 26, 29, 77
mertensii 14, 21, 22, 26, 28, 39, 42, 77, 147
mesoleucos 13
meyeri 15, 17, 77, 89, 149, 196, 197, 199
miliaris 14, 19, 21, 22, 23, 26, 27, 65, 66, 68, 69, 77, 92, 95, 102, 107, 112, 122, 138, 149, 151, 160, 162, 163, 166, 167, 168, 201
mitratus 13, 77, 148
modestus 89
multicinctus 14, 19, 20, 21, 26, 27, 61, 64, 65, 66, 68, 69, 70, 71, 72, 77, 80, 81, 82, 83, 84, 111, 112, 113, 114, 115, 116, 117, 118, 119, 120, 121, 122, 123, 125, 126, 127, 128, 129, 138, 139, 140, 160, 162, 163, 165, 166, 167, 168, 171, 172, 174, 175, 176, 177, 179, 180, 181, 182, 243, 244, 250
nigropunctatus 13, 77, 147
nippon 13, 88, 89, 96, 129, 195, 196, 197, 198, 199
ocellatus 13, 40, 44, 88, 89, 107, 132, 135, 136, 137, 138, 139, 148, 149, 161, 162, 165, 166, 167, 168
ocellicaudus 13, 23, 24, 25, 26, 28, 149
octofasciatus 14, 77, 95
oligacanthus 14, 15
ornatissimus 15, 17, 39, 41, 42, 65, 66, 69, 77, 80, 81, 82, 83, 89, 107, 149, 160, 162, 163, 164, 165, 166, 167, 168, 182, 196, 197, 199, 218
oxycephalus 13, 22, 23, 24, 26, 28

paucifasciatus 14, 21, 22, 26, 42, 49, 53, 54, 56, 57, 58, 66, 71, 77, 107, 147, 149
pelewensis 14, 19, 20, 21, 26, 27, 39, 41, 42, 77, 107, 112
plebeius 14, 16, 42, 77, 89, 95, 96, 144, 149, 197, 199, 201
punctatofasciatus 14, 19, 20, 21, 22, 26, 27, 77, 89, 104, 112, 196, 198, 199
quadrimaculatus 14, 22, 42, 61, 64, 65, 66, 68, 70, 71, 72, 77, 80, 83, 107, 149, 153, 154, 160, 162, 163, 164, 166, 167, 168
rafflesi 13, 42, 149, 196, 198, 199, 200
rainfordi 14, 42, 73, 77, 95, 96, 138
reticulatus 15, 17, 39, 41, 77, 107, 213, 214, 215, 217, 218
robustus 13
sanctaehelenae 10, 14, 21, 22, 23, 26, 27, 67, 77, 88
sedentarius 10, 14, 21, 22, 23, 26, 27, 89, 107, 132, 136, 138, 161, 162, 166, 167, 168
selene 13, 23
semeion 13, 39, 42, 149, 196, 197, 199, 200
semilarvatus 13, 147
smithi 14, 27, 77, 147
sp. 9, 10, 11, 13, 15, 23, 27, 34, 40, 65, 87, 88, 89, 90, 95, 96, 121, 131, 132, 136, 137, 138, 139, 143, 144, 145, 146, 147, 148, 149, 150, 152, 153, 155, 206, 240, 243
speculum 14, 42, 77, 95, 149, 151, 196, 197, 199
striatus 12, 40, 77, 107, 131, 132, 136, 137, 139, 161, 162, 166, 167, 168
tinkeri 13, 22, 27, 147, 148
triangulum 14, 77
trichrous 14, 17, 18, 19, 26, 27, 29, 39, 42
tricinctus 14, 77
trifascialis 14, 39, 41, 42, 49, 53, 54, 56, 57, 58, 64, 69, 70, 73, 77, 80, 81, 82, 83, 84, 95, 103, 104, 105, 107, 145, 149, 155, 160, 162, 165, 166, 167, 168, 186, 187, 188, 190, 191, 192, 197, 199, 201, 243, 244
trifasciatus 14, 16, 17, 26, 29, 32, 39, 41, 42, 43, 65, 66, 69, 73, 77, 80, 81, 82, 83, 103, 104, 105, 107, 160, 162, 163, 165, 166, 167, 168, 197, 199, 213, 214, 217, 221, 222, 223, 224, 227, 230, 231, 232, 235, 236, 237, 239, 241

ulietensis 13, 18, 19, 20, 26, 27, 39, 41, 42, 77, 196, 197, 199, 200
unimaculatus 14, 17, 18, 39, 42, 65, 66, 69, 70, 71, 73, 78, 89, 90, 95, 107, 138, 149, 160, 162, 163, 166, 167, 168, 196, 197, 198, 199, 200, 201
vagabundus 13, 23, 39, 41, 42, 78, 89, 95, 103, 104, 105, 106, 107, 149, 197, 199
weibeli 13, 77, 149
xanthocephalus 13, 78, 103, 104, 105, 106, 107, 147
xanthurus 14, 21, 22, 26, 28, 77
zanzibariensis 14, 16, 77, 149
Chaetodontops *collare* 206
flavirostris 206
lunula 206
sp. 13, 18, 23, 29, 148, 150, 206
Chagos-Laccadive ridge 26
Chelmon *marginalis* 12
mulleri 12
rostratus 12, 42, 78, 95
sp. 11, 12, 88, 89
Chelmonops *curiosus* 12
sp. 11, 12, 88
trucatus 12
Ciliary ganglion 223, 225, 229
Cirrhilabrus *flavidorsalis* 28
lubbocki 28
rubripinnis 28
Citharoedus *baronessa* 208
ornatissimus 206
reticulatus 206
sp. 11, 15, 16, 17, 89, 90, 147, 206
Cladistics 15, 29, 161
Cocos Keeling 26, 27
Coelenterates 102, 103
Cognitive maps 79, 84, 85
Coloration 145, 151
Common oculomotor nerve, see cranial nerve III
pillar 236
Competition interspecific
intraspecific 108
shelters 63
Competitive exclusion 16
Congrogadus *hierichthys* 28
Cooperation 61
Copepods 200
Coprophagy 200
Coradion *altivelis* 12, 196, 197, 199, 200
chrysozonus 12
melanopus 12
sp. 11, 12, 87, 88, 89, 90, 94, 97, 98
Coral abundance 173, 182, 190

assemblages 47, 53, 58
branching 47, 56, 57, 58, 59
browsers 47, 49, 56, 58
community 47, 56, 58
cover 56, 59, 69, 176, 188
distribution 47, 50, 51, 52
encrusting 57, 59
feeders 85, 101, 103, 107, 171, 187, 195, 199, 200, 202, 218
feeding 69, 189, 190
lipid 181
massive 57, 59
morphology 179
mucus 191
polyps 58, 81, 101, 103, 107, 173, 174, 175, 178, 179, 181, 182
reefs 67, 92, 94, 136, 171, 205, 235
scleractinian 168, 195, 197, 199, 200
Sea 92, 95, 97
skeleton 180
tissue 177
toxin 182
Corallivory 67, 73
Corallochaetodon austriacus 207
sp. 11, 14, 16, 17, 26, 207
Corpus glomerulosum 212
Coryphaena hippurus 208
Coscinarea monile 50
Courtship 125, 135, 136, 138
Cranial nerve III 221, 222, 227, 229, 231
nerve IV 221, 222, 224, 227, 228, 231
nerve V 227, 228
nerve VI 221, 222, 227, 229, 231
nerve VII 227, 228, 229
nerve VIII 229, 240
Crista 235, 236, 240
Cycloseris doderleini 50
patelliformis 50
Cynarina lacrymalis 52
Cyphastrea chalcidicum 51
microphthalma 51
serailia 51

Dascyllus aruanus 214, 215, 217
marginatus 63
Defense 176
Dendrophyllia arbuscula 52
Density compensation 33, 36, 37, 43, 45
stasis 43
Dentition 16, 159, 161, 162, 163, 165, 166, 167
Diencephalon 212, 214, 216
Diet 49, 182, 195, 196, 197, 200
Discochaetodon sp. 11, 14, 17, 29
Dispersal 16, 26, 27, 62, 65
Disruptive coloration hypothesis 143, 146, 155
Distribution patterns 47, 105
Divergent evolution 168
Diversity indices 56
Dorsolateral nucleus 212

Echinophyllia aspera 52
Echinopora fructiculosa 51
gemmacea 51
mammillosa 51
Ecsenius bandanus 28
sp. 28, 31
Eggs 117, 133, 134, 135, 136, 137, 139
pelagic 65, 87, 88
predators 66, 131, 137
El Nino 38
Encephalization 205, 208
Encounter rates 173
Endemism 24, 26, 28
Endolymph 241
Endolymphatic canal 235, 239, 240, 241
sac 235, 240
Energy content 171, 177, 182
maximizer 171, 172, 192
Engraulis mordax 122
Erythrastrea flabellata 51
Ethmoid 222
Ethmoidal plate 231
septum 225
Eupomacentrus jenkinsi 144
Exornator citrinellus 206, 218
kleinii 206, 213, 214, 215, 216, 217
multicinctus 206, 218
sp. 11, 13, 18, 19, 20, 27, 206
Extinction 16, 26, 27
Extraocular muscles 222
Eye 144, 147, 149, 150, 155, 221, 222
bars 144, 146, 147, 150, 155
camouflage 143, 149, 155
movements 179
Eyemasks 143, 155
Eyespots 143, 144, 145, 146, 147, 148, 149, 150, 151, 152, 153, 155

Facial nerve, see cranial nerve VII
False starts 133
Favia favus 51
laxa 51
pallida 51
rotumana 51
stelligera 51
Favites abdita 51
chinensis 51
complanata 51
flexuosa 51
melicerum 51
pentagona 51
peresi 51
Fecundity see spawning fecundity
Feeding behavior 96, 102, 183, 187, 188, 191, 192, 196, 205, 218
bites 176
ecology 171
groups 195, 196, 197
habits 159
preference 179, 180
rates 71, 72, 179, 187, 188, 189, 190, 191, 192
sponge 205, 218
Fish larvae 87, 88, 90, 93, 94, 95, 97
larval distribution 97
abundance 94
Fistularia petimba 216
Food preference 172, 179
resources 183
Foraging 70, 79, 172, 187
behavior 81, 171, 173, 174, 175, 178
ecology 183
model 171
paths 80, 82, 83, 173, 175
patterns 171
strategy 187, 192
Forcipiger flavissimus 12, 65, 66, 77, 95, 96, 144, 160, 161, 162, 163, 164, 165, 166, 167, 168, 194, 196, 198, 199, 207, 213, 214, 215, 217
longirostris 12, 78, 144, 160, 162, 163, 167, 168, 204, 207, 220
sp. 11, 12, 39, 41, 88, 89, 94, 98, 144, 150, 207, 216
Forebrain 208
French Polynesia 201
Fungia fungites 50
granulosa 50
klunzingeri 50
mollucensis 50
paumotensis 50
scruposa 50
scutaria 50

Galaxea fascicularis 51
Gamete 127, 137
release 133, 136
Ganglion 240
ciliaris 228
Generalists 65, 69, 71, 159, 168, 197
Genicanthus caudovittatus 206, 213, 214, 215, 217
sp. 206
Gobio gobio 230
Gonadosomatic index 112, 122, 126
Goniastrea aspera 51

edwardsi 51
pectinata 51
retiformis 51
Goniopora dijiboutiensis 51
  planulata 51
  savignyi 51
  somaliensis 51
  stutchburyi 51
Gonochaetodon baronessa 206
  sp. 11, 14, 206, 208
Great Barrier Reef 42, 87, 92, 94, 95, 97
Growth rates 98
Gulf of Aqaba 24
  Aden 24
Gut contents 173, 196
Gymnothorax marginatus 216, 217
  meleagris 151
Gyrosmilia interrupta 52

Habit formation 79
Haller's campanula 230
Halodule uninervis 48
Halophila stipulacea 48
Handling 173, 175
  costs 178
  time 171, 173, 174, 181, 182
Haplochromis sp. 165
Harems 61, 66, 72, 73
Hawaii 201
Hemitaurichthys multispinosus 12, 27
  polylepis 12, 68, 77, 196, 198, 199, 200, 207
  sp. 11, 12, 27, 69, 88, 89, 95, 207
  thompsoni 12, 77
  zoster 12, 77, 200
Heniochus acuminatus 12, 39, 42, 138
  chrysostomus 12, 39, 41, 42, 196, 198, 199, 207
  diphreutes 12, 77, 95·
  intermedius 12, 49, 53, 54, 56, 57, 77, 207, 208, 210, 213, 214, 215, 217
  monocerus 12, 39, 42, 196, 197, 199, 200, 207
  pleurotaenia 12
  singularis 12, 196, 197, 199, 207
  sp. 11, 12, 88, 89, 207
  varius 12, 196, 198, 199, 200, 207
Herpetoglossa simplex 50
Herpolitha limax 51
Heterochaetodon dolosus 207
  miliaris 211, 212, 214, 215, 217, 218
  quadrimaculatus 218
  sp. 207
  unimaculatus 218
Holacanthus arcuatus 206

sp. 90, 206, 217
tricolor 121
xanthotis 206, 213, 214, 217
Home range 64, 66, 218
Horizontal canal 236
Hydnophora exesa 51
  microconos 51
Hydroids 197
Hydro-Lab 132
Hypothalamus 212, 225, 227, 228, 229

Ictalurus nebulosus 231
Indian Ocean 26, 27, 28, 29
Indo-Pacific 9, 27, 28, 29, 33, 90
Intraocular muscles 229
Intruder 192
Iris 230
Isthmus of Panama 167
Iteroparity 62

Japan 195, 201, 202
Jaws 16, 163, 166
Johnrandallia nigrirostris 12, 40
  sp. 11, 12, 88, 89
Johnston Atoll 27

Kin selection 73

Labroides dimidiatus 214, 215, 217
  sp. 216
Labrus bergylta 212, 214, 215, 216, 217
  sp. 216
Labyrinth, see membranous labyrinth
Lagena 235, 239
Landmark orientation 84
Larvae, see Fish larvae
Lee Stocking Island 132, 134, 135
Lepidochaetodon quadrimaculatus 207
  sp. 11, 14, 17, 207
  unimaculatus 207
Leptastrea bottae 51, 176
  inaequalis 51
  purpurea 51, 176
Leptoseris explanata 50
  mycetoseroides 50
  yabei 50
Linophora auriga 207
  fasciatus 207
  sp. 207, 208
Lobophyllia corymbosa 52
  hemprichii 52
Lobus lateralis hypothalami 212
Locomotory aptitudes 217
Lunar periodicity 66, 131, 133, 137, 138, 140

Macula 236
  lagena 235, 239, 240
  neglecta 235, 239, 241
  sacculi 235, 239, 240
  utriculi 235, 240
Malay archipelago 26
Mating behavior 126
  systems 61, 67
Medulla oblongata 212, 216, 217, 227, 228, 229, 236, 240
Megaprotodon plebeius 207, 213, 214, 215, 216, 217
  sp. 11, 14, 207
  strigangulus 80
  trifascialis 187, 207
Melichthys niger 83, 137
Membranous labyrinth 235, 236
Mesencephalon 212, 216, 227, 228
Mesochaetodon melannotus 213, 214, 215, 216, 217
  sp. 207
  trifasciatus 207, 215, 216, 218
Metencephalon 212
Mexico 40
Midbrain 208
Millepora dichotoma 52
  exaesa 52
  platyphylla 52
Monogamy 61, 63, 71, 72, 111
Monophyly 22
Montastrea annuligera 51
  valenciennesi 51
Montipora danae 50
  erythraea 50
  informis 50
  monasteriata 50
  patula 188, 191
  sp. 70, 187, 188, 189, 190, 191, 192
  spongiosa 50
  spumosa 50
  tuberculosa 50
  venosa 50
  verrilli 188, 191
  verrucosa 50, 176, 188, 191
Moringua sp. 208
Morphospace 38, 45
Mortality 64
Mugil cephalus 121
Mullerian mimicry 143, 152, 155
Muscle inferior oblique 221, 222
  inferior rectus 221, 225
  lateral rectus 221, 225
  medial rectus 221, 225
  retractor bulbi 221, 232
  superior oblique 221, 222
  superior rectus 221, 223

Mycedium elephantotus 52
Myodome 221, 222, 223, 230
Myripristis parvidens 214, 215, 217
    sp. 26

Nalbantius sp. 207
    speculum 207, 213, 214, 215, 217
Nannocharax fasciatus 230, 231
Nematocysts 171, 174, 178, 181
Niche breadth 37, 45, 107
Nucleus geniculatus 205, 212
Nycthemeral rhythm 218

Ocellus 147, 150, 151, 152
Olfaction 216
Olfactory bulb 205, 212, 214, 216
Ontogeny 102, 105, 108
Oocytes 111, 112, 113, 114, 115, 116, 117,
    119, 120, 121, 122
Optic chiasma 224, 228
    system 205
    tectum 205, 208, 212
Optimal diet 172
    foraging 171
Orbit 222
Orientation 79, 84
Otoliths 96
Ovary 65, 111, 113, 114, 115, 120, 121, 135,
    136, 137
Oxychaetodon sp. 207
    ulietensis 207

Pacific 33, 166, 167
    Ocean 28, 29
Population 28
Pair bonding 67, 71
Pallium 205, 212
Parachaetodon ocellatus 90
    sp. 11, 15, 88, 89, 94, 97, 98
Parallel evolution 167
Paraphyletic 18, 29
Parasphenoid 223, 225
Paratethyan, see Tethyan seaway
Parental care 73
Patrolling 71, 190, 191, 192
Pavona cactus 50
    decussata 50
    divaricata 50
    explanulata 50
    maldivensis 50
    varians 50
Peripheral isolation 27
Persian Gulf 26
Phylogeny 159
Pielou evenness 48, 57
Piloting 84, 85

Place niche breadth 35, 45
Plagiotremus celebesensis 28
    iosodon 28
    spilistius 28
Planktivory 67, 73, 85, 98, 201
Plankton 61, 67, 201
Platygyra daedalea 51
    lamellina 51
Pleisiomorphy 17, 19, 21, 168
Pocillopora damicornis 50, 174, 178
    meandrina 70, 72, 171, 172, 173, 174, 175,
        176, 177, 178, 179, 180, 181, 182, 191
    sp. 57, 70, 179, 180, 181, 182, 187, 196
    verrucosa 50
Podabacia crustacea 51
Polychaetes 101, 103, 108, 197, 199, 200
Polygyny 62, 63, 72
Polypterus bichir 241
    senegalus 239, 240, 241
    sp. 241
Pomacanthus imperator 206
    maculosus 206
    semicirculatus 90, 206
    sexstriatus 206
    sp. 90, 206
    straitus 206
Porites columnaris 51
    compressa 70, 81, 171, 172, 173, 175, 176,
        177, 178, 179, 180, 181, 182
    lobata 51, 70, 72, 81, 171, 172, 173, 175,
        176, 177, 178, 179, 180, 181, 182
    lutea 51
    solida 51
    sp. 171, 173, 174, 175, 177, 178, 179, 180,
        181
    undulata 51
Poster coloration 143, 145, 146, 151, 153,
    155
Posterior canal 236
Predation 62, 64, 67, 73, 98, 137, 143, 144,
    145, 146, 148, 149, 150, 151, 152, 153,
    155
Profitability 171, 181, 182
Prognathodes aculeatus 12
    aya 12
    dichrous 12
    falcifer 12, 27
    guezei 12
    guyanensis 12
    guyotensis 12
    marcellae 12
    obliquus 12
    sp. 11, 12, 22, 27, 148
Prootic 225, 228, 229
Protandry 63
Psammocora explanulata 50

    haimeana 50
    profundacella 50
Pygoplites diacanthus 206
    sp. 206

Rabdophorus ephippium 207
    fasciatus 208, 210, 213, 214, 215, 217
    sp. 11, 12, 18, 23, 207
Ramus ciliaris brevis 229
    longus 228, 229
Reciprocal altruism 72
Recruitment 101, 102, 105, 106, 108, 109,
    120
Red Sea 24, 26, 42
Reef 52, 53
Reproduction 61, 111
Reproductive patterns 139
Rhombochaetodon mertensi 207
    paucifasciatus 207
    sp. 14, 20, 21, 22, 26, 28, 207
Roa excelsa 13
    jayakari 13
    modestus 13
    sp. 11, 13, 15, 22, 148
Roaops sp. 11, 13, 18, 21, 22
Route specific landmarks 79, 84, 85
Runula sp. 144

Sacculus 225, 227, 228, 229, 235, 236
Saccus vasculosus 212
Scarus rubroviolaceus 214, 215, 217
    sp. 216
Schooling 61, 67
Sea anemones 199, 200, 202
Search time 175
Seasonality 137, 138
Seriatopora hystrix 50
    sp. 57
Settlement 101
Sexual dichromatism 135
Shannon-Weaver index 48, 56, 102, 197,
    199, 200, 201, 202
Shelter 58, 63, 64, 73
Siderastrea savignyana 50
Social behavior 73, 218
    relationships 129, 218
    systems 67
Society Islands 27, 41
Solea lascaris 216
    vulgaris 236, 239
Spacial learning 79, 85
Sparisoma sp. 121
Spawning 73, 97, 119, 120, 121, 122, 123,
    125, 126, 127, 131, 134, 135, 136, 137,
    138, 139, 140
    ascent 136

behavior 66, 72, 129, 133, 135
fecundity 65, 71, 136, 139, 140
sites 131, 137
Specialists 65, 69, 70, 101, 102, 103, 107, 159, 168, 197, 201
Species recognition hypothesis 143, 145, 155
richness 33, 56, 58
Spermatozoa 114, 115
Sphyraena jello 217
Sponges 197, 218
Strongylochaetodon melannotus 207
sp. 207
Stylocoeniella guentheri 50
Stylophora pistillata 50
sp. 50, 57
Sulcus lateralis 21
ypsiliformis 212
Sun compass orientation 85
Swimming 240
Sympatry 9, 10, 11, 12, 15, 18, 24, 27, 29
Synapomorphy 167, 168

Tectum 212
mesencephali 215

Teeth, see dentition
Tegmentum 215, 227
Tela choroidea 212
Telencephalon 212, 214, 216
Temperature 138
Territory 70, 80, 83, 187, 188, 191, 192, 218
feeding 64, 71, 72, 176
intraspecific 65
size 189, 191, 192
Testis 111, 114, 115, 120
Tethyan seaway 9, 11, 27
Tetrachaetodon sp. 11, 14, 16, 29, 148
Tholichthys osseus 89, 90
sp. 87, 88, 90
Time budgets 191
Torus longitudinalis 215
Tridentiger trigonocephalus 227, 230, 231, 232, 236, 239
Trigeminal nerve, see cranial nerve V
Trochlea 224
Trochlear nerve, see cranial nerve IV
Trochlearis nerve, see cranial nerve IV
Truncus ciliaris 229
Tuamotu archipelago 41

Tubastrea diaphana 52
Tubipora musica 52
Turbinaria mesenterina 52

Utriculus 235, 236

Valvula cerebelli 205, 212, 216
Vas deferens 114, 115
Ventricles 216
Vicariance 166
Vision 221
Visiting 191
Visual abilities 216

Warning coloration 146, 155
Western Atlantic, see Atlantic

Xestospongia muta 133

Zoanthideans 199, 200
Zoogeography 98
Zooplankton 199, 200
feeders 195
Zooxanthellae 176